MTP International Review of Science

Amino Acids, Peptides and Related Compounds

MTP International Review of Science

Publisher's Note

The MTP International Review of Science is an important new venture in scientific publishing, which we present in association with MTP Medical and Technical Publishing Co. Ltd. and University Park Press, Baltimore. The basic concept of the Review is to provide regular authoritative reviews of entire disciplines. We are starting with chemistry because the problems of literature survey are probably more acute in this subject than in any other. As a matter of policy, the authorship of the MTP Review of Chemistry is international and distinguished; the subject coverage is extensive, systematic and critical; and most important of all, new issues of the Review will be published every two years.

In the MTP Review of Chemistry (Series One), Inorganic, Physical and Organic Chemistry are comprehensively reviewed in 33 text volumes and 3 index volumes, details of which are shown opposite. In general, the reviews cover the period 1967 to 1971. In 1974, it is planned to issue the MTP Review of Chemistry (Series Two), consisting of a similar set of volumes covering the period 1971 to 1973. Series Three is planned for 1976, and so on.

The MTP Review of Chemistry has been conceived within a carefully organised editorial framework. The over-all plan was drawn up, and the volume editors were appointed, by three consultant editors. In turn, each volume editor planned the coverage of his field and appointed authors to write on subjects which were within the area of their own research experience. No geographical restriction was imposed. Hence, the 300 or so contributions to the MTP Review of Chemistry come from many countries of the world and provide an authoritative account of progress in chemistry.

To facilitate rapid production, individual volumes do not have an index. Instead, each chapter has been prefaced with a detailed list of contents, and an index to the 10 volumes of the MTP Review of Organic Chemistry (Series One) will appear, as a separate volume, after publication of the final volume. Similar arrangements will apply to the MTP Review of subsequent series.

Butterworth & Co. (Publishers) Ltd.

Organic Chemistry
Series One
Consultant Editor
D. H. Hey, F.R.S.
Department of Chemistry
King's College, University of London

Volume titles and Editors

1 STRUCTURE DETERMINATION IN ORGANIC CHEMISTRY
Professor W. D. Ollis, F.R.S.,
University of Sheffield

2 ALIPHATIC COMPOUNDS
Professor N. B. Chapman,
Hull University

3 AROMATIC COMPOUNDS
Professor H. Zollinger, *Swiss Federal Institute of Technology, Zurich*

4 HETEROCYCLIC COMPOUNDS
Dr. K. Schofield, *University of Exeter*

5 ALICYCLIC COMPOUNDS
Professor W. Parker, *University of Stirling*

6 AMINO ACIDS, PEPTIDES AND RELATED COMPOUNDS
Professor D. H. Hey, F.R.S. and
Dr. D. I. John,
King's College, University of London

7 CARBOHYDRATES
Professor G. O. Aspinall, *Trent University, Ontario*

8 STEROIDS
Dr. W. F. Johns, *G. D. Searle & Co., Chicago*

9 ALKALOIDS
Professor K. Wiesner, F.R.S.,
University of New Brunswick

10 FREE RADICAL REACTIONS
Professor W. A. Waters, F.R.S.,
University of Oxford

INDEX VOLUME

Physical Chemistry
Series One
Consultant Editor
A. D. Buckingham
Department of Chemistry
University of Cambridge

Volume titles and Editors

1 **THEORETICAL CHEMISTRY**
Professor W Byers Brown, *University of Manchester*

2 **MOLECULAR STRUCTURE AND PROPERTIES**
Professor G. Allen, *University of Manchester*

3 **SPECTROSCOPY**
Dr. D. A. Ramsay, F.R.S.C., *National Research Council of Canada*

4 **MAGNETIC RESONANCE**
Professor C. A. McDowell, F.R.S.C., *University of British Columbia*

5 **MASS SPECTROMETRY**
Professor A. Maccoll, *University College, University of London*

6 **ELECTROCHEMISTRY**
Professor J. O'M Bockris, *University of Pennsylvania*

7 **SURFACE CHEMISTRY AND COLLOIDS**
Professor M. Kerker, *Clarkson College of Technology, New York*

8 **MACROMOLECULAR SCIENCE**
Professor C. E. H. Bawn, F.R.S., *University of Liverpool*

9 **CHEMICAL KINETICS**
Professor J. C. Polanyi, F.R.S., *University of Toronto*

10 **THERMOCHEMISTRY AND THERMODYNAMICS**
Dr. H. A. Skinner, *University of Manchester*

11 **CHEMICAL CRYSTALLOGRAPHY**
Professor J. Monteath Robertson, F.R.S., *University of Glasgow*

12 **ANALYTICAL CHEMISTRY —PART 1**
Professor T. S. West, *Imperial College, University of London*

13 **ANALYTICAL CHEMISTRY — PART 2**
Professor T. S. West, *Imperial College, University of London*

INDEX VOLUME

Inorganic Chemistry
Series One
Consultant Editor
H. J. Emeléus, F.R.S.
Department of Chemistry
University of Cambridge

Volume titles and Editors

1 **MAIN GROUP ELEMENTS—HYDROGEN AND GROUPS I–IV**
Professor M. F. Lappert, *University of Sussex*

2 **MAIN GROUP ELEMENTS—GROUPS V AND VI**
Professor C. C. Addison, F.R.S. and Dr. D. B. Sowerby, *University of Nottingham*

3 **MAIN GROUP ELEMENTS—GROUP VII AND NOBLE GASES**
Professor Viktor Gutmann, *Technical University of Vienna*

4 **ORGANOMETALLIC DERIVATIVES OF THE MAIN GROUP ELEMENTS**
Dr. B. J. Aylett, *Westfield College, University of London*

5 **TRANSITION METALS—PART 1**
Professor D. W. A. Sharp, *University of Glasgow*

6 **TRANSITION METALS—PART 2**
Dr. M. J. Mays, *University of Cambridge*

7 **LANTHANIDES AND ACTINIDES**
Professor K. W. Bagnall, *University of Manchester*

8 **RADIOCHEMISTRY**
Dr. A. G. Maddock, *University of Cambridge*

9 **REACTION MECHANISMS IN INORGANIC CHEMISTRY**
Professor M. L. Tobe, *University College, University of London*

10 **SOLID STATE CHEMISTRY**
Dr. L. E. J. Roberts, *Atomic Energy Research Establishment, Harwell*

INDEX VOLUME

Organic Chemistry
Series One

Consultant Editor
D. H. Hey, F.R.S.

MTP International Review of Science

Volume 6

Amino Acids, Peptides and Related Compounds

Edited by **D. H. Hey, F.R.S.** and **D. I. John**
King's College, University of London

Butterworths · London
University Park Press · Baltimore

THE BUTTERWORTH GROUP

ENGLAND
Butterworth & Co (Publishers) Ltd
London: 88 Kingsway, WC2B 6AB

AUSTRALIA
Butterworths Pty Ltd
Sydney: 586 Pacific Highway 2067
Melbourne: 343 Little Collins Street, 3000
Brisbane: 240 Queen Street, 4000

NEW ZEALAND
Butterworths of New Zealand Ltd
Wellington: 26–28 Waring Taylor Street, 1

SOUTH AFRICA
Butterworth & Co (South Africa) (Pty) Ltd
Durban: 152–154 Gale Street

ISBN 0 408 70280 X

UNIVERSITY PARK PRESS

U.S.A. and CANADA
University Park Press
Chamber of Commerce Building
Baltimore, Maryland, 21202

Library of Congress Cataloging in Publication Data

Hey, Donald Holroyde.
Amino acids, peptides, and related compounds.

(Organic chemistry, series one, v. 6) (MTP International review of science)
1. Amino acids. 2. Peptides. I. John, D. I., joint author. II. Title. [DNLM: 1. Amino acids.
2. Peptides. QU60 A517 1973]
QD251.2.074 vol. 6 [QD431] 547′.008s [547′.75]
ISBN 0–8391–1034–0 73–11000

MTP MEDICAL AND TECHNICAL PUBLISHING CO. LTD.
St. Leonard's House
St. Leonardgate
Lancaster, Lancs.
and
BUTTERWORTH & CO. (PUBLISHERS) LTD.

Filmset by Photoprint Plates Ltd., Rayleigh, Essex
Printed in England by Redwood Press Ltd., Trowbridge, Wilts
and bound by R. J. Acford Ltd., Chichester, Sussex

Consultant Editor's Note

The subject of Organic Chemistry is in a rapidly changing state. At the one extreme it is becoming more and more closely involved with biology and living processes and at the other it is deriving a new impetus from the extending implications of modern theoretical developments. At the same time the study of the subject at the practical level is being subjected to the introduction of new techniques and advancements in instrumentation at an unprecedented level. One consequence of these changes is an enormous increase in the rate of accumulation of new knowledge. The need for authoritative documentation at regular intervals on a world-wide basis is therefore self-evident.

The ten volumes in Organic Chemistry in this First Series of biennial reviews in the MTP International Review of Science attempt to place on record the published achievements of the years 1970 and 1971 together with some earlier material found desirable to assist the initiation of the new venture. In order to do this on an international basis Volume Editors and Authors have been drawn from many parts of the world.

There are many alternative ways in which the subject of Organic Chemistry can be subdivided into areas for more or less self-contained reviews. No single system can avoid some overlapping and many such systems can leave gaps unfilled. In the present series the subject matter in eight volumes is defined mainly on a structural basis on conventional lines. In addition, one volume has been specially devoted to methods of structure determination, which include developments in new techniques and instrumental methods. A further separate volume has been devoted to Free Radical Reactions, which is justified by the rapidly expanding interest in this field. If there prove to be any major omissions it is hoped that these can be remedied in the Second Series.

It is my pleasure to thank the Volume Editors who have made the publication of these volumes possible.

London D. H. Hey

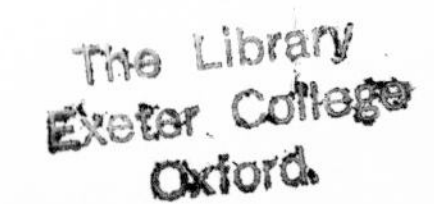

Preface

During the past 25 years progress in the chemistry of amino acids and peptides has been dominated by the objectives of structural elucidation and synthesis of peptides of interesting biological function. The establishment of the feasibility of sequencing and synthesis of these complex molecules, through the pioneering studies on insulin and oxytocin, has led in more recent years to spectacular advances in many areas and the first steps have already been taken into the field of protein synthesis. A new phase in the history of peptide chemistry is thus being opened up. In parallel with these developments significant progress is also being made in the elucidation of structure-activity relationships and in the implications of peptide conformation. The discovery of new amino acids and related compounds of natural origin, and the study of peptides of abnormal structure, in particular in the penicillin and depsipeptide areas, has attracted considerable attention.

In line with these recent developments three chapters in this volume have been devoted to those aspects of the chemistry of peptides which include structural elucidation and synthesis. A further two chapters have then been given to peptide conformation and structure-activity relationships, both areas which have derived considerable benefit from the greater availability of well-characterised model compounds. Other chapters cover amino acids, depsipeptides, and the penicillins and cephalosporins. The vast subject of protein and enzyme chemistry has been deliberately excluded from this volume. Limitations of space made this decision inevitable.

The editors wish to express their thanks to the authors who have contributed to this volume.

London

D. H. Hey
D. I. John

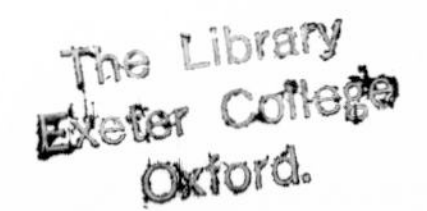

Contents

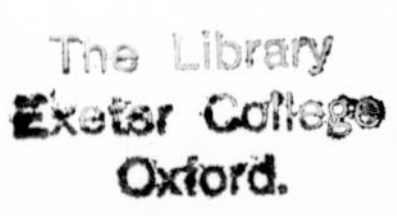

1
Amino Acids of Natural Origin

E. A. BELL
Kings College, University of London

1.1 INTRODUCTION

At the time of writing over 250 naturally occurring amino and imino acids have been isolated and their structures determined and there can be little doubt that many more remain to be discovered. Our knowledge of most of these compounds is of relatively recent origin and the advances in this area of biological chemistry spring directly from the development of chromatography and electrophoresis as rapid methods not only for the identification of known amino-acids in biological extracts but also for the detection of previously unknown compounds.

Of the known amino acids, 20 are found universally distributed as protein constituents in all forms of living organisms. A few others such as ornithine, homoserine and γ-aminobutyric acid are also frequently encountered as metabolic intermediates. The great majority of the natural amino acids enjoy a much more limited distribution however, and for the purposes of this chapter these will be referred to as the 'unusual' amino acids.

Most of the 'unusual' amino acids have been isolated from plants or micro-organisms, although various of them are known to occur in the animal kingdom. In plants, these compounds are usually found in the free state or combined as simple dipeptides, frequently as the γ-glutamyl peptides. In contrast, most of the 'unusual' amino-acids which have been found in micro-organisms occur in the combined form, frequently as constituents of antibiotic-type peptides which yield the 'unusual' amino-acids on hydrolysis.

1.2 POSSIBLE ORIGINS OF 'UNUSUAL' AMINO ACIDS

As yet, little is known either of the biosynthesis or fate of most amino acids. It has been suggested[1] that some of the 'unusual' amino acids may have arisen by the modification of biosynthetic pathways primarily concerned with the synthesis of 'usual' amino acids. It can be readily visualised, for example, that pipecolic acid could be formed if the enzymes normally concerned in the synthesis of proline had undergone a loss of substrate specificity. Similarly homoarginine might easily arise if lysine as well as ornithine was an acceptable substrate for a modified form of ornithine transcarbamylase in certain plant species. Certain of the 'unusual' amino acids have chemical structures which suggest that they are derived directly from one or other of the 'usual' amino acids by processes such as methylation or deamination; such modifications being possible either in the free amino acid or after it has been bound in a peptide or protein molecule. Many of the 'unusual' amino acids have very distinctive structures however, and it is difficult to imagine how these can arise by modification, either of familiar biosynthetic pathways or of more commonly occurring structures. It would seem then that these compounds arise in a variety of different ways; some owing their origins to modifications of familiar biosynthetic routes, and others to novel and sometimes unexpected pathways.

1.3 POSSIBLE BIOLOGICAL ROLES OF 'UNUSUAL' AMINO ACIDS

Just as it is not possible to generalise on the origins of all the various 'unusual' amino acids, so is it impossible to ascribe to them all a single biological role. The facts that many of the 'usual' amino acids are close chemical analogues of various of the 'usual' amino acids, and that certain of the 'unusual' amino acids or peptides containing them are toxic to different forms of life, suggest that some, at least, may function as protective agents in the organisms which synthesize them. It has for example been suggested[2] that L-dopa (L-3,4-dihydroxyphenylalanine) and 5-hydroxy-L-tryptophan may protect the seeds of *Mucuna* and *Griffonia* species from insect attack. It is also known that certain unusual plant amino acids are toxic to micro-organisms, fungi, other plants and mammals. It is possible then that some of these amino acids, which have no apparent role in the basic metabolic processes of the organisms which synthesise them, may yet be found to play a vital part in the organisms' overall economy, either by protecting them against potential predators or by discouraging competitors in their fight for survival.

1.4 RECENTLY DISCOVERED AMINO ACIDS

The rate at which new amino acids are being discovered has increased rather than decreased over the past few years. In the following sections, primary emphasis will be given to those new amino acids which have been reported in the literature during the years 1970 and 1971. Some reference will, however, be made to amino acids discovered previously when significant new information concerning their structures, biosynthesis or distribution has appeared in the years under review.

1.4.1 Neutral aliphatic amino acids

All of the newly-discovered amino acids in this group are of bacterial or plant origin. The biosynthesis of N^{β}-dimethyl-L-leucine (1), a constituent of etamycin, has been studied by Walker and his co-workers[3,4]. They have concluded that it is formed from L-leucine by methylation at both nitrogen and β-carbon atoms.

$$CH_3{\cdot}CH(CH_3){\cdot}CH(CH_3){\cdot}CH(NHCH_3){\cdot}CO_2H$$

(1)

Ornithine has been found to give rise to a new naturally occurring keto-amino acid, 2-amino-4-oxopentanoic acid (2) in *Clostridium sticklandii*[5], whilst a closely related unsaturated amino acid, L-2-aminopent-4-ynoic acid (3), has been isolated from a *Streptomycete* fermentation[6] and shown to act as an antimetabolite of L-methionine and L-leucine in *Bacillus subtilis*.

$$CH_3{\cdot}CO{\cdot}CH_2{\cdot}CH(NH_2){\cdot}CO_2H$$

(2)

$$HC{\vdots}C{\cdot}CH_2{\cdot}CH(NH_2){\cdot}CO_2H$$

(3)

Diaminopimelic acid (4) has long been known as a cell wall constituent of many micro-organisms. The configurations of two stereoisomeric forms of 2,6-diamino-3-hydroxypimelic acid (5), which also occur as microbial cell wall constituents, have recently been established[7] as the NH_2 *meso*, OH *threo* and the NH_2 L, OH *erythro* forms.

$$HO_2C{\cdot}CH(NH_2){\cdot}CH_2{\cdot}CH_2{\cdot}CH_2{\cdot}CH(NH_2){\cdot}CO_2H$$

(4)

$$HO_2C{\cdot}CH(NH_2){\cdot}CHOH{\cdot}CH_2{\cdot}CH_2{\cdot}CH(NH_2){\cdot}CO_2H$$

(5)

Another hydroxy-diamino-dicarboxylic acid, namely, 2,6-diamino-7-hydroxyazelaic acid (6), has been found as a constituent of the two antibiotic peptides edeine A and edeine B obtained from a strain of *Bacillus brevis*[8].

$$HO_2C{\cdot}CH(NH_2){\cdot}CH_2{\cdot}CH_2{\cdot}CH_2{\cdot}CH(NH_2){\cdot}CHOH{\cdot}CH_2{\cdot}CO_2H$$
(6)

The discovery of the unsaturated amino acid, β-methylene-L-(+)-norvaline (7), in the mushroom *Lactarius helvus* by Levenberg[9] has been followed by the isolation of the β-methylene-L-(+)-norleucine (8) from *Amanita vaginata*[10].

$$CH_3{\cdot}CH_2{\cdot}C(:CH_2){\cdot}CH(NH_2){\cdot}CO_2H$$
(7)

$$CH_3{\cdot}CH_2{\cdot}CH_2{\cdot}C(:CH_2){\cdot}CH(NH_2){\cdot}CO_2H$$
(8)

From yet another fungus, *Gibberella fujikuroi* have been isolated *N*-jasmonoyl-isoleucine (9) and *N*-dihydrojasmonoylisoleucine (10)[11].

O

CO·NH·CH(CO$_2$H)·CH(CH$_3$)·CH$_2$·CH$_3$

(9)

O

CO·NH·CH(CO$_2$H)·CH(CH$_3$)·CH$_2$·CH$_3$

(10)

In *Aesculus californica* (the Californian buckeye) the biosynthesis of the recently discovered 2-amino-4-methylhex-4-enoic acid (11) has been investigated. Isoleucine proved to be the best precursor of this amino acid and possible pathways from isoleucine to this compound and to homoisoleucine (12) are suggested[12, 13].

$$CH_3{\cdot}CH{:}C(CH_3){\cdot}CH_2{\cdot}CH(NH_2){\cdot}CO_2H$$
(11)

$$CH_3{\cdot}CH_2{\cdot}CH(CH_3){\cdot}CH_2{\cdot}CH(NH_2){\cdot}CO_2H$$
(12)

1.4.2 Hydroxy amino acids

Following the report of *N*-(2-hydroxyethyl)alanine (13) as a component of phospholipid from rumen protozoa[14], the lower homologue *N*-(2-hydroxy-ethyl)glycine (14) has been isolated from the seaweed *Petalonia jascia*[15].

$CH_3{\cdot}CH(NH{\cdot}CH_2{\cdot}CH_2OH){\cdot}CO_2H$ (13) $\qquad$ $CH_2(NH{\cdot}CH_2{\cdot}CH_2{\cdot}OH){\cdot}CO_2H$ (14)

From micro-organisms, the isolation of *threo*-L-α-amino-β,γ-dihydroxy-butyric acid (15)[16] and β-hydroxy-L-norvaline (16)[17] has been reported,

$$HO{\cdot}CH_2{\cdot}CHOH{\cdot}CH(NH_2){\cdot}CO_2H$$
(15)

$$CH_3{\cdot}CH_2{\cdot}CH(OH){\cdot}CH(NH_2){\cdot}CO_2H$$
(16)

while from the plant kingdom, γ,δ-dihydroxyisoleucine (17) has been found to occur in a peptide from *Amanita phalloides*[18]; γ-hydroxycitrulline (18), which was tentatively identified previously in seeds of *Vicia fulgens* and *V. unijuga*[19] has now been isolated from seeds of *V. faba*[20].

$$CH_2OH{\cdot}CHOH{\cdot}CH(CH_3){\cdot}CH(NH_2){\cdot}CO_2H$$
(17)

$$H_2N{\cdot}CO{\cdot}NH{\cdot}CH_2{\cdot}CHOH{\cdot}CH_2{\cdot}CH(NH_2){\cdot}CO_2H$$
(18)

1.4.3 Sulphur and selenium containing amino acids

The presence of *N*-formylmethionine (19) in proteins of *Clostridium pasteurianum rubredoxin*[21] and in the proteins of honeybee thorax[22] has been reported. The honeybee proteins also contain *N*-acetylmethionine (20).

$$CH_3{\cdot}S{\cdot}CH_2{\cdot}CH_2{\cdot}CH(NH{\cdot}OC{\cdot}H){\cdot}CO_2H$$
(19)

$$CH_3{\cdot}S{\cdot}CH_2{\cdot}CH_2{\cdot}CH(NH{\cdot}OC{\cdot}CH_3)CO_2H$$
(20)

The finding of L-homomethionine (21) in cabbage[23] adds to the list of free sulphur-containing amino acids in higher plants.

$$CH_3{\cdot}S{\cdot}CH_2{\cdot}CH_2{\cdot}CH_2{\cdot}CH(NH_2){\cdot}CO_2H$$
(21)

S-(2-Hydroxy-2-carboxyethyl)cysteine (22) has been identified in normal human urine[24].

$$HO_2C{\cdot}CHOH{\cdot}CH_2{\cdot}S{\cdot}CH_2{\cdot}CH(NH_2){\cdot}CO_2H$$
(22)

From the urines of patients suffering from cystathionuria and homocystinuria have been isolated several new sulphur containing amino acids[25–27]. These are *S*-(carboxymethyl)homocysteine (23), *S*-(2-hydroxy-2-carboxyethyl)homocysteine (24), *S*-(3-hydroxy-3-carboxy-n-propyl)cysteine (25), *S*-(3-hydroxy-3-carboxy-n-propylthio)homocysteine (26) and *S*-(2-hydroxy-2-carboxyethylthio)homocysteine (27).

$$HO_2C{\cdot}CH_2{\cdot}S{\cdot}CH_2{\cdot}CH_2{\cdot}CH(NH_2){\cdot}CO_2H$$
(23)

$$HO_2C{\cdot}CHOH{\cdot}CH_2{\cdot}S{\cdot}CH_2{\cdot}CH_2{\cdot}CH(NH_2){\cdot}CO_2H$$
(24)

$$HO_2C{\cdot}CHOH{\cdot}CH_2{\cdot}CH_2{\cdot}S{\cdot}CH_2{\cdot}CH(NH_2){\cdot}CO_2H$$
(25)

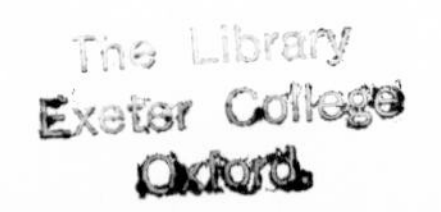

$$HO_2C{\cdot}CHOH{\cdot}CH_2{\cdot}CH_2{\cdot}S{\cdot}S{\cdot}CH_2{\cdot}CH_2{\cdot}CH(NH_2){\cdot}CO_2H$$
(26)

$$HO_2C{\cdot}CHOH{\cdot}CH_2{\cdot}S{\cdot}S{\cdot}CH_2{\cdot}CH_2{\cdot}CH(NH_2){\cdot}CO_2H$$
(27)

The toxic seleniferous plant *Morinda reticulata* has recently been shown by Peterson and Butler[28] to contain selenocystathionine (28).

$$HO_2C{\cdot}CH(NH_2){\cdot}CH_2{\cdot}CH_2{\cdot}Se{\cdot}CH_2{\cdot}CH(NH_2){\cdot}CO_2H$$
(28)

This amino acid has also been previously recognised[29] as the toxic constituent of *Lecythis ollaria* (monkey's coconut).

The biosynthesis of *Se*-methylselenocysteine (29) in *Astragalus bisculatus* has been studied by Chow and co-workers[30]. Their results suggest that this compound is synthesised by the same enzyme systems as are responsible for the synthesis of *S*-methylcysteine (30) which occurs in the same plant.

$$CH_3{\cdot}Se{\cdot}CH_2{\cdot}CH(NH_2){\cdot}CO_2H$$
(29)

$$CH_3{\cdot}S{\cdot}CH_2{\cdot}CH(NH_2){\cdot}CO_2H$$
(30)

1.4.4 Aromatic and heterocyclic amino acids

A number of cyclic guanidino amino acids have been isolated during the past few years from micro-organisms. These include, capreomycidine and epicapreomycidine, two epimers of α-(2-iminohexahydropyrimid-4-yl) glycine (31) obtained by hydrolytic degradation of the antibiotic capreomycin[31], and tuberactidine (32) which is produced on hydrolysis of tuberactinomycin[32].

H N NH NH $CH(NH_2){\cdot}CO_2H$
(31)

H HO N NH NH $CH(NH_2){\cdot}CO_2H$
(32)

L-Cyclohexa-1,4-diene-1-alanine (33), previously synthesised by Snow and co-workers[33] and shown to be an antimetabolite of phenylalanine in rat and *Leuconostoc dextranicum*, has now been isolated from a *Streptomycete*[34].

$CH_2{\cdot}CH(NH_2){\cdot}CO_2H$
(33)

HO $CH_2{\cdot}CH(NH_2){\cdot}CO_2H$ OH
(34)

Another recently discovered[35] natural antimetabolite of phenylalanine obtained from a *Streptomycete* is L-3-(2,5-dihydroxyphenyl)alanine (34).

The isolation of dimethylhistidine (35) from the Australian seaweed *Gracilaria secundata* has been reported[36].

$CH_3\overset{+}{N}$ — $CH_2CH(NH_2)CO_2^-$
N
CH_3

(35)

Among the higher plants, the leguminosae have proved the richest source of aromatic and heterocyclic amino acids. From leaves of *Vicia faba* has been isolated *N*-carbamoyl-2-(*p*-hydroxyphenyl)glycine (36)[37], and from *Pisum sativum* 2-(β-D-glucopyranosyl)-4-alanyl-3-isoxazolin-5-one (37)[38].

HO — $CH(NH{\cdot}CO{\cdot}NH_2)CO_2H$

(36)

O O NR
$CH_2{\cdot}CH(NH_2){\cdot}CO_2H$

R = α-D-glucosyl

(37)

Derivatives of tryptophan have figured prominently amongst the new amino acids from higher plants. Two unusual chlorinated tryptophans, N^α-carboethoxyacetyl-D-4-chlorotryptophan (38) and N^α-carbomethoxy-acetyl-D-4-chlorotryptophan (39), have both been obtained from immature seeds of *Pisum sativum*[39].

Cl
$CH_2{\cdot}CH(NH{\cdot}CO{\cdot}CH_2{\cdot}CO_2{\cdot}CH_2{\cdot}CH_3){\cdot}CO_2H$
N
H

(38)

Cl
$CH_2{\cdot}CH(NH{\cdot}CO{\cdot}CH_2{\cdot}CO_2{\cdot}CH_3){\cdot}CO_2H$
N
H

(39)

From the legume *Aotus subglauca* has been isolated[40] the methyl ester of *S*-(+)-N^β-methyltryptophan (40),

$CH_2{\cdot}CH(NH_2){\cdot}CO_2CH_3$
N
CH_3

(40)

and in the seeds of *Abrus precatorious*, the metho cation of the methyl ester of *N,N*-dimethyltryptophan (41) has been found to occur[41], although the corresponding anion was not identified.

$CH_2{\cdot}CH(\overset{+}{N}(CH_3)_3){\cdot}CO_2CH_3$
N
H

(41)

It has been known for a number of years that 5-hydroxy-L-tryptophan (42) occurs in certain plants. This amino acid has now been shown to exist, in concentrations equivalent to 6–10% of the fresh weight, in the seeds of the West African legume *Griffonia simplicifolia*[42].

HO — $CH_2{\cdot}CH(NH_2){\cdot}CO_2H$
N
H

(42)

In mammalian systems the synthesis of hydroxylated phenylalanines has continued to receive attention. The biosynthesis of L-dopa from *m*-tyrosine in rat liver and beef adrenal medulla has been reported[43]; in the medulla it has also been observed that *m*-tyrosine is formed from phenylalanine[44].

Pipecolic acid, an imino acid which is widely distributed in plants, has now been shown to arise from lysine in the mammalian liver[45].

1.4.5 Acidic amino acids

By hydrolysis of an antibiotic from a *Streptomycete*, Shoji and Sakazaki[46] obtained *threo*-β-hydroxy-L-glutamic acid (43) while β-methylaspartic acid (44), which occurs in the bound form in the antibiotic amphomycin, has also been shown to possess the L-*threo* configuration[47].

$$HO_2C{\cdot}CH_2{\cdot}CHOH{\cdot}CH(NH_2){\cdot}CO_2H$$

(43)

$$HO_2C{\cdot}CH(CH_3){\cdot}CH(NH_2){\cdot}CO_2H$$

(44)

The isolation of *erythro*-β-hydroxy-L-aspartic acid (45) from the unripe seeds of the legume *Astragalus sinicus*[48] has been reported.

$$HO_2C{\cdot}CH(OH){\cdot}CH(NH_2){\cdot}CO_2H$$

(45)

The amide pinnatanine has been obtained from *Staphylea pinnata* and its structure established as N^5-(2-hydroxymethylbutadienyl)*allo*(*threo*)-γ-hydroxy-L-glutamine (46)[49].

$$CH_2{:}CH{\cdot}C(CH_2OH){:}CH{\cdot}NH{\cdot}CO{\cdot}CHOH{\cdot}CH_2{\cdot}CH(NH_2){\cdot}CO_2H$$

(46)

1.4.6 Basic amino acids

The occurrence of α,β-diaminobutyric acid (47) in aspartocin was reported by Hausmann and co-workers[50] and two stereoisomers of the same basic amino acid, now found in another antibiotic amphomycin, have been identified as the L-*threo* and D-*erythro* forms[51,52].

$$CH_3{\cdot}CH(NH_2){\cdot}CH(NH_2){\cdot}CO_2H$$

(47)

The higher homologue, 2,4-diaminovaleric acid (48), has been isolated from *Clostridium sticklandii*[53].

$$CH_3{\cdot}CH(NH_2){\cdot}CH_2{\cdot}CH(NH_2){\cdot}CO_2H$$

(48)

A derivative of α,γ-diaminobutyric acid N^α-(6-methyloctanoyl)-α,γ-diaminobutyric acid (49) has been obtained from *Bacillus colistinus*[54], while

L-γ-oxalysine (50), a lysine antagonist in *E. coli*, has now been found as a naturally-occurring compound in two streptomycetes[55]. The racemic form of this amino acid was obtained synthetically a number of years ago[56].

$$H_2N{\cdot}CH_2{\cdot}CH_2{\cdot}CH[NH{\cdot}CO{\cdot}(CH_2)_4{\cdot}CH(CH_3){\cdot}CH_2{\cdot}CH_3]{\cdot}CO_2H$$

(49)

$$H_2N{\cdot}CH_2{\cdot}CH_2{\cdot}O{\cdot}CH_2{\cdot}CH(NH_2){\cdot}CO_2H$$

(50)

Both N^{ε}-trimethyl-L-δ-hydroxylysine (51) and the corresponding phosphate have been isolated from the cell walls of a diatom[57].

$$(CH_3)_3\overset{+}{N}{\cdot}CH_2{\cdot}CH(OH){\cdot}CH_2{\cdot}CH_2CH(NH_2){\cdot}CO_2^-$$

(51)

The biosynthesis[58], and physiological effects of the neurotoxic N^{β}-oxalyl-L-α,β-diaminopropionic acid (52) from *Lathyrus sativus* have been further investigated[59, 60].

$$HO_2C{\cdot}CO{\cdot}NH{\cdot}CH_2{\cdot}CH(NH_2){\cdot}CO_2H$$

(52)

Another recently discovered plant amino-acid[61] which is both hepatotoxic and teratogenic is indospicine (53). A total synthesis of the compound has now been described[62] and further work on its physiological effects reported[63].

$$H_2N{\cdot}C(:NH){\cdot}CH_2{\cdot}CH_2CH_2{\cdot}CH_2{\cdot}CH(NH_2){\cdot}CO_2H$$

(53)

The degradation of canavanine (54) to canalin (55) in plants has been studied by several groups of workers[64–66], and investigations of the fate of this amino acid in mammalian liver tissues have also been made[67].

$$H_2N{\cdot}C(:NH){\cdot}NH{\cdot}O{\cdot}CH_2{\cdot}CH_2{\cdot}CH(NH_2){\cdot}CO_2H$$

(54)

$$H_2N{\cdot}O{\cdot}CH_2{\cdot}CH_2{\cdot}CH(NH_2){\cdot}CO_2H$$

(55)

L-Homoarginine (56) previously known only in species of *Lathyrus* has been found in *Lotus helleri*, a species of a second legume genus[68].

$$H_2N{\cdot}C(:NH){\cdot}NH{\cdot}CH_2{\cdot}CH_2{\cdot}CH_2{\cdot}CH_2{\cdot}CH(NH_2){\cdot}CO_2H$$

(56)

Following the report of N^{ε}-trimethyl-lysine (57) in histones[69], many methylated derivatives of lysine and arginine have been found in animal proteins. The N^{ε}-monomethyl-lysine (58), as well as the trimethyl derivative, has been found in myosin[70–72] and in sheep plasma protein[73], while N^{ε}-dimethyl-lysine (59) has been reported to occur in actin from amoeba accompanied by traces of the mono- and trimethyl derivatives[74].

$$(CH_3)_3N^+(CH_2)_4CH(NH_2){\cdot}CO_2^-$$

(57)

$$CH_3{\cdot}NH{\cdot}(CH_2)_4{\cdot}CH(NH_2){\cdot}CO_2H$$

(58)

$$(CH_3)_2N{\cdot}(CH_2)_4{\cdot}CH(NH_2){\cdot}CO_2H$$

(59)

The presence of N^{ε}-mono-, di- and tri-methyl-lysines, N^G, N^G (60) N^G, N'^G-dimethylarginine (61) as well as the glucosylgalactosyl and galactosyl-δ-hydroxylysine have been detected in human urine[75].

$$(CH_3)_2N{\cdot}C(:NH){\cdot}NH{\cdot}(CH_2)_3{\cdot}CH(NH_2){\cdot}CO_2H$$

(60)

$$CH_3{\cdot}NH{\cdot}C(:NCH_3){\cdot}NH{\cdot}(CH_2)_3{\cdot}CH(NH_2){\cdot}CO_2H$$

(61)

Hypusine, a new basic amino acid identified as N^6-(4-amino-2-hydroxybutyl)-2,6-diaminohexanoic acid (62), has been reported to occur in bovine brain[76] and other mammalian organs as well as in human urine[77], while N^{ω}-methyl-arginine (63) has been found in nuclear protein from rat liver[78].

$$H_2N{\cdot}CH_2{\cdot}CH_2{\cdot}CH(OH){\cdot}CH_2{\cdot}NH{\cdot}(CH_2)_4{\cdot}CH(NH_2){\cdot}CO_2H$$

(62)

$$CH_3{\cdot}NH{\cdot}C(:NH){\cdot}NH{\cdot}(CH_2)_3{\cdot}CH(NH_2){\cdot}CO_2H$$

(63)

1.4.7 Imino acids

In *Santalum album* (Sandal), Kuttan and Radhakrishnan[79] have studied the biosynthesis of *cis*-4-hydroxy-L-proline (64) and have obtained evidence that it is formed by the direct hydroxylation of L-proline. The same workers have shown that *Santalum album* also contains *trans*-4-hydroxy-L-proline in the bound form[80].

The synthesis and x-ray analysis of *cis*-3,4-methano-L-proline (65), an imino acid newly isolated from *Aesculus parviflora*[81], has been carried out by Fujimoto and co-workers[82].

HOHC—CH$_2$
H$_2$C—N(H)—CH·CO$_2$H

(64)

H$_2$C
HC—CH
H$_2$C—N(H)—CHCO$_2$H

(65)

OH
N(H) CO$_2$H

(66)

Peganum harmala has proved to be a new source of L-(−)-4-hydroxypipecolic acid (66)[83], while 3-hydroxypipecolic acid (67) has been found in halophyte leaves, flowers and fruits[84].

The unsaturated imino acid baikiain (68), originally discovered in *Baikiaea plurijuga* (Rhodesian teak)[85], has now been identified in *Corallina officianalis* (an Australian seaweed)[36].

It has been shown that the lower homologue of proline, azetidine-2-carboxylic acid (69), which was originally isolated from *Convallaria majalis*

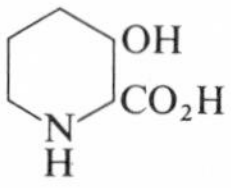

OH CO_2H N H
(67)

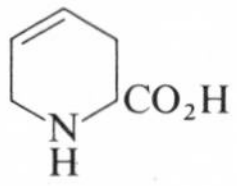

CO_2H N H
(68)

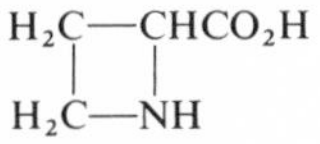

$H_2C—CHCO_2H$
$H_2C—NH$
(69)

$H_2C—CHCO_2H$
$H_2C—NCH_2{\cdot}CH_2{\cdot}CH_2{\cdot}NH{\cdot}CH(CO_2H){\cdot}CH_2{\cdot}CH(NH_2){\cdot}CO_2H$
(70)

(lily of the valley)[86] is derived from α,γ-diaminobutyric acid in *Delonix regia* (flamboyant tree)[87].

A second naturally occurring azetidine compound, nicotinamine 1-[3-(1-amino-1,3-dicarboxypropylamino)propyl] azetidine-2-carboxylic acid (70), has been found in tobacco leaves[88].

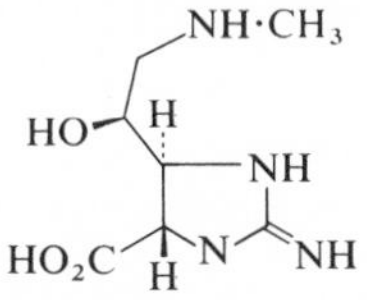

$NH{\cdot}CH_3$ HO H NH HO_2C H N NH
(71)

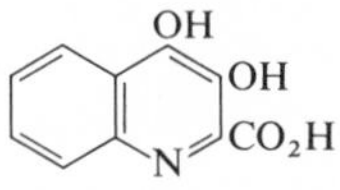

OH OH CO_2H N
(72)

From a microbial peptide has been isolated *N*-methylstreptolidine (71)[89] and from the sponge *Aplysina aerophoba* the isolation of 3,4-dihydroxy-quinoline-2-carboxylic acid (72)[90] has been reported.

1.4.8 Non α-amino acids

The hydrolysis of peptides from micro-organisms has given rise to a number of hydroxylated non-α-amino acids. From edienes A and B respectively, Hettinger and Craig[91] have obtained iso-serine (73) and β-tyrosine (74). The identification of δ-hydroxy-β-lysine (75) as one of the hydrolysis products of tuberactinomycin[32] has been reported, and 4-amino-2,3-dihydroxy-3-methyl-butanoic acid (76) has been isolated from the hydrolysate of carzinophilin[92].

An intermediate in 5-amino levulinate metabolism in *Rhodospirillum rubrum* has been identified as 5-amino-4-hydroxyvaleric acid (77).

Putreanine, *N*-(4-aminobutyl)-3-aminopropionic acid (78), which was first reported in the mammalian central nervous system[94], has since been detected in both human brain[95] and human urine[96]

$$H_2N{\cdot}CH_2{\cdot}CHOH{\cdot}CO_2H$$

(73)

$$HO-C_6H_4-CH(NH_2)\cdot CH_2\cdot CO_2H$$

(74)

$$H_2N\cdot CH_2\cdot CHOH\cdot CH_2\cdot CH(NH_2)\cdot CH_2\cdot CO_2H$$

(75)

$$H_2N\cdot CH_2\cdot C(CH_3)OH\cdot CHOH\cdot CO_2H$$

(76)

$$H_2N\cdot CH_2\cdot CHOH\cdot CH_2\cdot CH_2\cdot CO_2H$$

(77)

$$H_2N\cdot CH_2\cdot CH_2\cdot CH_2\cdot CH_2\cdot NH\cdot CH_2\cdot CH_2\cdot CO_2H$$

(78)

1.5 DISCUSSION

The development of chromatographic and electrophoretic techniques has led to the discovery, isolation and characterisation of over 250 naturally occurring amino acids, and to the detection of others which have not yet been isolated. As many of these compounds are of recent discovery, comparatively little is known of their biological significance, biosynthesis and ultimate fate in the organisms in which they are found. Nevertheless, this group of natural compounds is of great potential interest to workers in several fields of science.

Distribution studies of 'unusual' amino acids can for example provide information on the evolutionary relationships which exist between different species, genera and families of organisms. The presence of an 'unusual' amino acid in a restricted group of species, which are related in other respects, suggests strongly that these species are all descended from a common ancestral form in which the genome controlling the biosynthesis of that amino acid first arose. If the species have more than one 'unusual' amino acid in common and these compounds are of independent genetic origin, then the probability of common ancestry becomes almost certainty, as it is very unlikely that an identical combination of unusual genomes, responsible for an identical combination of 'unusual' amino acids, could have arisen more than once. In the plant kingdom, the 'unusual' amino-acids of the Leguminosae have been studied more thoroughly than those of other families and their phylogenetic significance has been discussed in a recent publication[97].

Many of the 'unusual' amino acids are close chemical analogues of various of the 'usual' amino acids, and it is not surprising to find that certain of the 'unusual' compounds act as antimetabolites when introduced into organisms to which they are normally foreign. This characteristic makes these 'unusual' amino acids of considerable interest both in the study of nutrition and medicine. In many parts of the world legume seeds (beans and peas) provide a major part of human dietary protein and the seeds of many legume species are rich in 'unusual' amino acids. The toxicity of some of these, as for example, canavanine, mimosine, α,γ-diaminobutyric acid and β-cyanoalanine to higher animals is well known and recent work, such as that on indospicine, suggests that other 'unusual' amino acids may be responsible for the toxic effects of certain plants eaten by man and animals. It is clear that the toxicity of an 'unusual' amino acid may be due to the competition between that compound and its 'usual' analogue, or it may be of a more complicated nature. Nevertheless, many of these compounds are a real, or potential hazard when they occur in plants or animals which are eaten by man and domestic animals[98, 99]. Conversely, an 'unusual' amino acid may be nutritionally valuable, L-homoarginine for example has been used to replace the essential amino acid lysine in the diets of experimental animals. The same amino acid will, however, inhibit growth in *Chlorella vulgaris*, *Escherichia coli* and *Staphylococcus aureus* suggesting that some of these compounds may find new roles as anti-microbial agents in the treatment of human disease. The medical importance of others as constitutents of antibiotics need scarcely be stressed.

One of the most interesting questions which the discovery of all these new amino acids raises is that of their role in the organisms which synthesise them. As has been suggested earlier, one role may be the protection of the organism against potential predators, and another, the discouragement of competitors in the same environment. Whatever the role or the target of a particular amino acid, it is clear that the species containing it would not have been selected for, if that particular amino acid had not conferred some overall advantage to that species. We may therefore expect that natural selection will have led to the perpetuation of species containing 'unusual' amino-acids, which are physiologically active in some, if not all, living systems.

While reports of toxicity and physiological activity of plant and animal materials are usually concerned with toxicity and physiological activity in man and higher animals, it is unlikely that he, or they, have exercised the environmental pressures which have led to the selection of most of the organisms which synthesise 'unusual' amino acids. A plant species containing a particular 'unusual' amino acid may have been selected because it is toxic to a particular species of predatory insect. That amino acid may, or may not, be toxic to man, to other insects, to other plants, or to micro-organisms. It may, however, play a vital role in the plant's strategy of survival, and a great deal of interesting work remains to be done, not only on the chemistry and biochemistry of these compounds in the organisms which contain them, but also on the interspecific chemistry which governs the relationships of different organisms, one to another, in the complicated and interwoven pattern of a living environment[100].

References

1. Fowden, L. (1965). *Biosynthetic Pathways in Higher Plants,* 73 (J. B. Pridham and T. Swain, editors). (London and New York: Academic Press)
2. Bell, E. A. and Janzen, D. H. (1971). *Nature (London),* **229,** 136
3. Walker, J. E., Bodanszky, M. and Perlman, D. (1970). *J. Antibiotics,* **23,** 255
4. Walker, J. E. and Perlman, D. (1971). *Biotech. Bioeng.,* **13,** 371
5. Tsuda, Y. and Friedmann, H. C. (1970). *J. Biol. Chem.,* **245,** 5914

6. Scannell, J. P., Preuss, D. L., Demny, T. C., Weiss, F., Williams, T. and Stempel, A. (1971). *J. Antibiotics,* **24,** 239
7. Perkins, H. R. (1969). *Biochem. J.,* **115,** 797
8. Hettinger, T. P. and Craig, L. C. (1970). *Biochemistry,* **9,** 1224
9. Levenberg, B. (1968). *J. Biol. Chem.,* **243,** 6009
10. Vervier, R. and Casimir, J. (1970). *Phytochemistry,* **9,** 2059
11. Cross, B. E. and Webster, G. R. B. (1970). *J. Chem. Soc. C,* 1839
12. Fowden, L. and Mazelis, M. (1971). *Phytochemistry,* **10,** 359
13. Boyle, J. E. and Fowden, L. (1971). *Phytochemistry,* **10,** 2671
14. Kemp, P. and Dawson, R. M. C. (1969). *Biochim. Biophys. Acta,* **176,** 678
15. Takagi, N., Hsu, H. Y. and Takemoto, T. (1970). *Yakugaku Zasshi,* **90,** 899 [*Chem. Abstr.,* **73,** 127 782u]
16. Westley, J. W., Preuss, D. L., Volpe, L. A., Demny, T. C. and Stempel, A. (1971). *J. Antibiotics,* **24,** 330
17. Godtfredsen, W. O., Vangedal, S. and Thomas, D. W. (1970). *Tetrahedron,* **26,** 4931
18. Wieland, T. and Fahrmeir, A. (1970). *Annalen,* **736,** 95
19. Bell, E. A. and Tirimanna, A. S. L. (1965). *Biochem. J.,* **97,** 104
20. Inatomi, H., Monta, Y., Ozawa, M., Suyama, Y. (1970). *Chem. Abstr.,* **72,** 18 541y
21. McCarthy, K. F. and Levenberg, W. (1970). *Biochem. Biophys. Res. Commun.,* **40,** 1053
22. Polz, G. and Dreil, G. (1970). *Biochem. Biophys. Res. Commun.,* **39,** 516
23. Suketa, Y., Sugii, M. and Suzuki, T. (1970). *Chem. Pharm. Bull.,* **18,** 249
24. Yao, K., Suzuki, M. and Ohmori, S. (1970). *Physiol. Chem. Phys.,* **2,** 195
25. Kodama, H., Yao, K., Kobayashi, K., Hirayama, K., Fujii, Y., Mizuhara, S., Haraguchi, H. and Hirosawa, M. (1969). *Physiol. Chem. Phys.,* **1,** 72
26. Kodama, H., Ohmori, S., Suzuchi, M. and Mizuhara, S. (1970). *Physiol. Chem. Phys.,* **2,** 287
27. Kodama, H., Ohmori, S., Suzuchi, M., Mizuhara, S., Oura, T., Isshiki, G. and Uemura, I. (1971). *Physiol. Chem. Phys.,* **3,** 81
28. Peterson, P. J. and Butler, G. W. (1971). *Aust. J. Biol. Sci.,* **24,** 175
29. Kerdel-Vegas, F., Wagner, F., Russell, P. B., Grant, N. H., Alburn, H. E., Clark, D. E. and Miller, J. A. (1965). *Nature (London),* **205,** 1186
30. Chow, C. M., Nigam, S. N. and McConnell, W. B. (1971). *Phytochemistry,* **10,** 2693
31. Bycroft, B. W., Cameron, D. and Johnston, A. W. (1971). *J. Chem. Soc. C,* 3040
32. Wakamiya, T., Shiba, T., Kaneko, T., Sakakibara, H., Take, T. and Abe, J. (1970). *Tetrahedron Letters,* 3497
33. Snow, M. L., Lauinger, C. and Ressler, C. (1968). *J. Org. Chem.,* **33,** 1774
34. Tamashita, T., Miyairi, N., Kunugita, K., Shimizu, K. and Sakai, H. (1970). *J. Antibiotics,* **23,** 537
35. Scannell, J. P., Pruess, D. L., Demny, T. C., Williams, T. and Stempel, A. (1970). *J. Antibiotics,* **23,** 618
36. Madgwick, J. C., Ralph, B. J., Shannon, J. S. and Simes, J. J. (1970). *Arch. Biochem. Biophys.,* **141,** 766
37. Eagles, J., Laird, W. M., Matai, S., Self, R., Synge, R. L. M. and Drake, A. F. (1971). *Biochem. J.,* **121,** 425
38. Lambein, F. and van Parijs, R. (1970). *Biochem. Biophys. Res. Commun.,* **40,** 557
39. Marumo, S. and Hattori, H. (1970). *Planta,* **90,** 208
40. Johns, S. R., Lamberton, J. A. and Sioumis, A. A. (1971). *Aust. J. Chem.,* **24,** 439
41. Ghosal, S. and Dutta, S. K. (1971). *Phytochemistry,* **10,** 195
42. Fellows, L. E. and Bell, E. A. (1970). *Phytochemistry,* **9,** 2389
43. Tong, J. H., D'Iorio, A. and Benoiton, N. L. (1971). *Biochem. Biophys. Res. Commun.,* **43,** 819
44. Tong, J. H., D'Iorio, A. and Benoiton, N. L. (1971). *Biochem. Biophys. Res. Commun.,* **44,** 229
45. Chamidi, H., Chou, W. and Kesner, L. (1971). *Biochem. Med.,* **5,** 56
46. Shoji, J. and Sakazaki, R. (1970). *J. Antibiotics,* **23,** 418
47. Bodanszky, M., Marconi, G. G. (1970). *J. Antibiotics,* **23,** 238
48. Inatomi, H., Inukai, F. and Murakami, T. (1971). *Chem. Pharm. Bull.,* **19,** 216
49. Grove, M. D., Daxenbichler, M. E., Weisleder, D. and Van Etten, C. H. (1971). *Tetrahedron Letters,* 4477
50. Hausmann, W. K., Borders, D. B. and Lancaster, J. E. (1969). *J. Antibiotics,* **22,** 257

51. Bodanszky, A. A. and Bodanszky, M. (1970). *J. Antibiotics*, **23,** 149
52. Strong, R. C., Bodanszky, A. A. and Bodanszky, M. (1970). *J. Antibiotics*, **23,** 257
53. Dyer, J. K. and Costilow, R. N. (1970). *J. Bacteriol.*, **101,** 77
54. Ito, M., Aida, K. and Uemura, T. (1970). *Agric. Biol. Chem.*, **34,** 476
55. Stapley, E. O., Miller, T. W., Mata, J. M. and Hendlin, D. (1970). *An. Inst. Farmacol. Espan.*, 1966–67 (Pub. 1970). **15-16,** 185
56. McCord, T. J., Ravel, J. M., Skinner, C. G. and Shive, W. (1957). *J. Amer. Chem. Soc.*, **79,** 5693
57. Nakajima, T. and Volcani, B. E. (1970). *Biochem. Biophys. Res. Commun.*, **39,** 28
58. Malathi, K., Padmanaban, G. and Sarma, P. S. (1970). *Phytochemistry*, **9,** 1603
59. Cheema, P. S., Padmanaban, G. and Sarma, P. S. (1970). *J. Neurochem.*, **17,** 1295
60. Cheema, P. S., Padmanaban, G. and Sarma, P. S. (1971). *J. Neurochem.*, **18,** 2137
61. Hegarty, M. P. and Pound, A. W. (1968). *Nature (London)*, **217,** 354
62. Culvenor, C. C. J., Foster, M. C. and Hegarty, M. P. (1971). *Aust. J. Chem.*, **24,** 371
63. Christie, G. S., de Munk, F. G., Madsen, N. P. and Hegarty, M. P. (1971). *Pathology*, **3,** 139
64. Rosenthal, G. A. (1970). *Plant Physiol.*, **46,** 273
65. Toepfer, R., Miersch, J. and Reinbothe, H. (1970). *Biochem. Physiol. Pflanz.*, **161,** 231
66. Whiteside, J. A. and Thurman, D. A. (1971). *Planta*, **98,** 279
67. Takahara, K., Nakanishi, S. and Natelson, S. (1971). *Arch. Biochem. Biophys.*, **145,** 85
68. Chwalek, B. and Przybylska, J. (1970). *Bull. Acad. Pol. Sci.*, **18,** 603
69. Hempel, K., Lange, H. W. and Birkofer, L. (1968). *Naturwissenschaften*, **55,** 37
70. Kuehl, W. M. and Adelstein, R. S. (1969). *Biochem. Biophys. Res. Commun.*, **37,** 59
71. Hardy, M., Harris, I., Perry, S. V. and Stone, D. (1970). *Biochem. J.*, **117,** 44p
72. Hardy, M., Harris, C. I., Perry, S. V. and Stone, D. (1970). *Biochem. J.*, **120,** 653
73. Weatherall, I. L. and Hadden, D. D. (1969). *Biochim. Biophys. Acta*, **192,** 553
74. Weihing, R. R. and Korn, E. D. (1970). *Nature (London)*, **227,** 1263
75. Kakimoto, Y. and Akazawa, S. (1970). *J. Biol. Chem.*, **245,** 5751
76. Shiba, T., Mizote, H., Kaneko, T., Nakajima, T., Kakimoto, Y. and Sano, I. (1971). *Biochim. Biophys. Acta*, **244,** 523
77. Nakajima, T., Matsubayashi, T., Kakimoto, Y. and Sano, I. (1971). *Biochim. Biophys. Acta*, **252,** 92
78. Paik, W. K. and Kim, S. (1970). *J. Biol. Chem.*, **245,** 88
79. Kuttan, R. and Radhakrishnan, A. N. (1970). *Biochem. J.*, **117,** 1015
80. Kuttan, R. and Radhakrishnan, A. N. (1970). *Biochem. J.*, **119,** 651
81. Fowden, L., Smith, A., Millington, D. S. and Sheppard, R. C. (1969). *Phytochemistry*, **8,** 437
82. Fujimoto, Y., Irreverre, F., Karle, J. M. and Karle, I. L. and Witkop, B. (1971). *J. Amer. Chem. Soc.*, **93,** 3471
83. Ahmad, V. U., Khan, M. A. (1971). *Phytochemistry*, **10,** 3339
84. Goas, G., Larker, M. F. and Goas, M. M. (1970). *Compt. Rend. Acad. Sci.*, Ser D., **271,** 1368
85. King, F. F., King, T. J. and Warwick, A. J. (1950), *J. Chem. Soc.*, 3590
86. Fowden, L. (1955). *Nature (London)*, **176,** 347
87. Sung, M. L. and Fowden, L. (1971). *Phytochemistry*, **10,** 1523
88. Noma, M., Noguchi, M., Tamaki, E. (1971). *Tetrahedron Letters*, 2017
89. Bordens, D. B., Sax, K. J., Lancaster, J. E., Hausmann, W. K., Mitscher, L. A., Wetzel, E. R. and Patterson, E. L. (1970). *Tetrahedron*, **26,** 3123
90. Fattorusso, E., Forenza, S., Minale, L. and Sodano, G. (1971). *Gazz. Chim. Ital.*, **101,** 104
91. Hettinger, T. P. and Craig, L. C. (1970). *Biochemistry*, **9,** 1224
92. Onda, M., Konda, Y., Omura, S. and Hata, T. (1971). *Chem. Pharm. Bull.*, **19,** 2013
93. Shigesada, K., Ebisuno, T. and Katsuki, H. (1970). *Biochem. Biophys. Res. Commun.*, **39,** 135
94. Kakimoto, Y., Nakajima, T., Kumon, A., Matsuoka, Y., Imaoka, N. and Sana, I. (1969). *J. Biol. Chem.*, **244,** 6003
95. Nakajima, T., Kakimoto, Y. and Sano, I. (1970). *J. Neurochem.*, **17,** 1427
96. Nakajima, T., Matsuoka, Y. and Akazawa, S. (1970). *Biochim. Biophys. Acta*, **222,** 405
97. Bell, E. A. (1971). *Chemotaxonomy of the Leguminosae*, 179 (J. B. Harborne, D. Boulter and B. L. Turner, editors). (London and New York: Academic Press)
98. Fowden, L., Lewis, D. and Tristram, H. (1967). *Advances in Enzymology*, **29,** 89

99. Thompson, J. F., Morris, C. J. and Smith, I. K. (1969). *Ann. Rev. Biochem.*, **38,** 137
100. Bell, E. A. (1972). *Phytochemical Ecology,* 163 (J. B. Harborne, editor). (London and New York: Academic Press)

2
Synthesis, Structural Properties and Reactions of Amino Acids

A. THOMSON
University of Lancaster

2.1 INTRODUCTION

This review covers papers in amino acid chemistry published since 1967 and covered by *Chemical Abstracts* before the end of 1971.

2.2 SYNTHESIS

2.2.1 General methods for α-amino acids

Throughout the period covered the predominant methods used for α-amino acids were the acetamidomalonate, hydantoin and Strecker syntheses; the azlactone route is now comparatively rare. The acetamidomalonate synthesis usually involves the reaction of an alkyl halide with the carbanion derived from diethyl acetamidomalonate, subsequent hydrolysis and decarboxylation giving the amino acid. Modifications of the technique are sometimes necessary, the most common being the use of formamidomalonate or of acetamidocyano-acetate. Hydantoin itself (1) is prepared by treatment of glycine with sodium cyanate, the active methylene can be condensed with aldehydes and the desired product obtained by subsequent reduction and hydrolysis. The Strecker method depends upon the hydrolysis of the amino-nitrile formed by reaction of an aldehyde or ketone with potassium cyanide and ammonium chloride.

Some new synthetic methods have been developed. Alkylation of the carbanion derived from ethyl *N*,*N*-bis(trimethylsilyl)glycinate gave up to

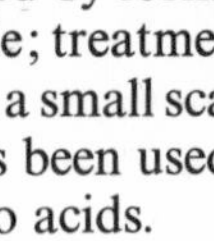

(1) 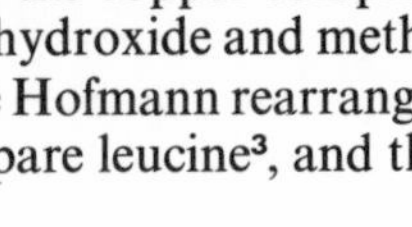 (2)

70% yields of amino acids after hydrolysis[1]. The methylene of glycine can also be activated by formation of the copper complex of its Schiff's base with salicylaldehyde; treatment with hydroxide and methyl iodide gave good yields of alanine on a small scale[2]. The Hofmann rearrangement of 2-isobutyl-acetoacetamide has been used to prepare leucine[3], and the method is applicable to other amino acids.

2.2.2 Asymmetric methods for α-amino acids

Since syntheses are usually aimed at optically active products there has been considerable interest in methods which incorporate the asymmetry during the synthesis and so remove the need for a wasteful and often tedious resolution. One method which has received much study involves the use of optically-active catalysts for hydrogenation of 2-acetamido-acrylic acids. The modification of the usual metal heterogeneous catalysts with L-amino acids or their polymers has given only poor results as far as asymmetry is concerned but a successful homogeneous catalyst has been found in the rhodium complex of the diphosphine ligand (2)[4]. When this complex was used as a catalyst in the hydrogenation of 2-acetamidocinnamic acid a 72% optical yield of *N*-acetyl-D-phenylalanine was obtained. Another approach depending upon reduction utilises the induced asymmetry of the C=N bond in the Schiff's base formed between an optically-active amine and an α-keto acid; the earlier work in this area has been reviewed[5]. The use of α-alkylbenzylamines or phenylglycine as the amine allows the alkyl groups to be removed from the nitrogen by hydrogenolysis so that the free, rather than *N*-alkylamino acid is the product. Optimal results (60–80% optical yield) for the preparation of alanine from pyruvic acid were obtained when α-alkylbenzylamines were used in apolar solvents[6], but the reaction was very temperature sensitive[7]. Unfortunately the degree of asymmetry incorporated decreased with increase in size of the side chain so that application of the method to aromatic amino acids does not appear promising. A model has been developed which allows selection of the particular benzylamine enantiomer required to produce a particular enantiomer of an amino acid[8].

Two other methods based on addition to Schiff's bases can be used. Substantially pure optical isomers of leucine and other amino acids have been obtained via the addition of hydrogen cyanide to the Schiff's base formed between an aldehyde and optically-active benzylamine, followed by hydrolysis and hydrogenolysis[9]. The D-isomer of α-methylbenzylamine gave D-amino acids. The addition of Grignard reagents to the Schiff's base formed between the menthyl ester of glyoxalic acid and an optically-active benzylamine has been studied[10]. The reagents derived from small alkyl groups were useless because they alkylated the nitrogen rather than the carbon of the C=N bond, but t-butyl or alkyl Grignards were successful. With organocadmium compounds *C*-alkylation occurred exclusively and optical yields up to 45% could be obtained. The complex formed between α-amino-α-methyl malonic acid and D-*cis*-α-dichloro-L,L-α,α′-dimethyltetraethylenetetraminocobalt(III) (3) decarboxylated stereospecifically to give the complex of L-alanine, but the amino acid was racemised by the procedures used to liberate it from the complex[11].

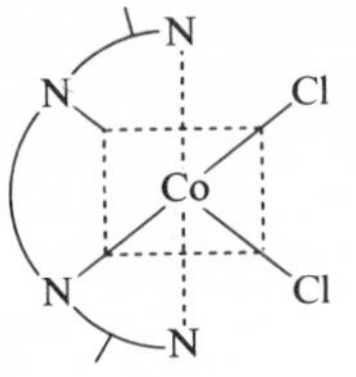

(3)

Most of the methods above suffer from the defect that the asymmetric inducer (e.g. the alkylbenzylamine) is lost during the synthesis and cannot be recovered; thus freedom from having to resolve the amino acid has only been achieved at the expense of continuously preparing and resolving the inducer. An ingenious synthesis has been developed by Corey which allows the inducer to be recovered and recycled[12]. The key reagent was the *N*-aminoindolenine (4) which reacted with an α-keto-acid ester to give the hydrazone (5). Cyclisation of this with sodium methoxide in benzene gave a lactone (6). The C=N bond of the lactone was reduced with aluminium amalgam and the product hydrogenolysed to give the amino acid ester (7). This ester on hydrolysis gave the indolenine (8), which could be recycled, and the free amino acid. D-Alanine prepared by this method had an optical purity of 80% which could be raised to 96% by using the modified reagent (9)[13]. The method works because the methylene in the lactone ring in (6) blocks the approach of the reducing agent to one side of the double bond; introduction of a methyl group at this position makes the blocking still more efficient.

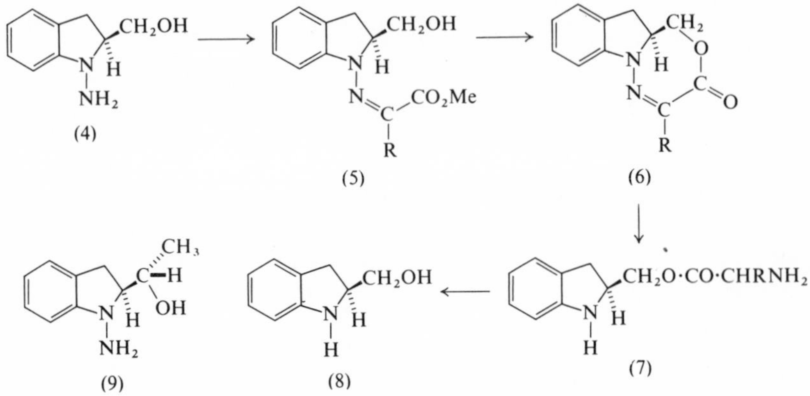

2.2.3 Resolution

The most significant advance of the period, so far as resolution was concerned, was the application of gas–liquid chromatography, but since this is not yet a preparative technique it will be covered in the analytical section of this review. Fractional crystallisation of a salt of an amino acid with an optically-active acid or base remains the preferred resolution method but the use of enzymes, particularly acylases, to hydrolyse preferentially a derivative of the L-isomer is becoming increasingly popular.

Some investigations have been made into the possibilities of separating racemates on ion-exchange resins containing asymmetric centres. The most successful of these used a copolymer of ethyl *N*-acryloyl-L-pyroglutamate and divinylbenzene as the resin and resolved basic amino acids up to 90% optical purity[14]. Simple attachment of amino acids to a chlorosulphonated polystyrene gave a material capable of separating DL-phenylglycine to 72% optical purity[15].

An interesting approach to resolution in which one enantiomer is converted to the other has been reported[16]. DL-t-Leucine was converted to its 2-trifluoromethyl-3-oxazolinone which reacted with dimethyl L-glutamate to give dimethyl *N*-trifluoroacetyl-L-t-leucine-L-glutamate, 60% of the original racemate being converted to L-t-leucine in the peptide.

2.2.4 Syntheses of particular types of amino acid

Syntheses will only be covered in this section if the method used was novel, or if the compound synthesised is interesting.

2.2.4.1 Protein amino acids

A synthesis of threonine by the base-catalysed reaction of ethanolamine with acetaldehyde in the presence of copper oxide has been described[17]. After initial oxidation of the ethanolamine to glycine, two molecules of acetaldehyde condensed with the copper–glycine complex to form the oxazolidine (10) which was isolated and the structure determined by X-ray crystallography. Hydrolysis of this intermediate gave threonine and *allo*-threonine in 6:4 ratio.

Two methods of converting serine to cysteine have appeared, but in both cases optically-active serine gives racemic cysteine. The reaction of ethyl serinate hydrochloride with thionyl chloride gave the β-chloroalaninate hydrochloride, which, after protection of the nitrogen, was treated with thioacetic acid; hydrolysis gave cysteine[18]. Alternatively the direct reaction

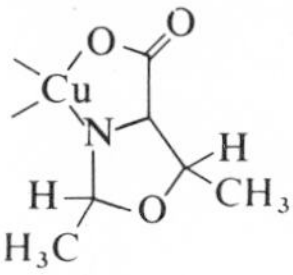

(10) 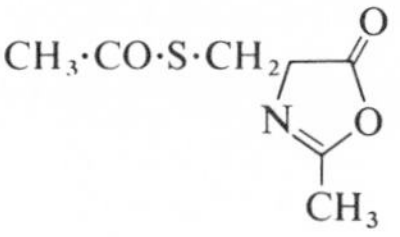 (11) 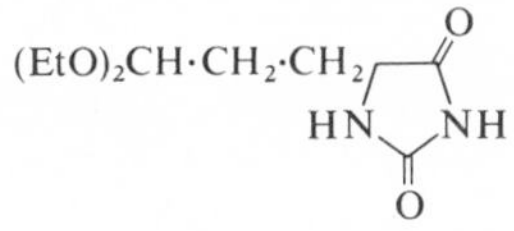 (12) 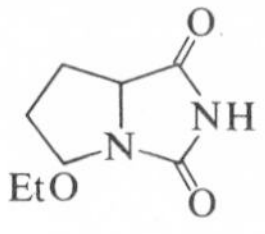 (13)

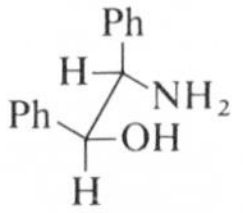

(14) 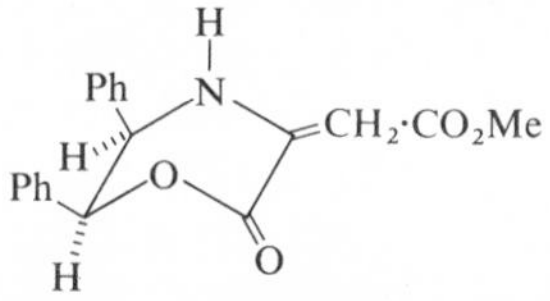 (15)

of serine with thioacetic acid and acetic anhydride gave the azlactone (11) which could be hydrolysed to cysteine[19]. A preparation of D-cysteine involved the addition of benzyl mercaptan to 2-acetamidoacrylic acid, the racemic mixture was resolved by reacting the *N*-acetyl-*S*-benzyl cysteine with aniline in the presence of papain; the L-isomer formed the anilide and could be removed[20].

Aromatic acids have been synthesised by reaction of the appropriate diazonium salt with acrylic acid derivatives in the presence of cuprous salts, giving an α-halogenoacid which was treated with ammonia to obtain the amino acid. Phenylalanine could be obtained in 50% yield from aniline by this method[21]. L-Phenylalanine has been prepared in 55% yield from the more commonly available L-tyrosine by hydrogenation of the *O*-tosyl derivative over Raney nickel. An interesting synthesis of L-tryptophan from acrolein used an Oxo reaction to prepare 4,4-diethoxybutyraldehyde which was condensed with hydantoin to form the hydantoin (12). This cyclised in solution to the *N*-alkylated hydantoin (13) which was used as a substrate for a Fischer indole synthesis. Tryptophan was obtained in 93% yield from (13)[22].

One optical isomer of 1,2-diphenylethanolamine (14) has been used as the starting material in a synthesis of L-aspartic acid[23]. Reaction with dimethyl acetylenedicarboxylate gave the lactone (15) which gave optically pure γ-methyl L-aspartate after catalytic reduction of the C=C bond over Raney nickel and hydrogenolysis. An industrial preparation of L-glutamic acid from acrylonitrile has been described in great detail and should prove useful as a means of supplementing the supply of this commercially useful amino acid[24].

2.2.4.2 *Other naturally-occurring α-amino acids*

The bacterial peptide component, α-amino-β-phenylbutyric acid, has been synthesised by a method which allowed determination of the relative configuration of the α- and β-carbons in the naturally occurring acid[25]. The good taste of mushrooms is imparted by tricholomic acid (16) which is also highly toxic to flies. Syntheses and resolution of both *threo* and *erythro*

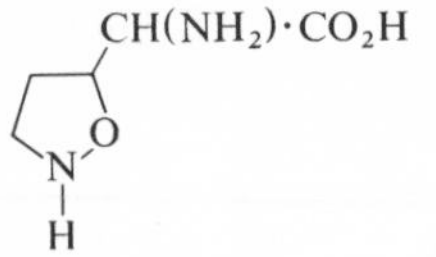

$CH(NH_2)\cdot CO_2H$ (isoxazolidinone ring, N–H)

(16)

$(OH)CH_2\cdot CH_2\cdot CMe_2\cdot S\cdot CH_2\cdot CH(NH_2)\cdot CO_2H$

(17)

$NH_2CH\cdot(CO_2H)\cdot CHMe\cdot S\cdot CH_2\cdot CH(NH_2)\cdot CO_2H$

(18)

isomers has shown that both biological activities are unique to the natural L-*erythro* isomer[26]. Two natural sulphur-containing amino acids have been synthesised by addition of thiols to double bonds. L-Felinine (17) from cat urine, was prepared by addition of cysteine to 2-methyl-4-hydroxybut-1-ene[27], and β-methyllanthionine (18), a bacterial peptide component, by conversion of DL-threonine to DL-β-methylcysteine which was added to 2-acetamidoacrylic acid[28].

2.2.4.3 *Aliphatic and alicyclic α-amino acids*

A new synthesis of 5-acylhydantoins, the precursors of β-keto-α-amino acids, has been developed which uses α-diketones as starting materials; modification of the β-keto group at the hydantoin stage would allow preparation of a variety of β-substituted amino acids[29]. Several aza-analogues of amino acids have been prepared as potential biological antagonists of the natural form; 4-aza-L-isoleucine was the only successful compound in a series of leucine and valine analogues[30], but 4-aza-L-lysine inhibited incorporation of lysine into RNA[31]. Other approaches to amino acid antagonists have involved fluorine substitution. If carboxyl groups are treated with sulphur tetrafluoride they are transformed into trifluoromethyl residues, and this method has been used to prepare ω,ω,ω-trifluoro-amino acids by treatment of the appropriate hydantoin[32]. In a preparation of β-trifluoromethylalanine, ethyl diazoacetate was treated with trifluoroacetic anhydride giving the β-keto-ester (19) which was reacted photochemically with acetonitrile to give the oxazole (20); catalytic reduction gave ethyl *N*-acetyl-β-trifluoromethylalaninate[33]. The method is applicable to the introduction of other β-perfluoroalkyl groups but β,β,β-trifluoroalanine must be prepared

$CF_3 \cdot CO \cdot C(N_2) \cdot CO_2Et$

(19)

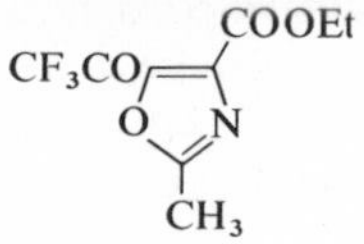

(20)

by a direct method from trifluoroacetaldehyde[34]. The conditions have been determined which optimise the yield of β-chloroalanine by direct chlorination of alanine[35].

A number of alicyclic amino acids have been prepared, often as possible phenylalanine analogues. The Birch reduction of phenylalanine itself gave 2-(1,4-cyclohexadienyl)alanine which was a successful antagonist of the parent acid[36]. A number of 2-amino-2-carboxybicycloheptanes have been prepared by the Diels–Alder reaction between cyclopentadiene and 2-nitroacrylic acids[37].

2.2.4.4 *Hydroxy substituted aliphatic α-amino acids*

The anion of ethyl *N,N*-bis(trimethylsilyl)glycinate reacted with non-enolisable aldehydes to give good yields of hydroxy acids, though ketones did not react under these conditions[38]. Base treatment of the copper complex of the glycine–pyruvate Schiff base and reaction with aldehydes gave complexes of β-hydroxy-α-amino acids from which the free acid could be liberated by precipitating the copper as sulphide. Because only mild bases were necessary to form the anion from the complex, a very wide variety of aldehydes could be used[39]. A novel preparation of β-hydroxyvaline, which is applicable elsewhere, used the reaction of *p*-nitrobenzenesulphonoxyurethane with ethyl β,β-dimethylacrylate in the presence of base to form the aziridine (21); subsequent acetolysis and hydrolysis gave the free amino acid[40].

Several syntheses of β-hydroxy-aspartic acid have appeared. In one of

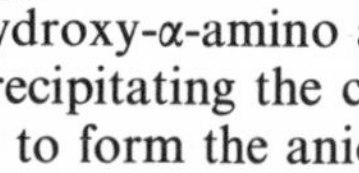

(21)

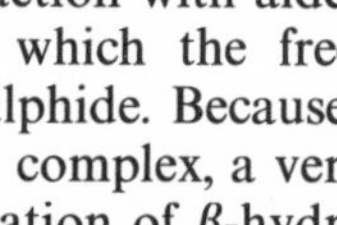

(22)

them ammonolysis of *cis*-epoxysuccinic acid gave *threo*-hydroxy-DL-aspartic acid while the *trans*-isomer gave a 2:1 mixture of *erythro* and *threo* isomers which could be separated by fractional crystallisation[41]. Alternatively β-furylserine has been prepared from glycine and furfural and treated with phosgene to form the cyclic urethane (22) which on oxidation with permanganate and hydrolysis gave the hydroxyaspartate[42].

2.2.4.5 α-Amino acids containing sulphur or selenium

A general synthesis has been reported for α-methyl substituted acids in which the hydantoin was formed from the addition product of benzylmercaptan and methylvinylketone. The method worked equally well with the selenium analogues and modifications to the α,β-unsaturated ketone allowed other acids to be prepared[43]. L-Lanthionine (23), which is formed from cystine residues when wool is treated with alkali, has been prepared by the selective desulphurisation of protected cystine with tris(diethylamino)phosphine[44]. The condensation of glyoxalic acid with cysteine gave the interesting cyclic

$S(CH_2{\cdot}CH(NH_2){\cdot}CO_2H)_2$

(23)

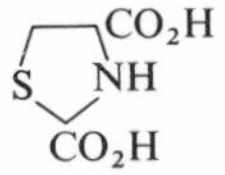

(24) [ring: S, NH; substituents CO_2H, CO_2H]

(25) [ring: S, S; $H_2N{\cdot}CH{\cdot}CO_2H$]

iminodiacid (24) which was an efficient radio-protective agent[45]; no activity was found for ω-dithiolanyl amino acids (25) which were prepared by a modified amidomalonate method[46]. The two sulphur enantiomers of *cis*-*S*-(β-styryl)-L-cysteine-*S*-oxides were prepared by oxidation of the addition product of L-cysteine and diphenylacetylene with a view to using them as possible substrates for onion enzymes, but no lachrymators were produced from them on testing[47].

2.2.4.6 Cyclic α-imino acids

3,4-Dehydroproline, a proline antagonist, has been prepared in good yield by reduction of pyrrole-2-carboxylic acid with hypophosphorous acid[48]. The L-enantiomer reacted with carbene (from diazomethane and cuprous chloride) to give 2,3-*cis*- and 2,3-*trans*-3,4-methylene-L-proline. The isomers were easily separated and the *cis* isomer was identical with a natural amino acid found in horse chestnuts[49]. Treatment of 3,4-dehydroproline with osmium tetroxide gave only 2,3-*trans*-3,4-*cis*-3,4-dehydroxyproline but permanganate gave both the 2,3-*trans*- and 2,3-*cis*-isomers[50]. A number of 4-halogeno-, 4-mercapto- and 4-amino-prolines have been prepared from 4-hydroxy-proline.

2.2.4.7 α-Amino acids containing aromatic residues

These compounds show a wide range of biological activity and, consequently, the majority of new amino acids synthesised fall into this class. The simplest activity is that of a competitive inhibitor of the corresponding protein amino acid and many compounds of this type have been prepared; as an example β-(2-fluoropyridyl)alanines were found to be efficient antagonists of phenylalanine in *E. coli*[51]. Alternatively the amino acid is used as a carrier to get useful biological activity into a particular part of the body; this technique is represented by the synthesis of *o*-boronyl-phenylalanine as a means of transporting a neutron capture agent for cancer therapy[52].

Three unusual syntheses have been reported within this class. In a preparation of 2-phenylaspartic acids the ethyl esters of *N*-chloroacetyl-2-

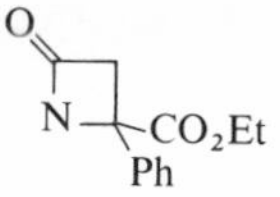

(26) $PhCHMe \cdot CH_2 \cdot O \cdot CONH_3$ (27) ⟶ (28)

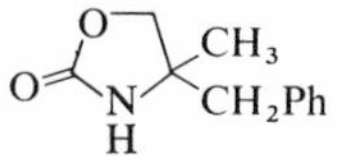

phenylglycines were cyclised in base to the corresponding β-lactams (26) which were easily hydrolysed to the acids[53]. A modification of the amidomalonate method has been used to prepare fluorotryptophan derivatives. Alkylation of formamidomalonate with piperidine and formaldehyde gave the Mannich base which reacted with fluoroindoles to give the tryptophan derivative after hydrolysis[54]. The method is capable of adaptation to a wide range of substituted tryptophans. A stereospecific photochemical synthesis[55] of *R*-α-methylphenylalanine began by reducing *S*-2-methyl-3-phenylpropionic acid to the corresponding alcohol, which was converted to the azidoformate derivative (27) which on irradiation gave the optically-active 2-oxazolidinone (28). The oxazolidinone was hydrolysed to the amino-alcohol and, with suitable protection, oxidised to the α-methylphenylalanine.

Several halogenophenylalanines have been investigated as hydroxylase inhibitors[56]. The highest activity *in vitro* was shown by 3-iodophenylalanine; the activity against tyrosine hydroxylase, but not that against phenylalanine hydroxylase, was increased by α-methylation of the amino acid. Unfortunately 3-iodophenylalanine is deiodinated *in vivo* rendering it useless as a phar-

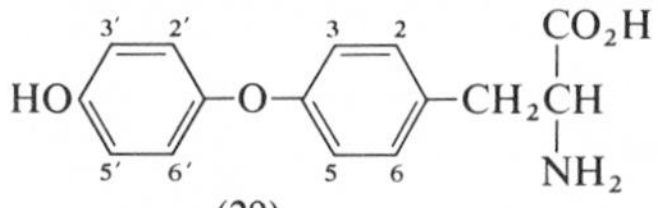

(29) 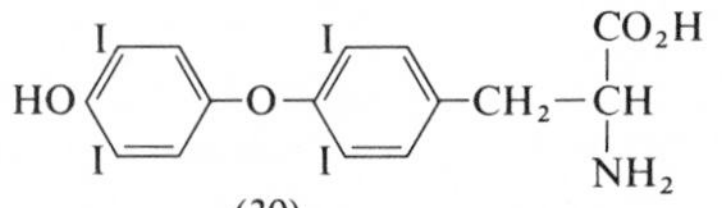 (30)

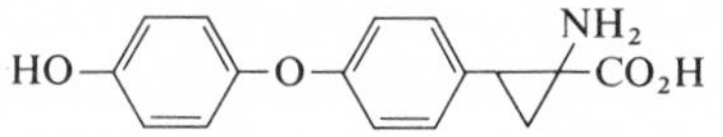

(31) 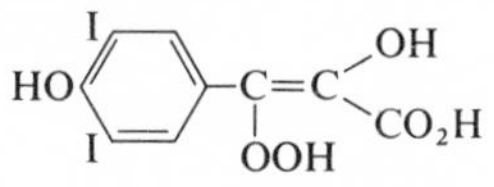 (32)

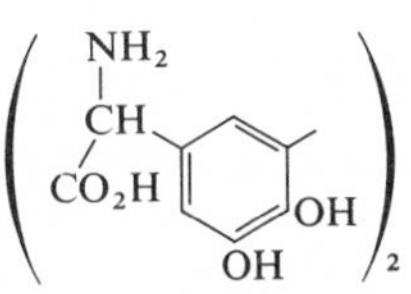

(33) 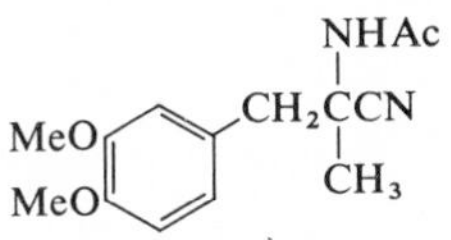 (34)

maceutical. In view of this a number of 3-alkyl derivatives have been tested but only the t-butyl compound showed even slight activity[56].

An area of intense activity has been that involving thyronine (29) and thyroxine (30); the normal method of preparing such compounds is by reaction of a di-(*p*-anisyl)iodonium chloride with an amino and carboxyl protected tyrosine derivative. Knowledge of the anti-thyroxine properties of 3,3′-di-iodo- and 3,3′5′-tri-iodo-thyronine has led to the synthesis of a number of other halogen derivatives but all had little activity. Other approaches have involved replacement of the 3′-iodine by an isopropyl group, a substitution which is known to give active compounds, and one compound with a potency greater than thyroxine has been found in 3,5-dibromo-3′-isopropyl-L-thyronine[57]. The insertion of isopropyl at other positions on the ring gives inactive compounds. Cyclopropylogues of thyronine (31) have been prepared by addition of carbenes to the exocyclic double bonds in the condensation products between aldehydes and azlactones; the 3′,5′-dibromo-3,5-di-iodo-derivative showed weak thyromimetic activity[58]. Thyroxine itself has been prepared in 18% yield by photo-oxidation of 4-hydroxy-3,5-di-iodophenyl-pyruvic acid to the hydroperoxide (32) followed by addition of 3,5-di-iodo-L-tyrosine in buffer[59].

Following the discovery that L-(3,4-dihydroxyphenyl)alanine (L-dopa) could be used in the treatment of Parkinsonism a number of preparations by standard methods were reported and patented, and some analogues were also synthesised. In an investigation of methyl substituted 2-hydroxyphenyl-alanines it was found that the 3-, 4- and 6-methyl derivatives would act as substrates for mammalian dopa decarboxylase but that the 5-methyl derivative was an efficient inhibitor[60]. A dopa dimer (33) has been synthesised from the biphenyldialdehyde by an azlactone method. The compound is as good a substrate for tyrosinase as L-dopa itself and is therefore believed to be an intermediate in the oxidation of dopa to melanin[61]. Several syntheses by standard methods of 2,4,5-trihydroxyphenylalanine, a centrally-active norepinephrine depleting agent, and of L-α-methyl-dopa, a blood pressure depressant, have been reported. One of the latter syntheses involved an interesting resolution step in that the nitrile intermediate (34) was resolved by selective crystallisation from DMSO, the unwanted enantiomer being racemised with sodium cyanide and recycled[62].

2.2.4.8 *Amino acids without α-amino groups*

An ingenious industrial preparation of β-alanine involved polymerisation of acrylamide followed by hydrolysis[63]. Substituted β-alanines have been prepared in high yield by heating pyrazolines with potassium hydroxide solution under pressure[64]. A possible analogue of cysteic acid, α-sulpho-β-alanine, has been prepared in 97% yield by direct treatment of β-alanine with oleum. The reaction probably involves the addition of sulphur trioxide to the enol form of the acid[65]. An interesting synthesis of ω-amino acids from macrocyclic ketones has been reported[66]. The ketone was converted to the tosyl derivative of its oxime which underwent a Beckmann rearrangement to the imino compound (35). This reacted with the enamine derived from cyclohexanone and morpholine to give the enamine (36) which hydrolysed in water to the cyclohexanone (37). Treatment of this with hydroxide ion caused ring opening to an ω-amino-7-keto-acid (38) which could be

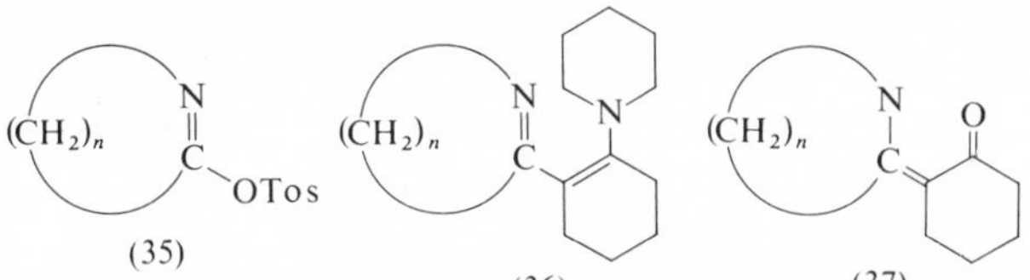

(35) (36) (37)

$NH_2(CH_2)_n \cdot CO \cdot (CH_2)_5CO_2H$

(38)

reduced to the required amino acid. The preparation worked well with ketones with ring sizes of from 6 to 12 carbon atoms. An examination of the distance between —COOH and $—NH_2$ groups in known inhibitors of fibrinolysis led to the development of rigid bicyclooctyl derivatives as inhibitors, the most powerful being 4-aminomethylbicyclo[2.2.2]octane-1-carboxylic acid[67].

2.2.4.9 N*-substituted amino acids*

A method which was used to prepare α-hydrazinohistidine stereospecifically from histidine itself could be applied to other amino acids[68]. The chloro acid was prepared, in 60 % yield, by treating histidine with nitrous acid and hydrochloric acid; the net reaction went with retention of configuration. The reaction of the chloro-acid with hydrazine had the kinetics expected for an S_N2 process and inversion therefore occurred at the last stage.

N-Methylamino acids have been prepared by treating an *N*-acylamino acid ester with methyl iodide and sodium hydride. Under suitable conditions the reaction was racemisation-free and no unmethylated material remained[69]. *N,N*-dimethylamino acids have been prepared by the catalytic reductive condensation of formaldehyde with the free acid[70]. The method worked well for most amino acids but aspartic acid underwent a Cope rearrangement instead of *N*-dimethylation. The dimethylamino acids were oxidised to the amine oxides which were decomposed by tosyl chloride in pyridine to aldehyde, carbon dioxide and dimethylamine[70].

The discovery that melphalan (L-4-bis(2-chloroethyl)aminophenylalanine) and its racemate sarcolysine were effective against some carcinomas has stimulated the synthesis of compounds containing bis(2-chloroethyl)amino- (= M) groups. Compounds of this type are believed to act as alkylating agents within the cell. The biological and chemical results up to the end of 1969 have been reviewed[71]. The M-group is usually introduced by treatment of a primary amine with ethylene oxide to give the bis(2-hydroxyethyl)amino compound, followed by reaction with thionyl chloride. Many compounds have been synthesised which contain the M-group within the amino acid unit and even the comparatively simple modification of sarcolysine itself by introducing another alkylating group in the form of the α-*N*-iodoacetyl derivative produced a material with twice the anti-carcinogenic activity of the parent compound[72]. A more normal way of attaching the M-group is to acylate the amino acid with an acid which already contains M; literally hundreds of derivatives of this type have been prepared. One of the more successful biologically was *N*-(*p*-M-phenylacetyl)glutamic acid[73].

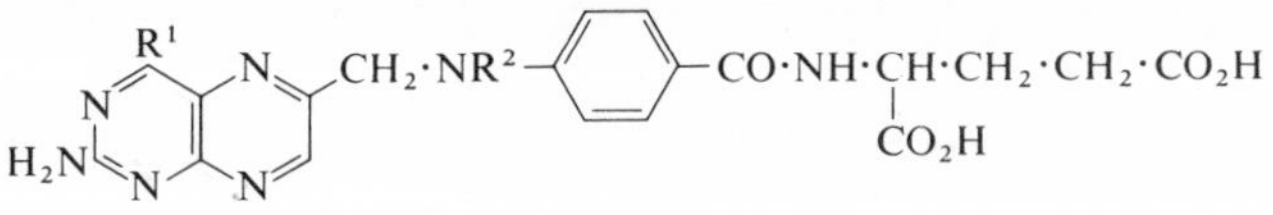

(39) $R^1 = NH_2$, $R^2 = Me$
(40) $R^1 = OH$, $R^2 = H$

Another approach to cancer therapy stems from the discovery that the folic acid analogue methotrexate (39) was effective against leukaemia. The replacement of the pteridine ring in folic acid (40) by diaminopyridines gave compounds with some activity as antagonists, but their replacement by purines destroyed the activity[74]. Compounds with other heterocyclic molecules in place of the pteridine system have been synthesised for testing. The replacement of the central phenyl ring in folic acid by pyridine gives biologically-inactive material. Modifications, such as fluoro-substitution, have also been made to the terminal glutamic acid residue in the hope of obtaining an efficient anticarcinogen.

2.2.5 Isotopically-labelled amino acids

A number of amino acids with a ^{14}C label in the carboxyl group have been synthesised by a Strecker synthesis using $K^{14}CN$, and for those with a label in the α-position [2-^{14}C]acetamidomalonate was usually used as the starting material. Carbon labels at other positions were introduced into the aldehyde or alkyl halide precursors of normal synthetic routes.

Tritiation or deuteration has been accomplished in several ways. Exposure of amino acids to tritium gas resulted in labelling at the α-position but in most cases this was accompanied by racemisation though D- and L-phthaloylglutamic acid in the solid state could be labelled stereospecifically[75]. Glutamine and asparagine have been labelled by reflux in tritiated ammonia. The incorporated tritium was not exchangeable in water and the method could presumably be used for other amino acids[76]. A more obvious method of incorporating a hydrogen label at the α-position is to decarboxylate the aminomalonate precursor of the acid in water containing the appropriate isotope; in pure heavy water 100% incoporation of deuterium is possible[77]. A transaminase has been used to incorporate tritium at the α-carbon of glutamic acid[78]. Enzymes have also been used to synthesise ^{15}N labelled compounds[79]. [^{15}N]Aspartic acid was prepared by the aspartase catalysed addition of $^{15}NH_3$ to fumaric acid. The labelled aspartate could then be used

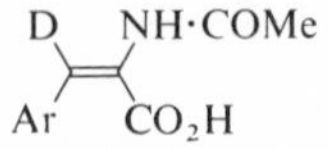

(41)

with a transaminase to prepare other [^{15}N]amino acids. Another direct way into these compounds is offered by the glutamate dehydrogenase catalysed exchange between glutamic acid and $^{15}NH_3$ [79]. Stereospecific labelling has been introduced into the β-methylene of phenylalanine by commencing an azlactone synthesis with the deuterated aldehyde[80]. The azlactone was hydrolysed to an acylaminocinnamic acid (41) which was catalytically hydrogenated with wholly *cis* addition. Subsequent resolution of the racemate gave the stereospecifically labelled compound.

2.2.6 Prebiotic synthesis

The investigation of the formation of amino acids under conditions which might have existed on a primitive earth has continued. There must now be little doubt that they could have been formed spontaneously, indeed given the results now obtained it would be a cause for surprise if they had never been formed at all. The action of heat, irradiation, high-energy discharge or shock waves in a mixture of methane, ammonia and water vapour have all produced most of the non-sulphur amino acids in small amounts and results of this type have now been obtained in many laboratories. Sulphur amino acids are immediately formed if ferrous sulphide or hydrogen sulphide is introduced into a vapour mixture of this type which is then irradiated or heated. A number of simple organic compounds such as ammonium fumarate or formamide have given amino acids on heat or irradiation followed by hydrolysis.

2.3 PHYSICAL PROPERTIES OF AMINO ACIDS

2.3.1 Absolute configuration and conformation

The use of protracted chemical interconversions to establish configuration is now rare. These were used in the case of (+)-α-methylphenylglycine and (+)-α-methylserine which were both established as belonging to the *S*-series by correlation with *R*-(−)-isovaline(α-ethylalanine)[81]. In some very simple cases chemical conversion is still the method of choice, for example the *p*-hydroxyphenylglycine derived from actinoidin has been assigned a D-configuration on the basis of its catalytic reduction to the known D-cyclohexylglycine[82]. In general though the possible use of such simple approaches is rare. An increasingly useful method of establishing the configuration of newly isolated natural α-amino acids is the use of stereospecific enzymes. There is a large number of hydrolases, transaminases and oxidases which are specific towards L-amino acids and for the D-isomers observation of oxidation catalysed by D-amino acid oxidase can be taken as proof of configuration. The latter enzyme has also proved useful in establishing the stereochemistry of asymmetrically tritiated glycines. x-Ray crystallography has been used occasionally but the labour involved in determining the absolute configuration by this technique is only worthwhile in very special cases. The sign of the Cotton effect at 225 nm in the optical rotatory dispersion spectrum is generally indicative of conformation but there are exceptions and the technique is not completely reliable, this problem is further discussed below.

The theoretical aspects of the conformation of amino acids in solution and in the solid state have been explored in some detail. In free zwitterionic amino acids it appears that the side chains have little effect on the stable conformation, the controlling factor being the electrostatic interaction between the two charged groups; detailed calculations which support this have been made for a number of amino acids[83]. *N*-Acetylamino acid amides have been studied as simple models of protein structures; the preferred conformation calculated by extended Hückel methods compared favourably with those actually found in peptides[84]. The proline ring has normally been regarded as a rigid structural unit in proteins, but some recent theoretical work on the energetics of its deformation suggests that it is not completely rigid and some small rotations are possible within it[85].

The major experimental methods of determining conformation are x-ray crystallography and n.m.r. spectroscopy and both these are covered in detail below. One interesting method of conformation determination was applied to 1-aminodecalin-1-carboxylic acids[86]. It was expected that equatorial groups would have different pK_a values to axial and that the differences would allow the compound with an equatorial amino group and axial carboxyl to be distinguished from that with the groups reversed. Unfortunately, the acids were not sufficiently soluble in water for reliable pK_a determination to be made and in aqueous acetone the results could not be interpreted. Use was therefore made of the pK_a change on going from 67% v./v. acetone to 25% v./v. acetone; the equatorial positions were assumed more accessible to solvent so that they should change pK_a more than the axial. A comparison of the results for the two isomers allowed their conformations to be established by this technique.

2.3.1.1 Crystallography

The crystal structures of many of the common amino acids, their salts and their hydrates have been determined. As would be expected the crystals show a strongly hydrogen-bonded lattice as well as the ionic interactions. Sometimes peculiarities arise in this hydrogen bond network; in cysteine hydrochloride for example the proximity of the groups suggests an S—H---Cl bond while one of the carboxyl oxygens is not hydrogen bonded at all[87]. Long aliphatic side-chains usually exist in a fairly random conformation, indeed one crystalline form of DL-α-aminobutyric acid contains three rotational isomers about the C_α—C_β bond within the unit cell[88], but in L-ornithine hydrochloride the aliphatic chain was planar and in its most extended form[89]. The β-synthesis and heavy-atom methods of structure determination have been compared in the analysis of L-arginine hydrobromide. β-Synthesis was found to reveal the unknown parts of the structure with greater strength than heavy-atom methods[90].

A sheet structure containing alternate D- and L-residues linked by hydrogen bonds has been found for *N*-acetyl-DL-leucine methylamide. The leucyl side chains are perpendicular to and form a hydrophobic region between the sheets. The crystal cleaved easily along this plane[91]. A survey has been made of proline conformations in published structures of amino acids, peptides and proteins, and it is established that the normal conformation of the proline ring has the α,β and δ carbons and the nitrogen atom coplanar with the γ-carbon puckered out of the plane[92].

2.3.2 Optical rotatory dispersion and circular dichroism

The existence of a strong positive band near 220 nm in the c.d. spectra of L-amino acids has been known for some time and has been used to determine absolute configuration. The c.d. spectra of a number of less common amino acids of known configuration have been measured as a test of the universality of the rule that a positive band indicates L-configuration[93]. In the few cases where the rule would have given the wrong answer, the molecule either contained an interfering chromophore, or else was subject to a severe steric constraint. A sector rule which relates the sign and amplitude of the band to the configuration has now been developed and accounts for some of the deviations[94]. It is assumed that the N—C—CO_2 atoms are coplanar in solution and the molecule is then arranged in the appropriate sectors with respect to this plane. The sector rule is particularly successful in accounting for the very low positive amplitude observed for proline and for the negative value for L-azetidinecarboxylic acid.

The transition responsible for the 220 nm band has been investigated in detail. An observation that, in addition to this strong positive band, L-amino acids had a weak negative c.d. band at 250 nm in water whose intensity enhanced as the solution was made more alkaline, led Anand and Hargreaves[95], to query the assignment of the 220 nm band to the n–π^* transition. On the normal order of such transitions in aliphatic systems one would expect the 250 nm band to be the n–π^* and the 220 nm band either the π–π^* or n–σ^* transition. This interpretation was opposed by others[96]. Lactic acid derivatives show similar c.d. spectra in which both bands are shifted in the same direction by temperature or by solvent change. In each case the direction observed supports the assignment of both bands to the n–π^* transition. The 250 nm band in amino acids is stronger than the equivalent band in the lactic acids which suggests that interaction between the oxygen or nitrogen lone-pair and the carboxyl may be responsible for producing it. This theory is supported by the disappearance of the band in acidified amino acid solutions and its intensification in basic solution.

The effect of aromatic substitution on the rotations and o.r.d. spectra of L-phenylalanines has been studied[97]. All showed positive Cotton effects at the 260 nm aromatic transition, but other large dispersion effects which were observed seemed to arise from the vicinal effect of the aromatic ring rather than from any induced asymmetry within it. A plot of molar rotation differences (relative to L-phenylalanine) against the Hammett σ was linear providing tyrosine derivatives were excluded. A comparison of the fine vibrational structure visible in the c.d. spectrum of phenylalanine with that visible in the u.v. shows that only non-totally symmetric vibrations of the aromatic ring will modify the c.d. spectrum[98]. A large increase in the total rotation was observed on cooling from room temperature to 77 K indicating the isolation of a particular isomer at the lower temperature. A detailed practical and theoretical study of the effect of conformation on the o.r.d. spectra of *o*-, *m*- and *p*-tyrosines has been made to further the use of o.r.d. in the determination of protein structures. The calculations, which were made using the Kirkwood oscillator theory, adequately explained the experimental results[99].

The rotational properties of the disulphide bond in cystine have also attracted interest because of their relevance to experimental studies on proteins. Studies of cystine in solution are difficult to interpret because of changes in the conformation about the S—S bond so that model compounds or cystine crystals have been used. Cyclic disulphides containing an asymmetric group showed similar solution spectra to that of crystalline cystine and the sign and position of the c.d. band seemed to reflect the spatial structure within the molecule, with a major contribution from the screw sense of the —S—S— bond[100].

Since one of the major uses of o.r.d. in amino acid chemistry lies in the determination of absolute configuration, attempts have been made to overcome the slight doubts which attach to the use of the 220 nm band alone by finding derivatives which have a well defined relationship between the spectrum and the absolute configuration. A number of *N*-acyl and hydantoin derivatives have their champions but the range of amino acids studied, using any of them, is not yet sufficient for certainty in the assignment of an unknown acid, although they may be of use for confirmation of a result by another method.

2.3.3 Nuclear magnetic resonance spectroscopy

The state of the art of n.m.r. as applied to amino acids and proteins has been critically reviewed and the observed chemical shifts tabulated[101]; the review is indispensable as a source of factual information. Since it appeared, the field has been extended by a study of other nuclei. In contrast to the complex proton spectra obtained with proteins, the ^{13}C spectra are very simple and individual carbon atoms can be picked out. As a prelude to study of larger molecules, the ^{13}C chemical shifts of the carbons have been measured for ^{13}C enriched amino acids which were prepared by growing algae on $^{13}CO_2$[102]. The conformation about the C_α—C_β bond can be measured by using ^{15}N—H vicinal coupling, the relation between dihedral angle and coupling constant has been determined and the method should be useful both for amino acids without an α-proton and for those where the α-proton resonance is masked by others in the ^1H n.m.r. spectrum[103].

In favourable cases, considerable information about solution conformations of amino acids and their derivatives can be obtained from the ^1H n.m.r. spectrum. The variation of the aliphatic proton resonances in the phenylalaninate ion with temperature and concentration indicated that the three classical staggered forms were the major conformations present[104]. Surprisingly, the least-favoured rotamer was the one which would be expected to be the most stable on steric grounds for the isolated anion. Solvation must therefore play a considerable part in the determination of conformation in aqueous solution. Similar solvent effects were found when the study was extended to tyrosine and histidine. A claim[105] made earlier that the occurrence of extra methyl peaks in DL-threonine and DL-valine arises because both exist in solution in two well defined conformations has been refuted[106]. The extra peaks for threonine arose through contamination by *allo*-threonine, whilst those for valine occur because the adjacent asymmetric centre makes the two methyl groups magnetically non-equivalent. In *N*-α-chloroacyl-valine esters, the separation between the methyl resonances of the isopropyl group increases with the size of the acyl group, indicating that the molecule is gradually frozen into a particular geometry by the applied steric constraint on the nitrogen[107].

Enantiomers of an α-amino acid methyl ester show different chemical shifts, particularly for the ester methyl and α-proton, if optically-active 2,2,2-trifluoro-1-phenylethanol is used as the solvent[108]. The small differences observed arise through the transient formation of complexes in the solution and structures have been suggested for these. The method is certainly very useful for determining optical purity, but a claim that it is a reliable method of determining absolute configuration must be proved by the investigation of more amino acids before it can be used for this purpose.

2.3.4 Other physical studies

Raman spectroscopy has been used to determine the solution conformation of α-aminobutyric acid; crystals containing known conformers were used as models[109]. The preferred conformation about the C_α—C_β bond was that with the most steric interaction in the isolated zwitterion with the methyl group lying between the ammonium and carboxylate groups; as in the case of phenylalanine referred to above solvation effects seem to account for this. Various considerations have suggested that a glycine molecule in isolation would exist as NH_2—CH_2CO_2H, rather than the zwitterion, and this has now been proved by sublimation of the compound into an argon matrix[110]. Infrared spectroscopy clearly demonstrated the existence of the neutral species.

The excited states of tryptophan and tyrosine continue to attract interest[111]. Quenching of the excited state occurs by simple interaction with the solvent but no exciplex seems to be involved. At temperatures where reorientation of the solvent is impossible, there is no temperature dependence of phosphorescence or fluorescence yield, but as soon as reorientation becomes possible, phosphorescence quenching begins, followed at slightly higher temperatures by fluorescence quenching. The quenching of tyrosine phosphorescence is reduced in deuterated solvents but the fluorescence quenching is unchanged. For tryptophan deuterated solvents *do* reduce fluorescence quenching.

2.4 ANALYSIS

The predominant methods used for analysis of amino acid mixtures were undoubtedly automatic ion-exchange chromatography, paper chromatography and electrophoresis. Many detailed modifications of these techniques have appeared aimed either at improving particular separations or at decreasing the time taken for an analysis. One problem which arises for workers with bacterial peptides is to distinguish D-amino acids when they are present. A simple method for this has now appeared[112] which involves treatment of the amino acid mixture with an L-amino acid carboxyanhydride; the resulting dipeptides can be separated by ion-exchange chromatography with sufficient resolution to detect 1 part of D-acid in the presence of 2000 parts of L-acid.

2.4.1 Gas–liquid chromatography

Development of this technique has continued but despite its obvious advantage of speed it is not used by most amino acid chemists. In part this is due to the wide distribution of automatic ion-exchange apparatus and simple human unwillingness to move from a known and proved method into the unknown. Undoubtedly though a major factor is the necessity for preparing derivatives and the difficulty of doing this reproducibly.

The first class of derivatives which has been used is the *N*-trifluoroacetylamino acid esters; n-butyl or n-amyl esters appear to be best as far as separation is concerned but preparation of these involves three steps. Direct esterification of amino acids with n-butanol or n-amyl alcohol is difficult because cysteine, threonine and histidine are insoluble in the alcohols. The methyl esters must therefore be made first and the butyl or amyl esters prepared by an alcoholysis reaction[113]. Trifluoroacetylation of the ester can be achieved, either with the anhydride or with 1,1,1-trichloro-3,3,3-trifluoroacetone, but if the former is used the mixture must have inorganic metal ions removed since a number of these interfere with the acylation process[114].

The other class of derivatives is prepared by persilylating the acid hydrochlorides using bis(trimethylsilyl)trifluoroacetamide or a related compound. Whilst only one reaction is involved, the hydrochlorides must be dry and care must be taken to see that all possible sites are silylated, partial silylation confuses the chromatogram. An advantage of these derivatives over the trifluoroacetyl ones is that arginine is made volatile. Silylation has to be used to make *N*-trifluoroacetyl-arginine butyl ester volatile. The problem with this technique is in getting the dry amino acid hydrochlorides in the first place.

Both types of derivatives are sensitive to water even when prepared, so care must be taken in handling them. Whilst the actual analysis is much quicker than ion-exchange chromatography, the time spent in preparing the derivatives by hand probably makes the total times for the analysis of a given peptide hydrolysate very similar. The method would become more popular for routine analysis if an automatic derivitiser, on the same lines as a protein sequenator, could be developed.

One very useful application of g.l.c. exists in its ability to separate enantiomers if optically-active stationary phases are employed. The commonest phases employed are *N*-trifluoroacetyl-L-valyl-L-valine cyclohexyl ester[115] and the equivalent derivative of L-phenylalanyl-L-leucine[116]. *N*-Trifluoroacetylamino acid isopropyl esters or other perfluoroacyl esters give excellent resolution of enantiomers on these columns. Long wall-coated columns are necessary for optimal separation and a satisfactory preparative technique has yet to appear.

2.4.2 Other methods

A study of the ^{19}F n.m.r. spectra of *N*-trifluoroacetylamino acids reveals that each protein amino acid imparts a different ^{19}F chemical shift to the trifluoroacetyl group[117]. This could be a useful method of analysis in certain cases but the sensitivity is low and considerably larger quantities than for chromatographic methods are required.

Enzyme methods for specific amino acids are becoming more popular. Acetyl-coenzyme A–D-amino acid *N*-acetyl transferase has been used to determine D-amino acids in the presence of their L-isomers[118]. The acetyl-coenzyme A was prepared *in situ* from acetyl phosphate and its consumption gave the total quantity of D-amino acids present.

2.5 REACTIONS OF AMINO ACIDS AND THEIR SIMPLE DERIVATIVES

2.5.1 Proton transfer and protonation

The rates of proton transfer between α-amino acids and water were measured by ultrasonic relaxation methods[119], and the rate of loss of protons from the ammonium group of the zwitterion was essentially constant at 2×10^{10} s^{-1} for all the acids studied. The differences in pK_A of this group in different acids derive entirely from different rates of protonation of the neutral amino groups. In superacids, every possible lone pair in the molecule is protonated; this does not induce any dehydration of α- or β-amino acids to oxocarbonium ion but dehydration does occur to some extent with γ- and completely with δ-amino acids[120].

The α-proton in α-amino acids exchanges with deuterium in deuterated acetic acid with simultaneous racemisation. The presence of a free amino group greatly facilitates racemisation but esterification of the carboxyl has little effect upon the rate. The rates observed for individual amino acids can be predicted by assuming that acetate ion acts as a base and that the side chains have their usual electronic effects. In alkaline heavy water, a free amino group or carboxyl group retards the rate of racemisation[121]. The rate of α-proton exchange of L-valine and L-alanine as the D- and L-cobalt complexes $[Co(en)_2\text{amino acid}]^{2+}$ are first order in both the complex and hydroxide ion, indicating a carbanion intermediate[122]. The D-isomer of the complex shows a preference for D- rather than L-valine at equilibrium (1.7:1 ratio).

2.5.2 Oxidation and reduction

The oxidation of an α-amino acid with silver(II) picolinate results in decarboxylation and deamination to the noraldehyde. Silver(II) oxide may also be used but in this case the aldehyde is further oxidised to the acid[123]. The reaction provides a facile method of degrading amino acids; the picolinate converts alanine to acetaldehyde in 30 min at 60 °C. Sarcolysine and other molecules containing bis-(2-chloroethyl)amino groups are easily oxidised in acidic solution, and radicals can be detected; the suggestion has been made that the anticancer activity of these molecules may be a function of this radical formation rather than their alkylating ability[124].

Cysteine reacts with *N*-methyl-*N*-nitrosotoluene-*p*-sulphonamide to give cystine rapidly and in virtually quantitative yield[125]. The *S*-nitroso derivative of cysteine has been proposed as an intermediate. Diethyldiazocarboxylate will oxidise a number of sulphur amino acids to sulphoxides but a proton donor in the form of a free carboxyl group is necessary for the reaction. Cysteine is oxidised to cystine in a few minutes and to cysteic acid in several days by this method[126]. The kinetics of the reduction of cystine to cysteine by sulphide ion were followed at pH 14 both polarographically and spectrophotometrically[127]. The reaction has two steps and the disulphide anion Cys—SS$^-$ is formed as an intermediate. The exchange reaction between cysteine and disulphides has been analysed in terms of two reactions, one between disulphide and the thiol anion, the other between the disulphide ion RSS$^-$ and the thiol anion[128]. With most disulphides the reaction is reversible but if thiamine is present as part of the disulphide the reverse reaction is blocked by formation of thiazolium ion.

2.5.3 Halogenation

Free radical chlorination of α-amino acids is easier in trifluoroacetic acid than in hydrochloric acid[129], and in time all the protons can be substituted except those at the α-position; glycine is inert under these conditions. The effect of organic solvent upon iodination of tyrosine and 3-iodotyrosine in water has been studied in order to help elucidate the biological reaction[130]. The effect of carrying out the tyrosine reaction in water containing increasing concentrations of methanol is amply explained by standard linear free energy solvent parameters, but in the case of 3-iodotyrosine the rate is not reduced as much as expected. It seems from this that apolar environments within the cell would favour di-iodination over monoiodination. The rate of iodination of thyronine (29) is only one-tenth that of tyrosine but iodination of the inner ring does not affect the reactivity of the outer, whilst iodination of the outer ring slows the rate still further.

2.5.4 Photochemical and radiolytic reactions

The exposure of amino acids to electromagnetic radiation usually results in the formation of radicals. In crystals of aliphatic α-amino acids at 77 K a radical anion is formed which deaminates on warming to 130 K giving the acyl radical $RCH{\cdot}CO_2H$. On further warming hydrogen abstraction occurs from the side chains of adjacent molecules in the crystal and a number of radicals is produced[131]. Similar initial processes occur in alkaline aqueous glasses but at the softening point the aminoacyl anion radical $NH_2\dot{C}R{\cdot}CO_2^-$ is formed by interaction of the acyl radical with amino acid molecules[132].

The aminoacyl anion radical and its protonated forms are also the primary products formed from the amino acid on γ-radiolysis of aqueous solutions; the abstracting species in this case appears to be the hydroxyl radical derived from water[133]. Other radicals detectable in the solution derive from the decarboxylation of the aminoacyl anion radical. The results of such experiments are very similar to those in which the hydroxyl radical is generated chemically from titanous chloride and hydrogen peroxide. Aromatic amino acids undergo photoionisation if irradiated in aqueous glasses[134].

The products finally obtained from photolysis of glycine in water are ammonia, acetic acid, glyoxalic acid and formaldehyde. The rate of their production is significantly increased by the addition of manganous or some other divalent metal ions, but slowed by addition of zinc or aluminium ions[135].

N-Chloroacetyl-*O*-methyl-L-tyrosine undergoes photorearrangement in aqueous solution to give the aza-azulene derivative (42); the reaction is

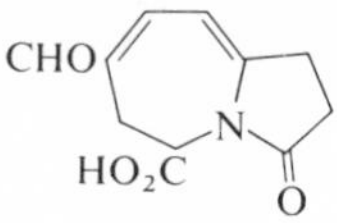

(42) 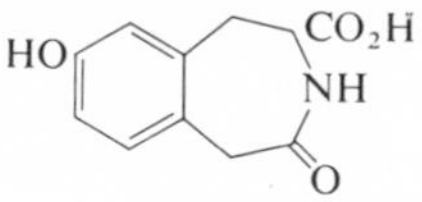 (43)

probably initiated by a homolysis of the C—Cl bond[136] and continues by a complicated radical arrangement. A similar initiation step occurs in the photolysis of *N*-chloroacetyl-tyrosine but here the radical substitutes *para* to the hydroxyl group to give the lactam (43)[137].

2.5.5 Hydrolysis of amino acid derivatives

Hydrolysis of *N*-benzoylglutamic acid in dilute acid solution occurs by intermediate formation of 2-pyrrolidone-5-carboxylic acid, but the hydrolysis of *N*-benzoylaspartic acid, which is faster, has β-benzoylaspartic anhydride as an intermediate rather than the corresponding azetidinone derivative[138].

Protonated amino acid ethyl esters are more rapidly hydrolysed in alkaline solution than the neutral species; rate increases of between 40 and 100 times occur on protonation[139]. These increases can be accounted for by electrostatic effects, and it may well be that many catalyses of amino acid esters by metal complexes are also due to this cause. However, it would be difficult to explain the observed 10^6 rate increase in the hydrolysis of cobalt-complexed isopropyl glycinate when compared with the free ester on this basis, and in this case at least temporary complexing of the ester carboxyl to the metal must occur[140]. An oxygen-isotope study on the hydrolysis of $[Co(en)_2(OH)(GlyOEt)]^{2+}$ ions showed that about half of the reaction can be accounted for by internal attack of the hydroxyl in the complex upon the ester carbonyl[141]. In none of the many different complexes studied was any evidence obtained of permanent coordination of the ester carbonyl group: only the free amino group is coordinated to the metal unless, as in cysteine, there is another possible ligand within the molecule allowing a chelate to be formed. Some evidence for the temporary interaction of the carbonyl with the metal contributing to catalysis is found in the hydrolysis of methyl L-histidinate which is catalysed more effectively by the nickel complex of D-histidine than by that of the enantiomer[142]. Molecular model building suggests that the ester carbonyl is better able to interact with the metal in the D-histidine complex.

2.5.6 Models of biological reactions

An n.m.r. study of the condensation of polyfunctional amino acids with pyridoxal in water to form Schiff's bases indicated that while hydroxy acids behave normally the cysteine base cyclised to a thiazolidine derivative, and that from histidine to a tetrahydropyridine[143]. 4-Trimethylammonium-salicylaldehyde is an excellent catalyst for the racemisation of glutamic acid; in the presence of copper, iron or aluminium ions racemisation yields over 80% can be obtained in 3 h at pH 10 and at 80 °C. When other substituted salicylaldehydes are used, in the absence of metal ions, a linear relationship is found between rate of racemisation and the Hammett σ-constant[144]. The phenolic hydroxyl in the 2-position is essential for reactivity in these compounds. *o*-Nitrosophenol also catalyses racemisation in the presence of metal ions; in this case an azo-compound takes the place of the Schiff's base[145].

Thyroxine (30) can be formed at room temperature by oxygenation of a mixture of 3,5-di-iodotyrosine, sodium glyoxalate and cupric acetate[146]; an initial transamination reaction to the corresponding pyruvic acid is followed by oxidative coupling of this to unreacted 3,5-di-iodotyrosine. Proline is hydroxylated in a mixture of the ferrous complex of EDTA, ascorbic acid and peroxide[147]. The function of the ascorbic acid is to generate hydroxyl radical as it is not essential and can be replaced by more iron.

2.5.7 Other reactions

Many of the general reactions of amino acids involve condensations with carbonyl compounds to produce heterocyclic compounds; out of the many only two of the more novel are given here. α-Amino acids undergo Strecker degradation when heated at a temperature sufficient to undergo decarboxylation with benzil and benzoin, giving as products 2,3,5,6-tetraphenylpyrazine and 2,3,4,5-tetraphenylpyrrole respectively[148]. Glutamic acid condenses with two molecules of diketene to give 3-acetyl-6-methyl(1*H*,3*H*)-pyridine-3,4-dione-1-acetic acid and similar *N*-substituted pyridine derivatives can be obtained from other amino acids. The amino acid esters react simply with diketene to give *N*-acetoacetyl-derivatives[149].

Amino acid amides are difficult to prepare by direct methods, but they can be prepared in good yield by reacting the amine with diethoxymagnesium and adding the resulting complex to a mixture of the amino acid ethyl ester and acetylacetone; hydrolysis of the resulting amide Schiff's base gives the product[150].

Thionyl chloride reacts rapidly with *erythro-N*-acyl-2-aryl serine esters (44) giving a *trans*-oxazoline (45) which opens slowly to give the *erythro*-2-chloro-2-arylalanine; the *threo*-isomer behaves similarly giving a *threo*-

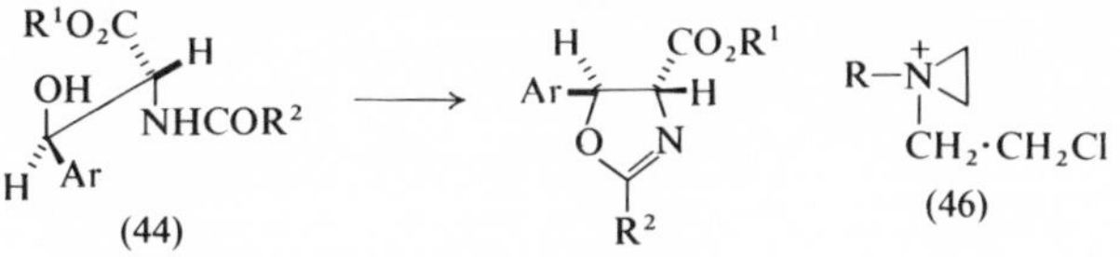

(44) (45) (46)

product if the aryl group is phenyl, but if it is nitrophenyl the reaction stops at the *cis*-oxazoline[151].

A kinetic study[152] using chlorambucil has shown that nucleophilic substitution of bis-(2-chloroethyl)amino compounds does involve the aziridinium ion (46). The attack of poor nucleophiles upon the aziridinium ion is slow enough for the step to be rate determining and second-order kinetics are observed, but with good nucleophiles the reaction increases in rate so that the formation of the ion becomes rate determining and the reaction first order. The change of order of reaction clearly indicates the presence of the aziridinium ion.

References

1. Ruhlmann, K. and Kuhrt, G. (1968). *Angew. Chem. Internat. Ed. Engl.*, **7,** 809
2. Nakahara, A., Nishikawa, S. and Mitani, J. (1967). *Bull. Chem. Soc. Jap.*, **40,** 2212
3. Yamamoto, M. and Oshima, K. (1968). *Japan Pat.*, 68, 12803
4. Kagan, H. B. and Dang, T. P. (1971). *Chem. Commun.*, 481
5. Klabunovskii, E. I. and Levitina, E. S. (1970). *Usp. Khim.*, **39,** 2154
6. Harada, K. and Matsumoto, K. (1967). *J. Org. Chem.*, **32,** 1794; (1968). *J. Org. Chem.*, **33,** 4467
7. Harada, K. and Yoshida, T. (1970). *Chem. Commun.*, 1071
8. Frainnet, E., Braquel, P. and Moulines, F. (1971). *Compt. Rend. Acad. Sci. Ser. C*, **272,** 1435
9. Patel, M. S. and Worsley, M. (1970). *Can. J. Chem.*, **48,** 1881
10. Fiaud, J. C. and Kagan, H. B. (1971). *Tetrahedron Letters*, 1019
11. Asperger, R. G. and Liu, C. F. (1967). *Inorg. Chem.*, **6,** 796
12. Corey, E. J., McCauley, R. J. and Sachdev, H. S. (1970). *J. Amer. Chem. Soc.*, **92,** 2476
13. Corey, E. J., Sachdev, H., Gougoutas, J. Z. and Saenger, W. (1970). *J. Amer. Chem. Soc.*, **92,** 2488
14. Suda, H., Hosono, Y., Hosokawa, Y. and Seto, T. (1970). *Kogyo Kagaku Zasshi*, **73,** 1250
15. Manecke, H. and Lamer, W. (1967). *Naturwissenschaften*, **54,** 647
16. Steglich, W., Frauendorfer, E. and Weygand, F. (1971). *Chem. Ber.*, **104,** 687
17. Maldonado, P., Richaud, C., Aune, J. P. and Metzger, J. (1971). *Bull. Soc. Chim. Fr.*,2933
18. Furuta, T. and Ishimaru, T. (1968). *Nippon Kagaku Zasshi*, **89,** 716
19. Rambacher, P. (1968). *Chem. Ber.*, **101,** 2595
20. Schoeberl, A., Rimpler, M. and Magosch, K. H. (1969). *Chem. Ber.*, **102,** 1767
21. Filler, R., White, A. B., Khan, B. T. and Gorelic, L. (1967). *Can. J. Chem.*, **45,** 329
22. Maeda, I. and Yoshida, R. (1968). *Bull. Chem. Soc. Jap.*, **41,** 2975
23. Vigneron, J. P., Kagen, H. and Horeau, A. (1968). *Tetrahedron Letters*, 5681
24. Watanabe, T. and Noyori, G. (1969). *Kogyo Kagaku Zasshi*, **72,** 1080; Kurokawa, H. and Noyori, G. (1967). *Kogyo Kagaku Zasshi*, **70,** 1360
25. Arold, H., Eule, M. and Reissmann, S. (1969). *Z. Chem.*, **9,** 447
26. Kamiya, T. (1969). *Chem. Pharm. Bull.*, **17,** 895
27. Schoeberl, A., Borcher, J. and Hantzsch, D. (1968). *Chem. Ber.*, **101,** 373
28. Belitz, H. D. (1967). *Tetrahedron Letters*, 749

29. Ramirez, F., Bhatia, S. and Smith, C. P. (1967). *J. Amer. Chem. Soc.,* **89,** 3030
30. McCord, T. J., Foyt, D. C., Kirkpatrick, J. L. and Davies, A. L. (1967). *J. Med. Chem.,* **10,** 353
31. Kolc, J. (1969). *Coll. Czech. Chem. Commun.,* **34,** 630
32. Babb, R. M. and Bollinger, F. W. (1970). *J. Org. Chem.,* **35,** 1438
33. Steglich, W., Heininger, H. U., Dworschak, H. and Weygand, F. (1967). *Angew. Chem. Internat. Ed. Engl.,* **6,** 807
34. Weygand, F., Steglich, W. and Fraunberger, F. (1967). *Angew. Chem. Internat. Ed. Engl.,* **6,** 808
35. Zaimia, T., Mitsusashi, K., Sasaji, I. and Asahara, T. (1970). *Kogyo Kagaku Zasshi,* **73,** 319
36. Snow, M. L., Lauinger, C. and Ressler, C. (1968). *J. Org. Chem.,* **33,** 1774
37. Kinoshita, M., Yanagisawa, H., Doi, S., Kaji, E. and Umezawa, S. (1969). *Bull. Chem. Soc. Jap.,* **42,** 194
38. Ruhlmann, K., Kaufmann, K. D. and Ickert, K. (1970). *Z. Chem.,* **10,** 393
39. Ichikawa, T., Maeda, S., Araki, Y. and Ishido, Y. (1970). *J. Amer. Chem. Soc.,* **92,** 5514
40. Berse, C. and Bessette, P. (1971). *Can. J. Chem.,* **49,** 2610
41. Okai, H., Imamura, N. and Izumiya, N. (1967). *Bull. Chem. Soc. Jap.,* **40,** 2154
42. Inui, T., Ohta, Y., Ujike, T., Katsura, H. and Kaneko, T. (1968). *Bull. Chem. Soc. Jap.,* **41,** 2148
43. Zdansky, G. (1968). *Ark. Kemi,* **29,** 47
44. Harpp, D. N. and Gleason, J. G. (1971). *J. Org. Chem.,* **36,** 73
45. Fourneau, J. P., Efimovsky, O., Gaignault, J. C., Jacquier, R. and Le Ridant, C. (1971). *Compt. Rend. Acad. Sci. Ser. C,* **272,** 1515
46. Mertes, M. P. and Ramsey, A. A. (1969). *J. Med. Chem.,* **12,** 342
47. Carson, J. F. and Boggs, L. E. (1967). *J. Org. Chem.,* **32,** 673
48. Corbella, A., Garibaldi, P., Jommi, G. and Mauri, F. (1969). *Chem. Ind. (London),* 583
49. Witkop, B., Fujimoto, Y., Irreverre, F., Karle, J. M. and Karle, I. C. (1971). *J. Amer. Chem. Soc.,* **93,** 3471
50. Robertson, A. V. and Simpson, W. R. J. (1968). *Aust. J. Chem.,* **21,** 769
51. Norton, S. J., Sullivan, P. T. and Sullivan, C. B. (1971). *J. Med. Chem.,* **14,** 211
52. Kuszewski, J. R., Lennarz, W. J. and Snyder, H. R. (1968). *J. Org. Chem.,* **33,** 4479
53. Martin, T. A., Comer, W. T., Combs, C. M. and Corrigan, J. R. (1970). *J. Org. Chem.,* **35,** 3814
54. Bentov, M. and Roffman, C. (1969). *Israel J. Chem.,* **7,** 835
55. Terashima, S. and Yamada, S. (1968). *Chem. Pharm. Bull.,* **16,** 2064
56. Counsell, R. E., Desai, P., Ide, A., Kulkarni, P. G., Weinhold, P. A. and Rethy, V. B. (1971). *J. Med. Chem.,* **14,** 789
57. Jorgensen, E. C. and Nulu, J. R. (1969). *J. Pharm. Sci.,* **58,** 1139
58. Pages, R. A. and Burger, A. (1967). *J. Med. Chem.,* **10,** 435
59. Matsuura, T., Omura, K. and Nishinaga, A. (1969). *Chem. Commun.,* 366
60. Bower, R. H. and Lambooy, J. P. (1969). *J. Med. Chem.,* **12,** 1028
61. Omote, Y., Fujinuma, Y. and Sugiyama, N. (1968). *Chem. Commun.,* 190
62. Reinhold, D. F., Firestone, R. A., Gaines, W. A., Chemerda, J. M. and Sletzinger, M. (1968). *J. Org. Chem.,* **33,** 1209
63. Szlompek-Nesteruk, D. and Znojek, L. (1967). *Przem. Chem.,* **46,** 218
64. Ioffe, B. V., Isidorov, V. A. and Stolyarov, B. V. (1971). *Dokl. Akad. Nauk. SSSR,* **197,** 91
65. Wagner, D., Gertner, D. and Zilkha, A. (1968). *Tetrahedron Letters,* 4479
66. Huenig, S., Graessmann, W., Meuer, V., Luecke, E. and Brenninger, W. (1967). *Chem. Ber.,* **100,** 3039
67. Loeffler, L. J., Britcher, S. F. and Baumgartner, W. (1970). *J. Med. Chem.,* **13,** 926
68. Sletzinger, M., Firestone, R. A., Reinhold, D. F., Rooney, C. S. and Nicholson, W. H. (1968). *J. Med. Chem.,* **11,** 261
69. Coggins, J. R. and Benoiton, N. L. (1971). *Can. J. Chem.,* **49,** 1968
70. Ikutani, T. (1971). *Bull. Chem. Soc. Jap.,* **44,** 271
71. Rasteikiene, L., Dagiene, M. and Knunyants, I. L. (1970). *Usp. Khim.,* **39,** 1537
72. Ciuslea, G. and Dragota, I. I. (1967). *Rev. Chim. (Bucharest),* **18,** 521
73. Karpavacius, K., Judickiene, A., Sadlauskaite, I., Kildisheva, O. V. and Knunyants, I. L. (1970). *Izv. Akad. Nauk. SSSR, Ser. Khim.,* 2150
74. Weinstock, L. T., Grabowski, B. F. and Cheng, C.-C. (1970). *J. Med. Chem.,* **13,** 995

75. Garnett, J. L., Law, S. W., O'Keefe, J., Halpern, B. and Turnbull, K. (1969). *Chem. Commun.*, 323
76. Bloss, K. (1969). *J. Label. Compounds,* **5,** 355
77. Thanassi, J. W. (1971). *J. Org. Chem.,* **36,** 3019
78. Wenzel, M., Bruehmuller, M. and Engels, K. (1967). *J. Label. Compounds,* **3,** 234
79. Zintel, J. A., Williams, A. J. and Stuart, R. S. (1969). *Can. J. Chem.,* **47,** 411
80. Kirby, G. W. and Michael, J. (1971). *Chem. Commun.,* 415
81. Takamura, N., Terashima, S., Achiwa, K. and Yamada, S. (1967). *Chem. Pharm. Bull.,* **15,** 1776
82. Lomakina, N. N., Zenkova, V. A. and Yurina, M. S. (1969). *Khim. Pri. Soedin.,* **5,** 43
83. Ponnuswamy, P. K. and Sasisekharan, V. (1971). *Int. J. Protein Res.,* **3,** 1, 9
84. George, J. M. and Kier, L. B. (1970). *Experientia,* **26,** 952
85. Nishikawa, K. and Ooi, T. (1971). *Progr. Theoret. Phys.,* **46,** 670
86. Chisholm, M., Cremlyn, R. J. W. and Taylor, P. J. (1967). *Tetrahedron Letters,* 1373
87. Ayyar, R. R. (1968). *Z. Kristallogr.,* **126,** 227
88. Ichikawa, T., Ittaka, Y. and Tsuboi, M. (1968). *Bull. Chem. Soc. Jap.,* **41,** 1027
89. Chiba, A., Ueki, T., Sasada, Y. and Kakudo, M. (1967). *Acta Crystallogr.,* **22,** 863
90. Chacko, K. K. and Mazumdar, S. K. (1969). *Z. Kristallogr.,* **128,** 315
91. Ichikawa, T. and Iitaka, Y. (1969). *Acta Crystallogr.,* **B25,** 1824
92. Balasubramian, R., Lakshminarayanan, A. V., Sabesan, M. N., Tegoni, G., Venkatesan, K. and Ramachandran, G. N. (1971). *Int. J. Protein Res.,* **3,** 25
93. Scopes, P. M., Fowden, L. and Thomas, R. N. (1971). *J. Chem. Soc. C.,* 833
94. Jorgensen, E. C. (1971). *Tetrahedron Letters,* 863
95. Anand, R. D. and Hargreaves, M. K. (1968). *Chem. Ind. (London),* 880
96. Barth, G., Voelter, W., Bunnenberg, E. and Djerassi, C. (1969). *Chem. Commun.,* 355; Toniolo, C. (1970). *J. Phys. Chem.,* **74,** 1390; Cymerman Craig, J. and Pereira, W. E. (1970). *Tetrahedron,* **26,** 3457
97. Blaha, K. and Fric, I. (1969). *Coll. Czech. Chem. Commun.,* **34,** 2852
98. Horwitz, J., Strickland, E. H. and Billups, C. (1969). *J. Amer. Chem. Soc.,* **91,** 184
99. Hooker, T. M. and Schellman, J. A. (1970). *Biopolymers,* **9,** 1319
100. Imanishi, A. and Isemura, T. (1969). *J. Biochem. (Tokyo),* **65,** 309
101. Rowe, J., Julian, M., Hinton, J. and Rowe, K. L. (1970). *Chem. Rev.,* **70,** 1
102. Horsley, W. J., Sternlicht, H. and Cohen, J. S. (1970). *J. Amer. Chem. Soc.,* **92,** 680
103. Lichter, R. L. and Roberts, J. D. (1970). *J. Org. Chem.,* **35,** 2806
104. Cavanaugh, J. R. (1968). *J. Amer. Chem. Soc.,* **90,** 4533
105. Aruldhas, G. (1967). *Spectrochim. Acta,* **A23,** 1345
106. Bak, B. and Nicolaisen, F. (1967). *Acta. Chem. Scand.,* **21,** 1980
107. Halpern, B., Westley, J. W. and Weinstein, B. (1967) *Chem. Commun.,* 160
108. Pirkle, W. H. and Beare, S. D. (1969). *J. Amer. Chem. Soc.,* **91,** 5150
109. Akimoto, T., Tsuboi, M., Kainosho, M., Tamura, F., Nakamura, A., Muraishi, S. and Kajiura, T. (1971). *Bull. Chem. Soc. Jap.,* **44,** 2577
110. Grenie, Y., Lassegues, J. C. and Garrigou-Lagrange, C. (1970). *J. Chem. Phys.,* **53,** 2980
111. Kuntz, E., Canada, R., Wagner, R. and Augenstein, L. (1968). *Molecular Luminescence Conference*, 551 (E. C. Lim, editor). (New York: Benjamin); Eisinger, J. and Navon, G. (1969). *J. Chem. Phys.,* **50,** 2069; Busel, E. P. and Burshtein, E. A. (1970). *Biofizika,* **15,** 993
112. Manning, J. M. and Moore, S. (1968). *J. Biol. Chem.,* **243,** 5591
113. Stalling, D. L., Gille, G. and Gehrke, C. W. (1967). *Anal. Biochem.,* **18,** 118
114. Gehrke, C. W. and Leimer, K. (1970). *J. Chromatogr.,* **53,** 195
115. Nakaparksin, S., Birrell, P., Gil-Av, E. and Oro, J. (1970). *J. Chromatogr. Sci.,* **8,** 177
116. Parr, W., Yang, C., Pleterski, J. and Bayer, E. (1970). *J. Chromatogr.,* **50,** 510
117. Sievers, R. E., Bayer, E. and Hunziker, P. (1969). *Nature (London),* **223,** 179
118. Schmitt, J. H. and Zenk, M. H. (1968). *Anal. Biochem.,* **23,** 433
119. Applegate, K. R., Slutsky, L. J. and Parker, R. C. (1967). *J. Amer. Chem. Soc.,* **90,** 6909
120. Olah, G. A., Brydon, D. L. and Porter, R. D. (1970). *J. Org. Chem.,* **35,** 317
121. Sato, M., Tatsumo, T. and Matsuo, H. (1970). *Chem. Pharm. Bull.,* **18,** 1794; *Yakagaku Zasshi,* **90,** 1160
122. Buckingham, D. A., Marzilli, L. G. and Sargeson, A. M. (1967). *J. Amer. Chem. Soc.,* **89,** 5133
123. Clarke, T. G., Hampson, N. A., Lee, J. B., Morley, J. R. and Scanlon, B. (1970). *J. Chem. Soc. C,* 815

124. Kozyrev, B. M. and Rivkind, A. I. (1968). *Dokl. Akad. Nauk. SSSR, Ser. Biol.*, **175**, 1396
125. Schulz, U. and McCalla, D. R. (1969). *Can. J. Chem.*, **47**, 2021
126. Axen, R., Chaykovsky, M. and Witkop, B. (1967). *J. Org. Chem.*, **32**, 4117
127. Angst, C. and Huegli, F. (1969). *Chimia*, **23**, 142
128. Nogami, H., Hasegawa, J., Ikari, N., Takeuchi, K. and Ando, K. (1970). *Chem. Pharm. Bull.*, **18**, 1091
129. Dixon, H. B. F., Kapeiskii, M. Y., Shlyapnikov, S. V., Oseledchik, V. S. and Turchin, K. F. (1967). *Zh. Obshch. Khim.*, **37**, 1237
130. Mayberry, W. E. and Hockert, T. J. (1970). *J. Biol. Chem.*, **245**, 697; *Endocrinology*, **86**, 225
131. e.g. Horan, P. K., Henrikson, T. and Snipes, W. (1970). *J. Chem. Phys.*, **52**, 4324
132. Sevilla, M. D. (1970). *J. Phys. Chem.*, **74**, 2096
133. Neta, P., Simic, M. and Hayon, E. (1970). *J. Phys. Chem.*, **74**, 1214
134. Finnstrom, B. (1971). *Photochem. and Photobiol.*, **13**, 375
135. Khenokh, M. A. and Bogdanova, N. P. (1968). *Dokl. Akad. Nauk. SSSR*, **180**, 492
136. Yonemitsu, O., Witkop, B. and Karle, I. L. (1967). *J. Amer. Chem. Soc.*, **89**, 1039
137. Yonemitsu, O., Tokuyama, T., Chaykovsky, M. and Witkop, B. (1968). *J. Amer. Chem. Soc.*, **90**, 776
138. Capindale, J. B. and Fan, H. S. (1967). *Can. J. Chem.*, **45**, 1921
139. Wright, M. R. (1967). *J. Chem. Soc. B*, 1265
140. Buckingham, D. A., Foster, D. M. and Sargeson, A. M. (1970). *J. Amer. Chem. Soc.*, **92**, 5701
141. Buckingham, D. A., Foster, D. M. and Sargeson, A. M. (1969). *J. Amer. Chem. Soc.*, **91**, 4102
142. Hix, J. E. and Jones, M. M. (1968). *J. Amer. Chem. Soc.*, **90**, 1723
143. Abbott, E. H. and Martell, A. E. (1970). *J. Amer. Chem. Soc.*, **92**, 1754
144. Ando, M. and Emoto, S. (1969). *Bull. Chem. Soc. Jap.*, **42**, 2624, 2628
145. Hirota, K. and Izumi, Y. (1967). *Bull. Chem. Soc. Jap.*, **40**, 178
146. Shiba, T., Kajiwara, M., Kato, Y., Inoue, K. and Kaneko, T. (1970). *Arch. Biochem. Biophys.*, **140**, 90
147. Hurych, J. (1967). *Hoppe. Seyler's Z. Physiol. Chem.*, **348**, 426
148. Al-Sayyab, A. F., Atto, A. T. and Sarah, F. Y. (1971). *J. Chem. Soc. C*, 3260
149. Kato, T. and Kubota, Y. (1967). *Yakugaku Zasshi*, **87**, 1219
150. Blazevic, K., Houghton, R. P. and Williams, C. S. (1968). *J. Chem. Soc. C*, 1704
151. Pines, S. H., Kozlowski, M. A. and Karady, S. (1969). *J. Org. Chem.*, **34**, 1621
152. Williamson, C. E. and Witten, B. (1967). *Cancer Res.*, **27A**, 33

3
The Structural Elucidation of Peptides

P. M. HARDY
University of Exeter

3.1 INTRODUCTION

In this chapter on structural elucidation, only primary structure has been covered. Considerations of space have limited the discussion of methods, but two areas of particular activity, the Edman method and mass spectrometry, have been dealt with in more detail than the other topics. Modification of amino acids other than those strictly related to primary structure determination are omitted entirely. Although a molecular weight of 10 000 is usually accepted as the dividing line between peptides and proteins, it has been found possible only to include peptides of up to *c.* 70 amino acid residues. Alamethicin and mycobacillin have been included in the section on cyclic peptides; as peptides antibiotics they have structural features other than those found in the other peptides considered, but several of the aspects of their structural elucidation are of rather general interest.

3.2 METHODS OF DETERMINING PRIMARY STRUCTURE

3.2.1 Amino acid determination

The separation of the 20 protein amino acids by g.l.c. on a single column has now been achieved. The amino acids were converted to their *N*-heptafluorobutyryl n-propyl esters and chromatographed using a temperature programme on a column impregnated with a siloxane (OV-1) as the stationary phase. Elution was complete with the emergence of the cysteine derivative after 43 min (Figure 3.1)[1]. After extensive studies in this field, Gehrke *et al.* have

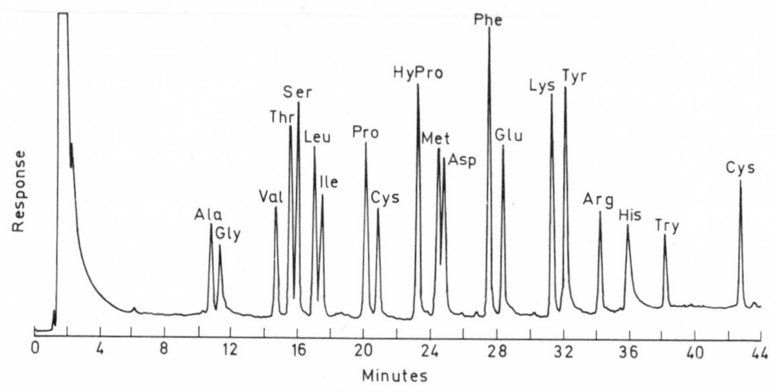

Figure 3.1 G.l.c. trace
(From Wayne-Moss *et al.*[1], reproduced by permission of Elsevier)

separated the same mixture of amino acids as their *N*-trifluoroacetyl n-butyl esters, but two columns were necessary for complete resolution of all components[2]. Tranexamic acid, *trans*-4-(aminomethyl) cyclohexane carboxylic acid, has been suggested as an internal standard for this type of amino acid analysis; it has the added advantage that it can be used to follow the performance of ion-exchange pretreatment of material from biological sources[2].

A fluorometric method of assaying amino acids has been developed and applied to the automated separation of amino acids by ion-exchange chromatography. This method involves the reaction of the amino acids with phenylacetaldehyde in the presence of ninhydrin, and is from 10 to a 100 times more sensitive than the ordinary ninhydrin colour reaction. Ammonia does not give a fluorescent derivative and since the reagents with which the amino acids are reacted do not themselves fluoresce it does not have the difficulties of *p*-dimethylaminonaphthalenesulphonyl (dansyl) derivatives as far as automation is concerned[3]. The normal ninhydrin reagent as used for amino acid assay either contains added hydrindantin, or it is produced *in situ*. In its absence α-amino acids produce little colour. However, the yellow-brown colour produced by the imino acids does not depend on the presence of hydrindantin and improved quantification of imino acids poorly resolved from other peaks on the amino acid analyser can be obtained by using hydrindantin-free ninhydrin[4].

The instability of asparaginyl and glutaminyl residues in peptides towards deamidation is reiterated in a recent paper. The use of phosphate buffers, high ionic strength, extremes of pH, and temperatures above 4 °C should be avoided if such decomposition is to be minimised. Even at pH 7 in a phosphate buffer of ionic strength 0.15 at 22 °C, H-Gly-Arg-Asn-Arg-Gly-OH deamidates with a half-life of one month[5].

3.2.2 Degradation by enzymes

An enzyme which is finding increasing application in preparing oligopeptides from proteins or larger peptides is thermolysin. It has found use, for example, in the preparation of peptides from bovine αsl-casein[6]. This is a phosphoprotein whose structural elucidation was hampered by its phosphoserine content. However, it proved possible to circumvent difficulties such as poor ninhydrin colour reactions, poor Edman degradation of phosphoserine residues and the proteolytic resistance of peptide bonds in the vicinity of phosphoserine residues by prior removal of the phosphate groups with alkaline phosphatase[6].

Cathepsin C cleaves peptides from the *N*-terminus, breaking them down into dipeptides. If the *N*-terminal amino acid is removed by one cycle of the Edman degradation, then subsequent treatment with this enzyme gives the complementary series of overlapping dipeptides. This interesting approach to structure determination, however, would only seem useful for small peptides[7]. A carboxy-peptidase from orange leaves[8] has been found to have a similar specificity to carboxypeptidase C from orange peel[9]. They both have the combined specificities of carboxypeptidases A and B, splitting both sides of proline residues, but failing to cleave glycyl–prolyl or prolyl–prolyl bonds. The enzyme from orange leaves sequentially releases the ten *C*-terminal amino acids of the B-chain of insulin[8].

Several ways of chemically modifying lysine residues so as to restrict cleavage by trypsin to arginyl peptide bonds are known. The more useful of these modifications are reversible and maleic and citraconic anhydride are two reagents of this type reacting with the ε-amino group. However, recent work on aldolase has shown that alkylation of the thiol groups of cysteine residues also occurs with these anhydrides[10]. Further work has now established that such a side-reaction can be avoided by using *exo-cis*-3,6-endoxo-Δ^4-tetrahydrophthalic anhydride, an anhydride in which the olefinic bond is not activated and hence not susceptible to nucleophilic addition of thiol groups[11].

3.2.3 Chemical cleavage at specific amino acid residues

Specific cleavages of this type have recently been reviewed[12]. The most widely used chemical cleavage is that at methionine using cyanogen bromide

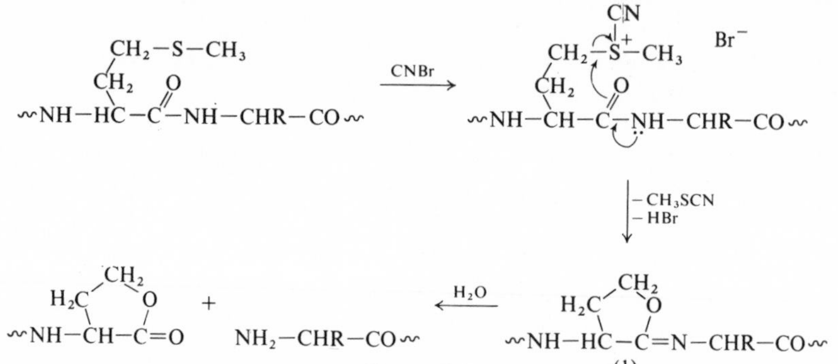

Figure 3.2

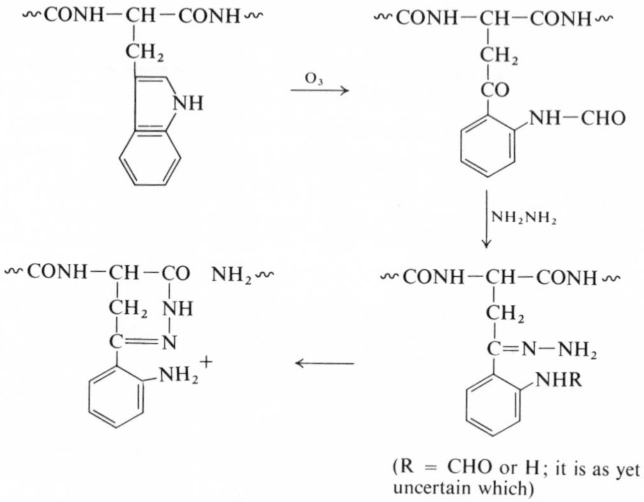

Figure 3.3

(Figure 3.2). Working under anhydrous conditions, it has now proved possible to isolate the imidate (1), previously postulated as an intermediate in this reaction, from *N*-acetyl-DL-methionylglycine. The rate of the cyclisation step for various compounds has been followed by g.l.c. analysis of the methyl thiocyanate evolved. In peptides where methionine is not a terminal residue the molecular weight of the substrate appears to have little influence on the reaction, but if the amino group of the methionine residue is free the reaction is slowed down[13].

New studies of the selective cleavage of tryptophyl links[14] and the bonds to cysteinyl residues[15] have been reported, but their scope has been limited so far to simple protected peptides. Tryptophyl residues react with ozone to give *N*-formyl-kyurenyl derivatives which are susceptible to hydrazine (Figure 3.3). Milder conditions for the rearrangement reaction (hydrazine acetate in glacial acetic acid) have been evolved which do not affect other peptide bonds. Leucine can be released from carbobenzoxy-L-alanyl-*N'*-formyl-L-kyurenyl-L-leucine in up to 87% yield under these conditions[14]. Serine and threonine residues in peptides are converted to oxazolidone derivatives by phosgene and subsequent treatment with alkali cleaves the peptide chain (Figure 3.4; for serine X = O). Cysteine residues have been found to undergo the same type of reaction (Figure 3.4; X = S). Yields of up

~CONH—CH—CONH~ ; CH_2 ; X —(i) $COCl_2$ (ii) heat→ ~CON—CH—CONH~ ; OC, CH_2 ; X

↓ OH^-

~COOH HN—CH—CONH~ ; OC, CH_2 ; X

Figure 3.4

to 80% of cleaved products were obtained from some di- and tri-peptides, although the 2-oxathiazolidine-4-carboxylic acid products were sometimes partly racemic. In the case of *N*-benzyloxycarbonyl-L-phenylalanyl-L-cysteinyl-glycine methyl ester the method failed; for some reason the *S*-chlorocarbonyl derivative decomposes rather than cyclising[15].

3.2.4 Chemical methods of sequencing

3.2.4.1 The Edman degradation

The partial determination of the amino acid sequences of proteins using the automated Edman machine or sequenator has now become a familiar sight in papers on primary structure. Both Niall[16] and Li[17] have used such machines to advantage in revising the structure of human growth hormone. These studies stemmed from the observation that there is a remarkable similarity between the primary structures of human growth hormone and human chorionic somatomammotropin which can be enhanced by repositioning a pentadecapeptide in the former. The two sequenator degradations showed that sequence 1–25 was in fact that predicted from the comparative homologies[16,17], and one of the tryptic peptides of the hormones cleaved by cyanogen bromide had been wrongly positioned in the original structural assignment.

The flavodoxins, proteins which can replace ferredoxin in certain anaerobic bacteria, have been found well suited to sequenator studies. That from *Clostridium pasteurianum* gave good results over 51 cycles and is suggested as an alternative to myoglobin for test runs of the instrument[18]. Although Tanaka *et al.*[18] state that for a variety of proteins they have been averaging about thirty residues in sequenator runs, in some proteins at least double this number can be achieved. The first 66 residues of the 84 residue chain of bovine parathyroid hormone, whose action is reciprocal to that of calcitonin (see later), have been successfully identified by this method[18], and partial sequence determinations of 67 and 50 positions have been reported for two smaller than usual immunoglobin light chains[19]. In the latter determinations two modifications to the original Edman–Begg design of sequenator were incorporated and dithiothreitol was added to the solvents. This permits the protein to be applied to the machine with its disulphide bonds intact, which is often advantageous as proteins are more soluble in their native form than after their disulphide bonds have been cleaved. The thiazolinones were not converted to thiohydantoins but hydrolysed directly to the amino acids either by hydriodic acid or sodium hydroxide–dithionate[20].

Automated Edman degradation has now been reviewed[21]. The operation of the sequenator relies on the insolubility of the protein during the extraction procedure and short peptide chains are too soluble to be successfully degraded. Attempts to overcome this limitation by coupling the peptide to a solid support have been made and an automated apparatus using this principle has now been described in detail[22]. Reagents are pumped through a column containing the resin-bound peptide, and it proved possible to degrade a 0.24 μmol sample of the 21 residue insulin A-chain within three days with an average degradative yield of 94% per cycle. However, under the conditions used aspartic acid blocks further degradation, presumably through its cyclic imide, and the phenylthiohydantoin of glutamic acid cannot be detected because it remains bound by its γ-carboxyl group to the resin[22].

The use of paper as a support for the Edman degradation of peptides has also been revived. A system of analysis of the liberated phenylthiohydantoins by t.l.c. permits an octapeptide to be sequenced using 30–40 nmol, and even on a 1 nmol scale the first two amino acids can be identified[23]. The Edman degradation can be carried out on peptides ionically bound to alumina. After protecting the free α-amino group as its t-butoxycarbonyl (BOC) derivative, the peptide α-carboxyl is linked to the aminomethyl group of 4′-hydroxy-2′-aminomethylazobenzene-4-sulphonic acid (2). The coloured

$$HO_3S-C_6H_4-N{=}N-C_6H_3(CH_2NH_2)-OH$$

(2)

peptide derivatives can be fixed as their anions on acidic alumina and after removal of the BOC-group with trifluoroacetic acid, the usual Edman degradation cycle can be carried out. The method has so far only been tested successfully on simple peptides without functional groups in the side-chains[24].

As far as the manual Edman degradation is concerned, the dansyl method

of detection of the freshly-exposed *N*-terminal amino acid is still the most widely used in the sequencing of novel peptides. It does not require expensive apparatus, like the method using a paper support described above. However, much ingenuity continues to be expended in the search for improvements to the method. The use of polyamide-coated glass plates for the t.l.c. separation of phenylthiohydantoins (PTHs) has been described[25], but other papers ring the changes in the structure of the isothiocyanate used for derivatisation. The use of methyl isothiocyanate and the detection by mass spectrometry of the methylthiohydantoins is discussed towards the end of this section, and g.l.c. has also been used to identify these derivatives, both in the free[26] and trimethylsilylated forms[27]. 1-Naphthyl isothiocyanate has the advantage as a degrading agent that the thiohydantoin produced is fluorescent, enhancing the sensitivity of the method[28].

The use of 4-isothiocyanatobenzenesulphonic acid (IBSA) has been suggested to reduce the side-reactions occurring through incomplete formation of the phenylthiocarbamyl peptide. Any peptide remaining unreacted after this first stage can in this case readily be separated from the derivative by chromatography in 0.1 M hydrochloric acid on DEAE-Sephadex. The 4-sulphophenylthiohydantoins can subsequently be identified by t.l.c.[29]. Reaction with IBSA[30] and other more highly sulphonated aromatic isothiocyanates[31] has also been used to improve the hydrophilicity of tryptic peptides in which the predominance of hydrophobic side-chains present solubility difficulties as the peptides are too easily removed during the thiazolinone step of the Edman degradation. The ε-amino-group of the *C*-terminal lysine residue is coupled to the hydrophilic isocyanate in a pretreatment step. In the case of the peptide H-Phe-Phe-Tyr-Pro-Thr-Lys-Ala-OH the normal Edman procedure gave only a 5% yield of phenylalanine-PTH after one cycle, but after pretreatment with IBSA 22% of alanine-PTH was liberated after six cycles[30].

The PTH-derivatives of cysteine and cystine are too unstable to be useful for identification purposes. Although cysteic acid and S-carboxymethyl-cysteine form stabler PTHs, their properties are such that they require special treatment in the sequenator routine. However, some *S*-alkylcysteines form PTHs which are suitable for such routine t.l.c. identification, and conditions for the alkylation of cysteine residues in proteins without appreciably affecting amino acids other than methionine have been described[32].

Mass spectrometry (m.s.) is becoming a valuable supplementary tool in the Edman degradation. If *p*-bromophenylisothiocyanate is used for forming the phenylthiocarbamyl derivative, the *p*-bromo-PTHs subsequently liberated can be simply identified by m.s. as the characteristic fragment and molecular ions are emphasised by a double peak of mass difference $\Delta m/e = 2$ caused by the isotope ratio of the bromine. Sequences of up to 13 residues were determined in this way by the manual Edman degradation of peptides derived from porcine pancreatic trypsin inhibitor[33]. This method was developed by Weygand and Obermeier, and tested using 16 cycles of the Edman degradation on glucagon; unequivocal assignment of glutamic and aspartic acids and their γ- and β-amides are possible by this technique[34].

One way of applying the Edman degradation to protein mixtures has now been fully described. Methyl isothiocyanate is used for derivative formation, and the methylthiohydantoins liberated at the end of each cycle, in contrast to PTHs, all give good molecular ions. The volatility of methyl isothiocyanate also removes the necessity of extracting excess reagent with solvent. When used with protein mixtures, a standard cocktail of ^{15}N-enriched amino acids as their methylthiohydantoin derivatives is added before each cycle. At the end of the cycle the extracted methylthiohydantoins are examined by m.s. and those released from the proteins are identified by the changes occurring in the $^{14}N:^{15}N$ ratios. The mass spectrometer is temperature programmed to volatilise the hydantoins sequentially into the ion beam; the temperatures for maximal ion current range from 70–80 °C for valine to 220–230 °C for histidine. If the peptide chains in the mixture are present in other than equimolar proportions the relative abundances of the liberated methylthiohydantoins can be used to assign them to the various components of the mixture. In this way the *N*-terminal decapeptide sequences of a mixture of 5 µmol of ribonuclease and 2.5 µmol of lysozyme could be correctly deduced[35].

The volatility of dansyl amino acids is unexpectedly high, and further esterification of these compounds for m.s. analysis is normally unnecessary. Characteristic fragments which give information on the structures of the dansylated molecules are, however, rare, and unequivocal identification of the usual protein amino acids by low-resolution studies is not possible. Molecular ions can be observed except in the cases of dansyl histidine and dansyl cystine, the latter being cleaved to dansyl cysteine[36].

3.2.4.2 *Sequencing from the* C*-terminus*

Work aimed at improving the Stark method of sequence determination from the *C*-terminal amino acid[37] has been reported. The method involves the formation of an acetylated peptide thiohydantoin and its subsequent cleavage

$$\sim CONH{-}CHR^1{-}CONH{-}CHR^2{-}CO_2H$$

$$\downarrow Ac_2O{-}NCS^-$$

$$\sim CONH{-}CHR^1{-}CO{-}\underset{\text{ring: } N{-}CHR^2{-}CO{-}NH{-}C(=S){-}}{N}$$

$$\downarrow CH_3{\cdot}CO{\cdot}NHOH \text{ or } 12M\ HCP$$

$$\sim CONH{-}CHR^1{-}CO_2H \; + \; \underset{\text{ring: } NH{-}CHR^2{-}CO{-}NH{-}C(=S){-}}{CHR^2{-}CO}$$

Figure 3.5

into the free thiohydantoin of the original *C*-terminal amino acid and the acetylated peptide in which a new residue has become *C*-terminal (Figure 3.5). Quantitative formation of acetyl amino acid hydantoins can be achieved in acetyl chloride–trifluoroacetic acid solution after the initial treatment with thiocyanic acid. Under these conditions *C*-terminal proline residues are converted directly to the thiohydantoin, no hydrolytic step being required[38]. This contrasts with the inertness of proline under the original conditions, which was associated with its inability to form an oxazolinone. Greater selectivity in the cleavage of peptide thiohydantoins is possible by the use of either 0.5 M triethylamine[38] or the ion-exchange resin Amberlite IR–120 (H^+ form) at room temperature for 2–6 h[39]. Using the latter method successive recoveries of the amino acid residues from the *C*-terminus of ribonuclease A were 85, 80 and 80% respectively for three cycles of the method[39]. On the fourth cycle recovery was down to 10%, but this residue was aspartic acid which is known to be resistant to this method of degradation[37].

The resistance of proline to oxazolinone formation also impedes the selective tritium-labelling method of *C*-terminal amino acid determination. The conditions which are normally successful (reaction with tritium oxide and acetic anhydride in the presence of pyridine) fail when proline is at the *C*-terminus. It has, however, been reported that in hot acetic anhydride–acetic acid mixtures *C*-terminal proline does undergo tritiation. Side reactions during this process have now been noted; some non-terminal residues undergo tritium incorporation, the imidazole ring protons of histidine being generally susceptible to exchange under these conditions[40].

3.2.5 Mass spectrometry

Mass spectrometric studies of peptide primary structures really began in 1965 when it proved possible to determine the sequence of the naturally-occurring nonapeptide fortuitine by this method using only a few micrograms of material[41]. Since that time advances in techniques and spectral interpretation have been steady rather than spectacular and the papers published in the last two years continue this trend. The advantage of the method lies not so much in its sensitivity, for the combination of the Edman method of removing the *N*-terminal amino acid with the detection of the freshly-exposed *N*-terminal amino acid as its dansyl derivative requires rather less material, but more in it being a single operation rather than one which has to be applied repetitively. Its chief limitation for routine use lies in the restriction of the sequence size which can be determined to about ten residues. As a result of this limitation, the peptides whose structures have been determined for the first time by m.s. have tended to be of the (cyclic) peptide antibiotic type rather than linear peptides or protein fragments, although of course small synthetic peptides have been widely used for evaluating fragmentation patterns and chemical modifications[42]. However, methods of application to protein fragments are being actively investigated, as will be seen in the discussion that follows.

3.2.5.1 Chemical modification

Permethylation of peptides is normally carried out prior to mass spectral studies. The technique using methyl-iodide, dimethyl sulphoxide, and sodium hydride[43] has gained widest acceptance for this purpose, and it can be applied successfully on a very small scale[44]. Although it improves volatility by preventing hydrogen bonding between amide groups and stabilises those fragmentation pathways containing sequence information, it can cause complications through the quaternisation of free amino groups, methionine, cysteine and histidine residues. Reaction with amino groups can be avoided by acetylating before methylating and it has proved possible in the case of some simple peptides at least to permethylate without affecting a methionine residue by careful control of the conditions[45]. Peptides containing histidine may be studied after treatment with diethyl pyrocarbonate. At pH 8.0 in aqueous solution this reagent ethoxycarbonylates free amino groups, the thiol group of cysteine, the phenolic hydroxyl of tyrosine and the imidazole nitrogen of histidine. Further treatment with excess diethyl pyrocarbonate splits off formate from the imidazole ring, converting it into a 1,2-bis-(ethoxycarbonyl-amino)-ethylene group. After acidification, the modified peptide can be extracted with ethyl acetate and after permethylation is suitable for sequence studies[46]. During permethylation, *C*-alkylation of glycine, tryptophan, and histidine residues has been reported in some cases[47].

Difficulties with the sulphur containing amino-acids can be circumvented by catalytic desulphurisation. Milder conditions than those originally recommended for this purpose using Raney nickel are now advocated, cysteine peptides losing their sulphur more rapidly than methionine ones[48]. The use of *N*-acyl peptide esters may also be advantageous, as these are less adsorbed by the nickel. Since the desulphurisation of cysteine gives rise to alanine, it may be desirable in peptides already containing alanine to use Ni–D_2 to label the alanine residues arising from cysteine with deuterium[49]. A catalyst which is easily prepared from nickel chloride and sodium borohydride has also proved successful for desulphurisation[50]. Where methionine is the only sulphur-containing amino acid, temporary protection as the sulphoxide during permethylation can be useful[51].

Although decomposition of S-methylcysteinyl sulphonium iodide and methionylsulphonium iodide to dehydroalanine and C-vinylglycine respectively occurs in the mass spectrometer, the spectra obtained after permethylation of peptides containing cysteine and methionine are complex and difficult to interpret[49, 52]. Use of a limited amount of methyl iodide during methylation affords *S*-methylcysteine from cysteine and useful sequence data can be obtained from peptides containing this moiety[52]. Arginine has proved to be the most difficult of the protein amino acids as far as mass spectrometry is concerned. The mass spectra of acetylated and permethylated arginine-containing peptides contain sequence information only from the *N*-terminus to the amino acid preceding the first arginine residue. Two ways of converting the guanido function into more amenable derivatives are commonly used, conversion to ornithine with hydrazine or formation of a heterocycle with β-diketones. A recent paper reports that *N*-acetyl peptides whose arginine residues have been modified to N^{δ}-*Z*-(4,6-dimethyl)pyrimidyl ornithine with acetylacetone can be permethylated under conditions that do not cause quaternisation of the heterocycle[53].

3.2.5.2 Fragmentation patterns and their interpretation

The combination of high-resolution mass spectrometry and computer analysis for sequence determination is finding less favour these days. The relatively high cost, the loss of sensitivity in the high-resolution mode and the relative success of low-resolution studies have been stressed recently[54]. Readily interpretable low-resolution spectra have also been obtained from simple peptides by using chemical ionisation instead of electron impact to generate the ions. In this method the sample is ionised by collision with carbonium ions derived from methane and the lower temperature involved limits unwanted cleavages and promotes a more uniform intensity of those fragmentations useful for sequencing. As little as 2 nmol of penta-alanine was required to obtain a good spectrum of the highest sequencing ions[55].

An investigation of the appearance in the mass spectra of certain peptides of ions of mass $(M+14)$ has shown that intermolecular methylation is occurring. Both histidine and tryptophan residues are capable of abstracting a methyl carbonium ion from the carboxymethyl group of a second molecule of the peptide; the mechanism of this reaction is illustrated for histidine in Figure 3.6. In molecules containing two histidine residues ions of mass

~NH—CH—CO~ | CH_2 ... PEPTIDE → ~NH—CH—CO~ | CH_2 ... N^+ CH_3 → ~NH—CH—CO~ | CH_2 ... N CH_3 + H^+

Figure 3.6

$M+28$ have been observed. The carboxymethyl group may not be the sole source of CH_2 units as small amounts of molecular ions of mass $(M+14)$ have been observed in such peptides as Dec-His-Ala-Pro-Val-OBu^t. The process is in all cases temperature dependent, rarely being observed below 200 °C[56].

3.2.5.3 *Application to proteins and mixtures of peptides*

The permethylation reaction normally carried out prior to mass-spectrometric analysis has been cunningly utilised to isolate the derivatised *N*-terminal segment of a protein. The protein is first acetylated, digested with a protease and then permethylated. Although the *N*-terminal *N*-acetyl peptide is chloroform soluble, the other peptides will be water-soluble quaternary salts as the α-amino groups exposed by the proteolytic cleavage will be quaternised on permethylation. Thus a simple solvent extraction procedure enables the *N*-terminal sequence to be subsequently obtained mass spectrometrically from as little as one milligram of protein[57].

The first application of mass spectrometry to the determination of the amino acid sequences of peptides derived from a protein was in 1969, when Geddes *et al.*[58] examined four peptides from a chymotryptic digest of silk fibroin. Three groups have now reported on their attempts to sequence the individual peptide chains in an unseparated mixture of oligopeptides such as might be obtained by further proteolytic digestion of a protein fragment. All three emphasise the necessity in this connection of partial vaporisation studies, the changes in relative abundances of peaks at varying temperatures serving to distinguish which peaks arise from the same oligopeptide[54, 59, 60]. Roepstorff *et al.*[59] found it possible to deduce the sequences of a mixture of such peptides as Ac-Ala-Leu-Phe-Gly-OMe and Ac-Phe-Gly-Leu-Ala-OMe using low-resolution mass spectra, but knowledge of the amino acid composition of the mixture was essential. McLafferty *et al.*[60] used a computer-assisted high-resolution system, and were able to correctly sequence a mixture of four components (Ac-Gly-OMe, Ac-Gly-Ala-Leu-OMe, Ac-Val-Gly-Gly-OMe, and Ac-Met-Phe-Gly-OMe). Morris *et al.*[54] examined a mixture of peptides obtained by partial fractionation of a tryptic digest of pepsin. The mixture contained three components (H-Val-Gly-Leu-Ala-Pro-Val-Ala-OH, H-Ala-Asn-Asn-Lys-OH and H-Gln-Tyr-Tyr-Thr-Val-Phe-Asp-Arg), and the structure of these three peptides was correctly deduced except that the Asp-Arg terminus of the heptapeptide was not identified. However, no attempt had been made to modify the arginine residue. This was a low resolution study, and the authors feel that any ambiguities that may arise in mixture analysis by this method may be resolved by isotopic labelling and individual manual high-resolution measurements. They further propose a general strategy for the application of mass spectrometric sequencing to proteins. The protein is first cleaved by a specific protease such as chymotrypsin and the mixture of peptides produced fractionated on Sephadex G.25. Pre-calibration with standards enables those peptides fragments of molecular weight <1000, which are directly suitable for m.s. sequencing, to be located. Fractions of higher molecular weight are subjected to further proteolysis until peptides of ten residues or less are obtained. Each oligopeptide is subjected to electrophoresis and cysteine, methionine, histidine and arginine containing peptides are detected by staining or radio-active labelling. Peptides containing these residues are treated separately according to their special requirements, but the remainder can be simply *N*-acetylated, permethylated and subjected to mass spectrometric analysis[54].

3.3 NEW PRIMARY STRUCTURES DETERMINED

3.3.1 Small peptides

3.3.1.1 Pituitary and hypothalamic peptides

Notable advances have been made recently in the characterisation of peptides from pituitary and hypothalamic sources. The melanocyte stimulating hormone (MSH) from the dogfish *Squalus acanthias* has been shown to be H-Ser-Met-Glu-His-Phe-Arg-Trp-Gly-Lys-Pro-Met-OH. Half the *C*-terminal α-carboxyl groups are amidated. This sequence contains the heptapeptide core Met-Glu-His-Phe-Arg-Trp-Gly common to natural peptides with MSH activity. The close similarity of this material to α-MSH (the latter has four additional amino-acid residues at the *N*-terminus) suggest that α-MSH is the more primitive melanin dispersing hormone. β-MSH-like peptides appear to be absent from dogfish pituitaries[61].

Hypophysectomised rats have been found unable to acquire conditional avoidance responses, e.g. learning to jump a barrier in response to a buzzer signal to avoid an electric shock. However, when such animals are treated with porcine pituitary extracts their responses resemble those of normal rats. This observation has led to the isolation and identification of the active factor as desglycinamide-[lysine-8]-vasopressin (3). It is possible that one or both of the acidic amino acids may be amidated. This octapeptide has no hormonal activity, but can readily be prepared from [lysine-8]-vasopressin by incubating it with trypsin; it probably arises naturally by this route[62].

The most important small peptides whose sequences have been announced recently are two releasing factors found in the hypothalamus. Porcine hypothalami have yielded an octapeptide which controls the secretion of growth hormone from the anterior pituitary[63]. Successive dansyl-Edman degradations established the amino-acid sequence H-Val-His-Leu-Ser-Ala-Gly-Glu-Lys-Glu-Ala-OH. The same source has yielded another peptide of this size which stimulates the release of both luteinising hormone and follicle-stimulating hormone. The sequence of this peptide proved rather more difficult to determine. Only a small quantity (<200 nmol) was available, and the

S——————S
|　　　　　　|
H-Cys-Tyr-Phe-Glu-Asp-Cys-Pro-Lys-OH

(3)

TL　TL　TL
↓　↓　↓
Glu-His-Trp-Ser-Tyr-Gly-Leu-Arg-Pro-Gly-NH_2
↑　↑
CT　CT

(4)

EDMAN　CP　EDMAN
H-Arg-Pro-Lys-Pro-Gln-Gln-Phe-Phe-Gly-Leu-Met-NH_2
↑
CT

CP = carboxypeptidase

(5)

N-terminus is blocked as pyroglutamyl (initially indicated by mass spectrometry), ruling out simple application of the Edman method. Enzymic degradation with thermolysin (TL) and chymotrypsin (CT) produced a series of peptides with suitable overlaps (except for one position), and sequencing these established the structure (4; in this and later structures the vertical arrows indicate points of enzymic cleavage)[63]. Attempts to open the pyroglutamyl residue with alkali to give a glutamyl peptide were unsuccessful, but a subsequent paper describes the use of pyrrolidone carboxylyl peptidase to achieve this end, and Edman degradation then confirmed the sequence deduced from the enzymic fragments[64].

Details of the structure determination of thyrotropin releasing hormone have now appeared[65]. Although only a tripeptide, this material occurs in such minute quantities that to obtain enough for structural studies required a

250 000 porcine hypothalami. Even then the actual amino acid content of the purified material (active dose 1 μ *in vivo*) was only 50–60% due to contamination by plasticisers etc. during chromatography. After cleavage with *N*-bromosuccinimide and subsequent dansylation and hydrolysis, dansyl proline was formed. This is indicative of a histidyl–prolyl linkage. Mild alkaline hydrolysis of the hormone enabled glutamic acid to be identified as *N*-terminal, but the native hormone showed no free amino group. Pyroglutamyl hydrazide was detected after hydrazinolysis, but no free proline was liberated. Mass spectral studies of the permethylated peptide confirmed the structure as pyroglutamyl-histidyl-prolinamide. There was evidence in the fragmentation pattern for the —$CONMe_2$ grouping at the carboxyl end of the proline, and after methanolysis a peak due to proline methyl ester could be observed. The presence of a substituent at either end of the molecule cannot be entirely excluded on the degradative evidence, and it is necessary to assume bound acetate to account for a methyl signal in the n.m.r. spectrum and a peak at 59 mass units in the mass spectrum of the native hormone[65]. However, synthetic pyroGlu-His-Pro-NH_2 [66, 67] cannot be distinguished from the natural hormone.

In 1931 von Euler and Gaddum found that both brain and intestine contain a substance that stimulates contraction of the isolated rabbit jejunum and causes transient hypotension when injected intravenously into anaesthetised rabbits[68]. Many workers have attempted to isolate the active component in the 40 years following and bovine hypothalamus has at last yielded the pure peptide, which has come to be known as substance P. The chemical and enzymic degradations and the sequence deduced from these for this undecapeptide are outlined in (5). Removal of the arginine in the first cycle of the Edman degradation of the intact peptide leaves a very hydrophobic compound, and subsequent solvent extraction had to be kept to a minimum to avoid large losses of material. This desarginyl peptide also adhered strongly to glass surfaces[69]. Substance P has the same C-terminal tripeptide amide sequence as physalaemin and eledoisin. These peptides are of the same size, but possess N-terminal pyroglutamyl residues.

3.3.1.2 *Bradykinin and its potentiators*

Five bradykinin potentiating peptides have been isolated from the venom of the Japanese snake *Agkistrodon halys blomhoffii*, and the sequences of three of these, potentiators B (6), C (7)[70], and E (8)[71], have been established. The terminal amino-acids are such that the peptides are resistant to Edman degradation, aminopeptidase, and carboxypeptidase, and the high proline content reduces their susceptibility to endopeptidases. Bacterial protease cleaves B at the glycyl and lysyl peptide bonds, and the identity of the liberated pyroglutamylglycine was established by comparison with synthetic material. Four cycles of the Edman degradation on the central peptide extended the known sequence to position 6. Collagenase was found to attack potentiator B at three peptide bonds. The most useful peptide from this digest, 7–11, whose sequence was determined by four cycles of the Edman method, probably arose from clostridiopeptidase A contamination of the collagenase. Potentiator C is highly resistant to attack by pepsin, thermolysin, and Nagarse. The protease from *S.griseus* cleaves only one bond, but this proved sufficient to establish the sequence[70]. Potentiator E has a lysine residue at position 2 which can be attacked by trypsin, liberating a nonapeptide. Edman degradation of this and a *C*-terminal fragment of it split off by bacterial protease (see (8)) enabled the primary structure to be deduced[71]. A pentapeptide (9) with bradykinin potentiating activity has also been isolated from *Bothrops jacara* venom. Cleavage at the lysyl bond with trypsin gave a tripeptide

BP CA

Glu-Gly-Leu-Pro-Pro-Arg-Pro-Lys-Ile-Pro-Pro-OH
1 2 3 4 5 6 7 8 9 10 11
CA = collagenase, BP = bacterial proteinase

(6)

BP

Glu-Gly-Leu-Pro-Pro-Gly-Pro-Pro-Ile-Pro-Pro-OH

(7)

T

Glu-Lys-Trp-Asp-Pro-Pro-Pro-Val-Ser-Pro-Pro-OH

T = trypsin BP

(8)

Glu-Lys-Trp-Ala-Pro-OH

(9)

sequenced by the Edman method. Although the intact pentapeptide is resistant to pyrrolidase carboxylyl peptidase, the pyroglutamyllysine liberated by trypsin was slowly split by this enzyme to give free lysine. Carboxypeptidase B gave the same products[72].

A new plasma kinin from the turtle *Pseudemys scripta elegans* is thought to be [threonine-6]-bradykinin (H-Arg-Pro-Pro-Gly-Phe-Thr-Pro-Phe-Arg-OH), although so far only its amino acid composition is known. This is the first example of a naturally occurring modified bradykinin[73].

3.3.2 Larger peptides

3.3.2.1 Clupeine Z

The partial sequence of clupeine Z, one of the three main components of herring sperm protamine, was determined in 1970 from studies on peptides obtained by tryptic digestion[74]. Of the 31 amino acid residues of clupeine Z, 21 are arginine residues and endopeptidases other than trypsin were found to be unsuitable for fragmentation into peptides overlapping those produced by trypsin. This situation has led to the application of a chemical method of cleavage at the three serine residues[75]. It is possible to induce an N → O peptidyl shift by treating the peptide with concentrated sulphuric acid for 3 days at 25 °C, and selective cleavage at the *O*-peptidyl bonds occurred on exposure to 6 M hydrochloric acid at 20 °C or by alkaline hydrolysis after acetylation. Acetylation was necessary as a prior step in the latter case to prevent the reversion to *N*-peptide bonds that occurs in alkaline solution. Sequence studies on the four peptides produced enabled clupeine Z to be formulated as H-Ala-Arg_4-Ser-Arg_2-Ala-Ser-Arg-Pro-Val-Arg_4-Pro-Arg_2-Val-Ser-Arg_4-Ala-Arg_4-OH [75]. This is the first determination of the complete sequence of a basic nuclear protein and it shows none of the simple periodic nature once suggested for such compounds. Thermolysin does not attack peptide bonds to the amino groups of arginine or lysine residues and it is suggested that this enzyme may be better than the chemical method for fragmenting basic peptides in a complementary way to trypsin[75].

3.3.2.2 *Pro-insulin C-peptides*

The structure of C-peptide, the peptide which connects the A and B chains of insulin in its biosynthetic precursor pro-insulin, was first described for the pig[76]. In this work and in subsequent studies on bovine C-peptide the sequences were determined on the pro-insulins. The C-peptides, however, survive in the pancreas after cleavage to insulin, and rather larger quantities can be isolated from this source. Studies on uncombined bovine C-peptide[77,78] have confirmed its identity with the sequence proposed for the central segment of pro-insulin, and parallel studies on human C-peptide have revealed the amino acid sequence of this third variety[79,80]. The sequences are compared in Figure 3.7*. The isolation of human C-peptide was complicated by the limited amount of material available and the lack of anti-serum or biological activity to distinguish the peptide. However, material more than 70% pure was obtained using an isolation procedure based on that devised for porcine C-peptide. It is suggested that the heterogeneity is partly due to the action of pancreatic carboxypeptidases. Porcine and bovine C-peptides have resistant C-terminal sequences (-Pro-Pro-Gln-OH) but human C-peptide (-Ser-Leu-Gln-OH) is susceptible to attack.

Some differences of opinion exist as to the positions of glutamine or glutamic acid residues in human and porcine C-peptides, and the arrangement at positions 18 and 20 is thought by Salokangas *et al.*[78] to be the reverse of the original assignment. There seems to be a considerable degree of freedom in the structural requirements for C-peptide and c.d. studies indicate that the connecting chain in pro-insulin is largely random coil in nature[81]. Rhesus monkey C-peptide differs from its human counterpart only in the replacement of one proline by leucine[80].

3.3.2.3 *Calcitonin*

The structure of the hormone porcine calcitonin (see Figure 3.8) was first announced in 1968, and this peptide has excited a great deal of interest

*In this and subsequent figures the sequences are arranged to show maximum homology.

Man:* H-Glu-Ala-Glu-Asp-Leu-Gln-Val-Gly-Glx-Val-Glx-Leu-Gly-Gly-Gly-Pro-Gly-Ala-Gly-Ser-Leu-Gln-Pro-Leu-Ala-Leu-Glu-Gly-Ser-Leu-Gln-OH
Pig:† H-Glu-Ala-Glx-Asn-Pro-Gln-Ala-Gly-Ala-Val-Glu-Leu-Gly-Gly-Gly-Leu-Gly-Gly- - - - -Leu-Gln-Ala-Leu-Ala-Leu-Glu-Gly-Pro-Pro-Gln-OH
Ox:‡ H-Glu-Val-Glu-Gly-Pro-Gln-Val-Gly-Ala-Leu-Glu-Leu-Ala-Gly-Gly-Pro-Gly-Gly-Gly- - - - - - - - - -Ala-Leu-Glu-Gly-Pro-Pro-Gln-OH
10 20 30

* Ko *et al.*[143] give Gln at 9 and Glu at 11, but Steiner *et al.*[77] Glu at 9 and Gln at 11
† Chance *et al.*[76] give Glu at 3, but Ko *et al.*[143] Gln at 3
‡ Salokangas *et al.*[78] give Ala at 18 and Gly at 25

Figure 3.7 Comparison of peptide sequences

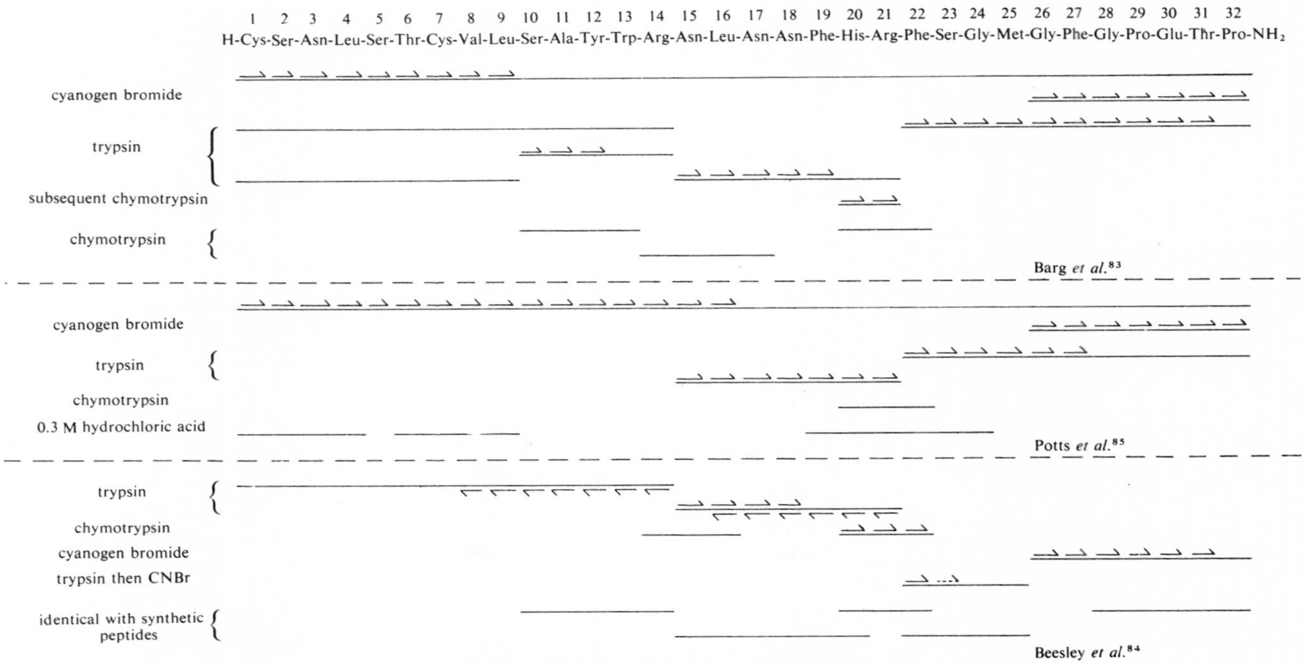

Figure 3.8 Comparison of peptide sequences

because of its potential therapeutic use. It is a dotriacontapeptide which inhibits the action of osteoclasts, the cells which resorb bone. The compound is unique among peptides in that its sequence has been established by four groups working independently, and with the publication in 1970 of the degradative details of the last of these structural determinations, it is interesting to compare these four approaches to the sequencing of a medium-sized peptide. There are two arginine residues in the calcitonin molecule conveniently located so that treatment with trypsin cleaves the chain into three fragments not too dissimilar in size. The CIBA group determined the sequence by studies on the tryptic fragments alone[82]. Early in tryptic hydrolysates two peptides were isolable, 1–14 and 15–32, but after longer incubation 15–32 was cleaved at position 21. Since the peptide corresponding to 22–32 contained no arginine it must be the *C*-terminal fragment and since 1–14 was split off while the *C*-terminus was still bound to the central peptide it must be *N*-terminal. Thus the relative positions of the tryptic fragments was established without recourse to other degradations to obtain overlapping peptides.

Using the dansyl–Edman combination, it proved possible to determine directly the sequence of peptide 1–14. With native calcitonin no PTH-derivatives appear at residues 1 and 7 as the disulphide bridge prevents their release, but performic oxidised material gives cysteic acid derivatives. Tryptophan was not positively identified at position 13 and was positioned by difference. The sequence of peptide 15–21 was obtained by a combination of stepwise determinations using Edman at the *N*-terminus and carboxypeptidase C at the *C*-terminus. Peptide 22–32 was sequenced by the Edman alone as it was resistant to carboxypeptidase. In order to confirm the *C*-terminal sequence, this latter peptide was split with chymotrypsin to liberate the *C*-terminal pentapeptide; a high concentration of enzyme was required to effect this. After 2 cycles of the Edman degradation, the resulting tripeptide was indistinguishable from synthetic H-Glu-Thr-Pro-NH_2 [82].

The major peptides and sequence studies involved in the other three structural determinations are outlined in Figure 3.8. Barg *et al.*[83] found it possible to carry out nine cycles of the Edman degradation on the intact peptide. In addition to splitting calcitonin at arginines 14 and 21, trypsin also furnished this group with a peptide 10–14 through fission at leucine 9. Sequences 10–12, 15–19 and 22–31 were established on tryptic peptides by the Edman method and the proline at 32 on a heptapeptide produced by cyanogen bromide cleavage. As in the Beesley *et al.*[84] and Potts *et al.*[85] determinations the major problem was isolating peptides with overlaps 14–15 and 21–22. Sequence 20–21 was established on a chymotryptic fragment of 15–21 and overlap peptides for 12–15 and 21–22 were obtained from chymotryptic digests of calcitonin itself to enable the complete sequence to be deduced[83]. Threonylprolinamide is unstable towards dioxopiperazine formation, complicating the sequencing at the *C*-terminus[82, 83].

Potts *et al.*[85] successfully carried out 16 cycles of the Edman degradation on calcitonin and seven cycles on both the *C*-terminal heptapeptide (obtained by cyanogen bromide cleavage) and the central tryptic fragment 15–21. Six Edman cycles on the *C*-terminal tryptic fragment sufficed to establish an overlap with the cyanogen bromide *C*-terminal fragment, and the sequence could then only be written in one way even though there was no overlap at

21–22. Overlap peptides for this region were in fact obtained from chymotryptic and partial acid hydrolysates (see Figure 3.8). Fragments of the *N*-terminal portion of the molecule were obtained for confirmation by degrading [^{14}C]iodoacetic acid treated calcitonin and isolating the labelled products. In all this group studied 59 peptide fragments, enough to be able to obtain two independent solutions for the amino acid sequence. Cleavage of the native hormone with sodium in liquid ammonia was found to give a high yield of prolinamide[85].

Beesley *et al.*[84] only report studies defining the sequence 8–32. The sequence 8–21 was largely established by following carboxypeptidase A and B digests of two tryptic peptides, while 26–32 was found by seven cycles of the Edman degradation on the *C*-terminal cyanogen bromide peptide. A peptide bridging the gap 22–25 was isolated from the *C*-terminal tryptic fragment after further treatment with cyanogen bromide and its sequence was confirmed by comparison with synthetic phenylalanylserylglycylhomoserine. Four other synthetic peptides were prepared for comparison with ones obtained by degradation of calcitonin (see Figure 3.8). One chymotryptic fragment 10–13 contains a Tyr–Trp bond stable to chymotrypsin; this was confirmed by studies on the synthetic tetrapeptide[84].

3.3.2.4 *Bombinin*

A peptide with haemolytic activity whose structure (10) bears some resemblance to melittin has been isolated from the defensive secretion of a European toad. This tetracosapeptide, bombinin, contains three lysine residues and so is split into four fragments by trypsin. Chymotrypsin attacks bombinin at the single phenylalanine residue and also breaks the chain at leucine residues 5, 11, 14 and 18 to give peptides conveniently overlapping those from

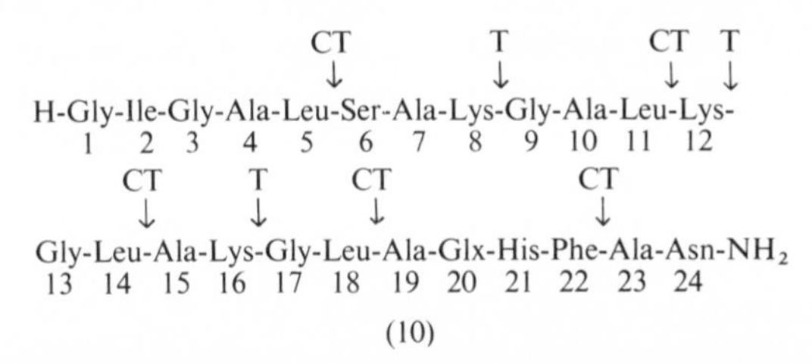

(10)

trypsin. A tripeptide sequence Gly-Ala-Leu and a tetrapeptide sequence Lys-Gly-Leu-Ala occur twice in the chain, somewhat complicating the determination of the primary structure. However, it proved possible to unravel the sequence, although the blocking group of the glutamic acid residue γ-carboxyl has not yet been identified[86].

3.3.2.5 *Gastrointestinal peptides*

Details of the structure determination of porcine secretin have now been published[87]. The sequence itself (Figure 3.9) was announced several years ago, but publication of the details was withheld pending the development of a more satisfactory method for the separation of tryptic peptides. Synthetic peptides were prepared in some instances to check sequence assignments, e.g. both H-Leu-Leu-Gln-Gly-Leu-Val-NH_2 and H-Leu-Gln-Leu-Gly-Leu-Val-NH_2 were synthesised to check the *C*-terminal sequence; only the former behaved identically to one of the tryptic fragments on further degradation with subtilisin[87]. Secretin is produced by the duodenum and glucagon by the pancreas. These two peptides show a degree of homology and the *N*-terminal region of a rather longer peptide discovered more recently, 'gastric inhibitary polypeptide'[88], is obviously more closely related to glucagon (Figure 3.9). This potent inhibitor of gastric acid secretion has been isolated from a preparation of the gastro-intestinal hormone cholecystokinin-pancreozymin[87]. A decapeptide isolated from human gastric juice has the sequence H-Leu-Ala-Ala-Gly-Lys-Val-Glu-Asp-Ser-Asp-OH, but its function is uncertain[89].

3.3.2.6 Trypsin inhibitors

The amino acid sequences of several peptides from animal sources which act as trypsin inhibitors have been determined since 1966, but in general there is little resemblance between them. Recently two further compounds with this type of activity have been sequenced. Porcine pancreatic trypsin inhibitor was rendered sensitive to tryptic digestion by reduction and carboxymethylation, and nine fragments were isolated. These were sequenced using the *p*-bromophenylisothiocyanate variation of the Edman method (see section on methods). Overlap peptides were obtained by restricting tryptic cleavage to the two arginine residues, which was accomplished by modifying the lysine residues with methylisothiocyanate[90]. The sequence determined differs at twelve positions from that found by Greene *et al.*[91] for bovine pancreatic inhibitor (Figure 3.10). One interesting difference lies in the active sites of these peptides. In the bovine variety the reactive site has been located at the Arg—Ile (18–19) bond because it undergoes limited specific site cleavage by trypsin resin and catalytic quantities of trypsin at acid pH, but in the porcine peptide a Lys—Ile bond is involved. A second trypsin inhibitor occurs in porcine pancreas, but its sequence is identical except for the loss of the *N*-terminal tetrapeptide Thr-Ser-Pro-Gln [92].

The trypsin inhibitor of *Arachis hypogaea* seeds occurs as a tetramer of sub-units of 48 residues, but it bears little resemblance to the pancreatic materials other than the possession of five disulphide bridges[93].

3.3.2.7 Snake toxins

Snake venoms are mixtures of proteins with enzymic activity and of basic peptides which are toxins or neurotoxins. The primary structures of some of these toxins from the families *Elapidae* (mainly cobra species) and *Hydropheidae* (sea snakes) have recently been determined; most of these structures were published in 1970 or 1971. Of the 13 toxins whose sequences have been

Glucagon: H-His-Ser-Gln-Gly-Thr-Phe-Thr-Ser-Asp-Tyr-Ser-Lys-Tyr-Leu-Asp-Ser-Arg-Arg-Ala-Gln-Asp-Phe-Val-Gln-Trp-Leu-Met-Asn-Thr-OH (10, 20, 29)

Gastric inhibitory polypeptide: H-Tyr-Ala-Glu-Gly-Thr-Phe-Ile-Ser-Asp-Tyr-Ser-Ile-Ala-Met-Asp-Lys-Ile-Arg-Gln-Gln-Asp-Phe-Val-Asn-Trp-Leu-Leu-Ala-Gln-Gln*

Secretin: H-His-Ser-Asp-Gly-Thr-Phe-Thr-Ser-Glu-Leu-Ser-Arg-Leu-Arg-Asp-Ser-Ala-Arg-Leu-Gln-Arg-Leu-Leu-Gln-Gly-Leu-Val-OH

-* sequence continues: Lys-Gly-Lys-Lys-Ser-Asp-Trp-Lys-His-Asn-Ile-Thr-Gln-OH

Figure 3.9 Peptide sequence

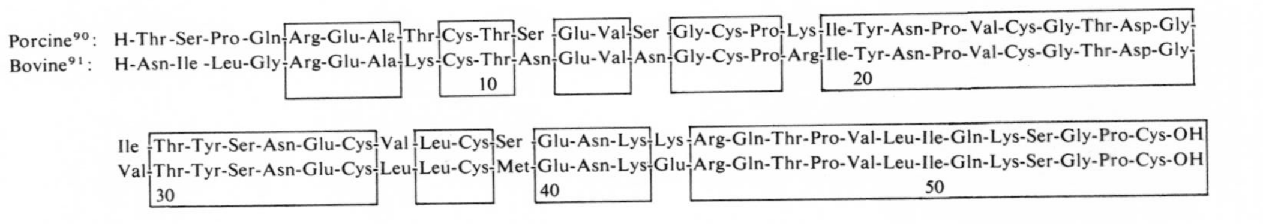

Figure 3.10 Peptide sequence

established, eight contain either 61 or 62 amino acid residues, about half of which are common to all the members (Figure 3.11; cardiotoxin is shown for comparison). In particular, there is a heptapeptide sequence which is invariant and which contains two of the eight cysteine residues. A comparison of the sequences shows that in the peptides with 61 residues it is the residue at position 19 of the 62 residue sequence which is missing. None of the peptides contain alanine, and methionine only occurs at the *N*-terminus of toxin *β*. The sea snake peptides erabutoxins a and b differ in sequence at only one position, 26, and this is explicable by a single-base replacement in the triplet code[94].

The pairing of the eight cysteine residues has been determined only for cobratoxin (11) and erabutoxin a, but there is little reason to suppose that

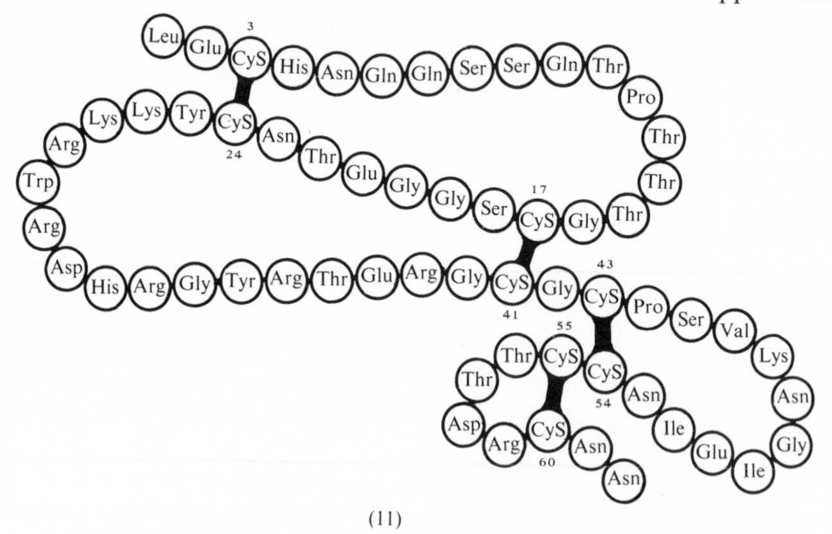

(11)

From Nakai, Sasaki and Hayashi[107], reproduced by permission of Elsevier.

the other members outlined in Figure 3.11 will differ from this. The single tryptophan residues of cobratoxin[108] and erabutoxin a [104] have been shown to be essential to the toxic action of the peptides. Six of the seven *γ*-carboxyl groups of cobratoxin will react with methyl glycinate in the presence of 1-ethyl-3-dimethylaminopropylcarbodi-imide with retention of toxicity. The seventh *γ*-carboxyl can be modified also in 5 M guanidine hydrochloride solution, but the peptide is then devoid of activity. Radiotracer studies with methyl [^{14}C]glycinate have located the buried glutamic acid residue at position 21 [103].

The sequences of two toxins from different species (cardiotoxin from the Formosan cobra and cytotoxin I from the Indian cobra) which contain 60 amino acid residues are very similar in sequence (Figure 3.12) but are quite different to the toxins shown in Figure 3.11. The cysteine residues, however, are rather similarly disposed. If they are paired up in the same fashion as cobratoxin, then only one loop (3–24 as opposed to 3–21) differs in size. Cytotoxin I is so named because of its high toxicity to Yoshida sarcoma and Ascites hepatoma cells. This toxicity disappears on reduction and either

(1) Erabutoxin a:[94] Arg-Ile-Cys-Phe-Asn-Gln-His-Ser-Ser-Gln-Pro-Gln-Thr-Thr-Lys-Thr-Cys-Pro-Ser-Gly-Ser-Glu-Ser-Cys-Tyr-Asn-Lys-
(2) Erabutoxin b:[94] Arg-Ile-Cys-Phe-Asn-Gln-His-Ser-Ser-Gln-Pro-Gln-Thr-Thr-Lys-Thr-Cys-Pro-Ser-Gly-Ser-Glu-Ser-Cys-Tyr-His-Lys-
(3) Toxin α*:[95] Leu-Glu-Cys-His-Asn-Gln-Gln-Ser-Ser-Gln-Pro-Pro-Thr-Thr-Lys-Thr-Cys-Pro- - Gly-Glu-Thr-Asn-Cys-Tyr-Lys-Lys-
(4) Cobratoxin:[96] Leu-Glu-Cys-His-Asn-Gln-Gln-Ser-Ser-Gln-Thr-Pro-Thr-Thr-Thr-Gly-Cys-Ser-Gly-Gly-Glu-Thr-Asn-Cys-Tyr-Lys-Lys-
(5) Toxin α†:[97] Leu-Gln-Cys-His-Asn-Gln-Gln-Ser-Ser-Gln-Pro-Pro-Thr-Thr-Lys-Thr-Cys-Pro- - Gly-Glu-Thr-Asn-Cys-Tyr-Lys-Lys-
(6) Toxin II:[98] Leu-Glu-Cys-His-Asn-Gln-Gln-Ser-Ser-Gln-Pro-Pro-Thr-Thr-Lys-Ser-Cys-Pro- - Gly-Asp-Thr-Asn-Cys-Tyr-Asn-Lys-
(7) Toxin β:[99] Met-Ile-Cys-His-Asn-Gln-Gln-Ser-Ser-Gln-Arg-Pro-Thr-Ile-Lys-Thr-Cys-Pro- - Gly-Glu-Thr-Asn-Cys-Tyr-Lys-Lys-
(8) Toxin IV:[98] Leu-Glu-Cys-His-Asn-Gln-Gln-Ser-Ser-Gln-Thr-Pro-Thr-Thr-Gln-Thr-Cys-Pro- - Gly-Glu-Thr-Asn-Cys-Tyr-Lys-Lys-
(9) Cardiotoxin:[105] Leu-Lys-Cys-Asp-Lys-Leu-Val-Pro-Leu-Phe-Tyr-Lys-Thr-Cys-Pro-Ala-Gly-Lys-Asn-Leu-Cys-Tyr-Lys-Met-Phe-Met-Val-
10 20

(1) Gln-Trp-Ser-Asp-Phe-Arg-Gly-Thr-Ile-Ile-Glu-Arg-Gly-Cys-Gly-Cys-Pro-Thr-Val-Lys-Pro-Gly-Ile-Lys-Leu-Ser-Cys-Cys-Glu-Ser-Glu-Val-
(2) Gln-Trp-Ser-Asp-Phe-Arg-Gly-Thr-Ile-Ile-Glu-Arg-Gly-Cys-Gly-Cys-Pro-Thr-Val-Lys-Pro-Gly-Ile-Lys-Leu-Ser-Cys-Cys-Glu-Ser-Glu-Val-
(3) Val-Trp-Arg-Asp-His-Arg-Gly-Thr-Ile-Ile-Glu-Arg-Gly-Cys-Gly-Cys-Pro-Thr-Val-Lys-Pro-Gly-Ile-Lys-Leu-Asn-Cys-Cys-Thr-Thr-Asp-Lys-
(4) Arg-Trp-Arg-Asp-His-Arg-Gly-Tyr-Arg-Thr-Glu-Arg-Gly-Cys-Gly-Cys-Pro-Ser-Val-Lys-Asn-Gly-Ile-Glu-Ile-Asn-Cys-Cys-Thr-Thr-Asp-Arg-
(5) Arg-Trp-Arg-Asp-His-Arg-Gly-Ser-Ile-Thr-Glu-Arg-Gly-Cys-Gly-Cys-Pro-Ser-Val-Lys-Lys-Gly-Ile-Glu-Ile-Asn-Cys-Cys-Thr-Thr-Asp-Lys-
(6) Arg-Trp-Arg-Asp-His-Arg-Gly-Thr-Ile-Ile-Glu-Arg-Gly-Cys-Gly-Cys-Pro-Thr-Val-Lys-Pro-Gly-Ile-Asn-Leu-Lys-Cys-Cys-Thr-Thr-Asp-Arg-
(7) Arg-Trp-Arg-Asp-His-Arg-Gly-Thr-Ile-Ile-Glu-Arg-Gly-Cys-Gly-Cys-Pro-Ser-Val-Lys-Lys-Gly-Val-Gly-Ile-Tyr-Cys-Cys-Lys-Thr-Asp-Lys-
(8) Gln-Trp-Ser-Asp-His-Arg-Gly-Ser-Arg-Thr-Glu-Arg-Gly-Cys-Gly-Cys-Pro-Thr-Val-Lys-Pro-Gly-Ile-Lys-Leu-Lys-Cys-Cys-Thr-Thr-Asp-Arg-
(9) Ala-Thr-Pro-Lys-Val-Pro-Val-Lys-Arg- - -Gly-Cys-Ile-Asp-Val-Cys-Pro-Lys-Ser-Ser-Leu-Val-Leu-Lys-Tyr-Val-Cys-Cys-Asn-Thr-Asp-Arg-
30 40 50

(1) Cys-Asn-Asn
(2) Cys-Asn-Asn
(3) Cys-Asn-Asn
(4) Cys-Asn-Asn
(5) Cys-Asn-Asn
(6) Cys-Asn-Asn
(7) Cys-Asn-Arg
(8) Cys-Asn-Asn
(9) Cys-Asn
60

* from *Naja nigricollis* † from *Naja haje haje*; identical to toxin δ from *Naja nivea*[101]
The erabutoxins are from *Laticauda semifasciata*, cobratoxin from *Naja naja atra*, toxin β from *Naja nivea*, and toxins II and IV from *Hemachatus haemachatus*.

Figure 3.11 Peptide sequence

carboxymethylation or aminoethylation, showing that the disulphide bridges are essential to the activity[100]. Partial sequences have been determined for two other toxins (CM-XI and CM-XII) from the Indian cobra[106]. Comparison of the sequences 1–27 of CM-XI and 1–28 of CM-XII (as determined by Edman degradation of the intact toxins) with cardiotoxin and cytotoxin I shows many similarities (Figure 3.12), although the amino acid compositions of the former peptides indicate that they probably contain 61 or 62 amino acid residues.

The three remaining toxins whose primary structures have been established fall into a third structural category (Figure 3.13). Toxin α of *Naja nivea*[99] and toxin A of *Naja naja*[107] both contain 71 residues and differ in sequence at

Cardiotoxin[105]: H-Leu-Lys-Cys-Asn-Lys-Leu-Val-Pro-Leu-Phe-Tyr-Lys-Thr-Cys-Pro-Ala-Gly-Lys-Asn-Leu-
Cytotoxin I[100]: H-Leu-Lys-Cys-Asn-Lys-Leu-Ile-Pro-Leu-Ala-Tyr-Lys-Thr-Cys-Pro-Ala-Gly-Lys-Asn-Leu-
CM-XI[106]: H-Leu-Lys-Cys-Asn-Lys-Leu-Ile-Pro-Leu-Ala-Tyr-Lys-Thr-Cys-Pro-Tyr-Gly-Lys-Asn-Leu-
CM-XII[106]: H-Leu-Lys-Cys-Asn-Lys-Leu-Val-Pro-Leu-Phe-Tyr-Lys-Thr-Cys-Pro-Ala-Gly-Lys-Asn-Leu-
10 20

Cys-Tyr-Lys-Met-Phe-Met-Val-Ala-Thr-Pro-Lys-Val-Pro-Val-Lys-Arg-Gly-Cys-Ile-Asp-
Cys-Tyr-Lys-Met-Tyr-Met-Val-Ser-Asn-Lys-Thr-Val-Pro-Val-Lys-Arg-Gly-Cys-Ile-Asp-
Cys-Tyr-Lys-Met-Tyr-Met-Val-
Cys-Tyr-Lys-Met-Tyr-Met-Val-Ala-
30 40

Val-Cys-Pro-Lys-Ser-Ser-Leu-Val-Leu-Lys-Tyr-Val-Cys-Cys-Asn-Thr-Asp-Arg-Cys-Asn-OH
Val-Cys-Pro-Lys-Asn-Ser-Leu-Val-Leu-Lys-Tyr-Glu-Cys-Cys-Asn-Thr-Asp-Arg-Cys-Asn-OH
50 60

Figure 3.12 Peptide sequence

Bungarotoxin[108]: H-Ile-Val-Cys-His-Thr-Thr-Ala-Thr-Ile-Pro-Ser-Ser-Ala-Val-Thr-Cys-Pro-Pro-Gly-Glu- Asn-Leu-Cys-Tyr-Arg-
Toxin A[107]: H-Ile-Arg-Cys-Phe-Ile-Thr-Pro-Asp-Val-Thr-Ser-Lys-Asp- - - -Cys-Pro-Asn-Gly-His- Val- - -Cys-Tyr-Thr-
Toxin α[99]: H-Ile-Arg-Cys-Phe-Ile-Thr-Pro-Asp-Val-Thr-Ser-Gln-Ala- - - -Cys-Pro-Asp-Gly-His- Val- - -Cys-Tyr-Thr-
10 20

Lys-Met-Trp-Cys-Asp-Ala-Phe-Cys-Ser-Ser-Arg-Gly-Lys-Val-Val-Glu-Leu-Gly-Cys-Ala-Ala-Thr-Cys-Pro-Ser-
Lys-Thr-Trp-Cys-Asp-Gly-Phe-Cys-Ser-Ile-Arg-Gly-Lys-Arg-Val-Asp-Leu-Gly-Cys-Ala-Ala-Thr-Cys-Pro-Thr-
Lys-Met-Trp-Cys-Asp-Asn-Phe-Cys-Gly-Met-Arg-Gly-Lys-Arg-Val-Asp-Leu-Gly-Cys-Ala-Ala-Thr-Cys-Pro-Lys-
30 40 50

Lys-Lys-Pro-Tyr-Glu-Glu-Val-Thr-Cys-Cys-Ser-Thr-Asp-Lys-Cys-Asn-His-Pro-Pro-Lys-Arg-Gln-Pro-Gly-OH
Val-Arg-Thr-Gly-Val-Asp-Ile-Gln-Cys-Cys-Ser-Thr-Asp-Asp-Cys-Asp-Pro-Phe-Pro-Thr-Arg-Lys-Arg-Pro-OH
Val-Lys-Pro-Gly-Val-Asn-Ile-Lys-Cys-Cys-Ser-Arg-Asp-Asn-Cys-Asn-Pro-Phe-Pro-Thr-Arg-Lys-Arg-Ser-OH
60 70

Figure 3.13 Peptide sequence

17 positions although the 10 cysteine residues are in identical positions. α-Bungaratoxin from *Bungarus multicinctus* possesses 74 residues[108], but if deletions are assumed at positions 14, 15 and 22, 34 residues are homologous with toxins β and A and the cysteines fit in correctly. The cysteine pairings of toxin α have been determined and are shown in Figure 3.14. This toxin has relatively few thermolysin labile bonds and three of the five disulphide bridges could be positioned directly from the amino acid composition of the separated cystine-containing thermolysin peptides (Figure 3.14, peptides I, II and III). The other two disulphide bridges were in one peptide (peptide IV) and to position them required cleavage between the adjacent cystine residues at positions 56 and 57. This peptide proved resistant to cleavage by subtilisin, pepsin, or pronase, but after treatment with trypsin had largely cleaved the

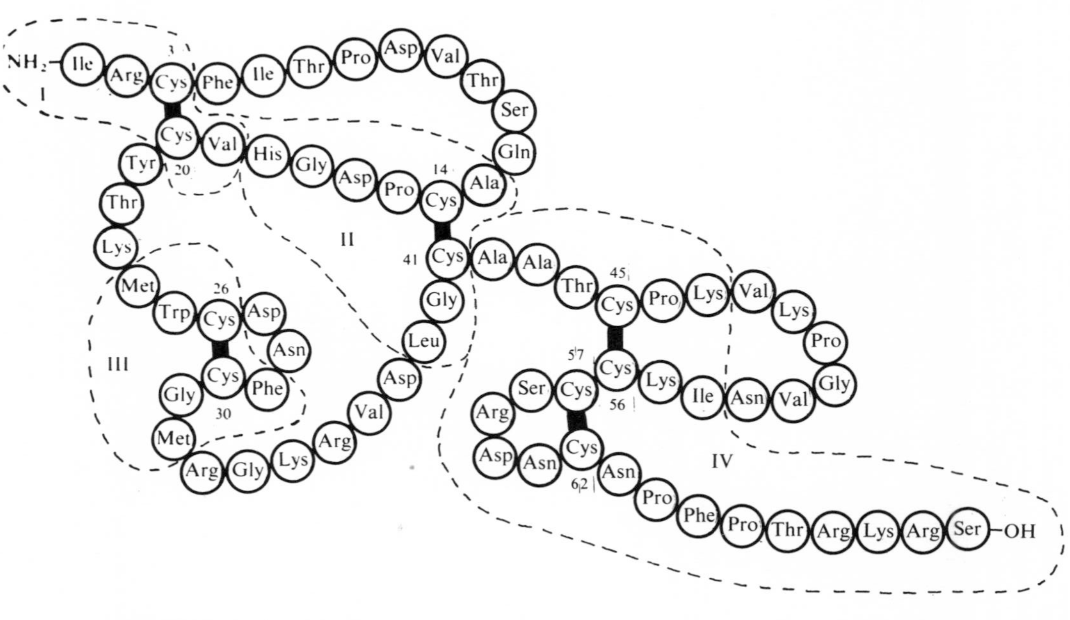

Figure 3.14 Structure
(From Botes[99], reproduced by permission of the American Society of Biological Chemists)

small loop at arginine 59, prolonged digestion with pronase gave a mixture of peptides. From this mixture the isolation of a peptide with the composition Cys_2Lys_2Pro enabled a link between 45 and 56 to be assigned and by difference 57 must be linked to 62[99].

Cobratoxin is not attacked by pepsin and neither trypsin or chymotrypsin, alone or in combination, cleaves to give single cystine peptides. The cysteine pairings in this case were successfully located by studying fragments produced by acid protease A and separated by high-voltage electrophoresis on paper. The five cystine containing peptides were oxidised with performic acid and amino acid analyses were carried out after a further electrophoretic separation. As in toxin α, one peptide contained two cystine residues owing to the contiguous positions of two cystines at 54 and 55 and required further treatment. Partial acid hydrolysis at 37 °C with 5 M sulphuric acid in 50% acetic acid gave a mixture of 18 products, the largest of which was a heptapeptide. Amino acid analysis of the cysteic derivatives of these enabled the 43–54 and 55–60 bridges to be established (11)[109]. An identical pattern of disulphide bridges has also been established for erabutoxin a [110]. Both toxin α and cobratoxin contain a loop with a disulphide bridged ring the same size as ones contained in insulin, oxytocin and vasopressin, but the biological significance of this remains to be explored.

In the determination of the sequences of these snake toxins a few anomalies are worth noting. Chymotrypsin was found to cleave the *Hemachatus* toxins[98] and cobratoxin[96] at both threonine and glutamine residues, while the Lys—Pro bond at 47–48 in erabutoxin b proved resistant to trypsin[94]. In toxin A, application of the tritiation method of determination of the *C*-terminal amino acid indicated an alanine residue. However, hydrazinolysis liberated proline and this was confirmed by the sequencing of tryptic and chymotryptic peptides. Only two alanine residues are present, and these were unequivocally located at positions 42 and 43. This mystery remains unsolved[102]. The Edman method proved ineffective on a hexapeptide H-Ser-Ser-Leu-Val-Leu-Lys-OH (45–50) tryptic fragment of cardiotoxin and its sequence was determined by use of carboxypeptidase A and thermolysin[105].

Of the two toxins of rather different types isolated from the Formosan cobra, cobratoxin is the major toxic component, causing peripheral respiratory paralysis. Cardiotoxin produces cardiovascular changes, and is responsible for local necrotic lesions[105].

3.3.2.8 Viscotoxins

The amino acid sequence of viscotoxin A3 (Figure 3.15), one of the principal pharmacologically active peptides from the European mistletoe *Viscum Album L.*, was established in 1968. Seven tryptic and five chymotryptic peptides are obtained as primary digestion products and these give complete overlaps for the whole molecule. The further degradation of the tryptic peptides is also outlined in Figure 3.15. The cleavage of the Thr–Cys bond at 25–26 by chymotrypsin is unusual, and the failure of this enzyme to split the peptide bond after residue 43 is indicative of a 43–44 sequence of Tyr-Pro rather than Pro-Tyr. Enzymic degradations were carried out after performic acid oxidation of viscosin A3, and during this oxidation considerable conversion of the tyrosine residues to 3-chlorotyrosine occurred. In peptide fragments containing the Ile-Ile sequence the normal conditions for complete acid hydrolysis of peptides only liberated one free residue of isoleucine[111].

More recently the arrangement of the disulphide bridges in viscotoxin A3 has been investigated. A link between cysteine residues 16 and 26 was established, but so far it has proved impossible to selectively cleave the peptide

T T CT T CT T

H-Lys-Ser-Cys-Cys-Pro-Asn-Thr-Thr-Gly-Arg-Asn-Ile-Tyr-Asn-Ala-Cys-Arg-Leu-Thr-Gly Ala-Pro-Arg-

10 (Thr) (Gly) (Gly-Ser) 20

CT T CT T

Pro-Thr-Cys-Ala-Lys-Leu-Ser-Gly-Cys-Lys-Ile-Ile-Ser-Gly-Ser-Thr-Cys-Pro-Ser-Tyr-Pro-Asn-Lys-OH

(Glu-Arg) (Ser) (Ala)

30 40

Figure 3.15 Viscotoxin A_3. The different amino acids of viscotoxin B are given in brackets

bond between the adjacent cysteine residues 3 and 4 in fragments which contain the intact disulphide bridges. It is clear that residues 3 and 4 are not bridged to each other, but there remain two possible arrangements of the bridges, as shown in (12)[112]. The sequence of a peptide from the same source,

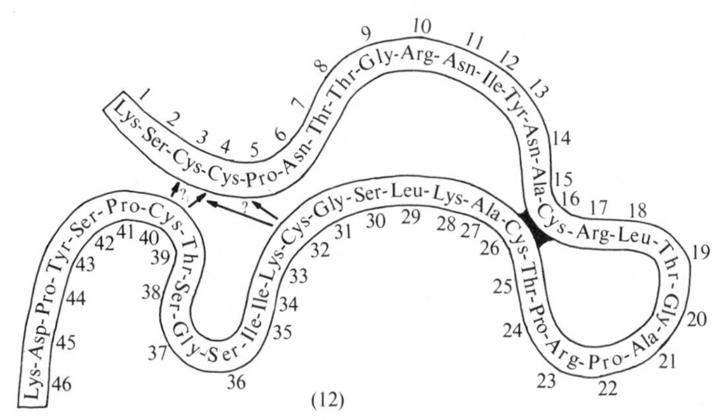

From Samuelsson and Pettersson[112], reproduced by permission of Munksgaard

viscotoxin B, has been found to differ from that of A3 at eight positions, principally in the central region (Figure 3.15)[113].

3.3.3 Cyclic peptides

3.3.3.1 Cyclotetrapeptides

Naturally occurring cyclic tetrapeptides are rare. The first totally α-linked compound of this type to be identified, fungisporin, was obtained as a crystalline sublimate on the destructive distillation of *Penicillium* and *Aspergillus* spores. Although originally formulated as a cyclic octapeptide, mass spectrometric work and synthesis have now established it as cyclo (D-Val–L-Val–D-

Phe–L-Phe)[114]. More recently tentoxin, a metabolite of *Alternaria tenuis Auct.* which produces chlorosis of many dicotyledonous plant species, has been assigned a cyclotetrapeptide structure. Resonance from all its 30 protons are evident in the n.m.r. spectrum, which supports a cyclic structure and the presence of a PhCH:C or PhC:CH grouping. Acid hydrolysis gives glycine, leucine, *N*-methylalanine and methylamine. Hydrogenation of tentoxin gives a mixture of dihydro, hexahydro, and octahydro products and acid hydrolysis of this mixture gave a compound which was chromatographically indistinguishable from *N*-methyl-β-cyclohexylalanine. This established the presence of a residue of *N*-methyl-αβ-dehydrophenylalanine, which accords with the observed u.v. spectrum. The sequential arrangement of this unsaturated amino acid with respect to the glycine, *N*-methylalanine, and leucine residues has not yet been determined and the configurations remain undefined[115].

A β-linked cyclic tetrapeptide has been isolated from the lichen *Roccella canariensis Darb.* Acid hydrolysis yields L-proline and D-β-amino-β-phenylpropionic acid. The mass spectrum and the symmetrical nature of the n.m.r.

NH–CH–CH_2 / CO Ph CO / N N / CO Ph CO / CH_2–CH–NH

(13)

CH_2–CO–NH / Ph–CH CH–Ph / NH CH_2 / CO CO / N / N–CO

(14)

spectrum favour (13), but (14) cannot be ruled out[116]. No cyclic peptides larger than a dioxopiperazine have previously been isolated from lichens.

3.3.3.2 *Mycobacillin*

Mycobacillin is an acidic macrocyclic peptide from *Bacillus subtilis B_3* whose amino acid sequence was defined some 10 years ago. Recent work has now established the configuration of its component amino acids and that it is not wholly α-linked[117]. Titration shows that this tridecapeptide contains seven free carboxyl groups, which corresponds to its known content of five aspartic acid and two glutamic acid residues. Treatment of mycobacillin with *N*-bromosuccinimide liberates carbon dioxide in an amount corresponding to the decarboxylation of two α-carboxyl groups. Oxidation of this antifungal antibiotic with sodium hypobromite followed by acid hydrolysis liberates succinic acid. This is indicative that at least one of the two glutamic acid residues is γ-linked, but this reaction proved unsuitable for distinguishing between one and two γ-linked species. However, hydrazinolysis of the intact macrocycle liberated only the α-hydrazide of aspartic acid and only the γ-hydrazide of glutamic acid. This evidence for two γ-linkages was confirmed by examination of the acid hydrolysate of the lithium borohydride reduction product of the heptamethyl ester of mycobacillin. α-Amino-γ-hydroxybutyric acid and γ-amino-δ-hydroxyvaleric acid were produced in a molar ratio of 5:2, which corresponds to the aspartic and glutamic acid proportions.

The location of D-amino acid residues was determined by enzymic methods on 15 peptides produced by partial acid hydrolysis, of the cyclic peptide. Each peptide was subjected to total acid hydrolysis and the hydrolysates were analysed both before and after incubation with D-amino-acid oxidase. Since aspartic and glutamic acids are resistant to this enzyme, parallel experiments were carried out with L-glutamate decarboxylase. This enzyme also possessed L-aspartate decarboxylase activity, and the appearance of alanine coupled

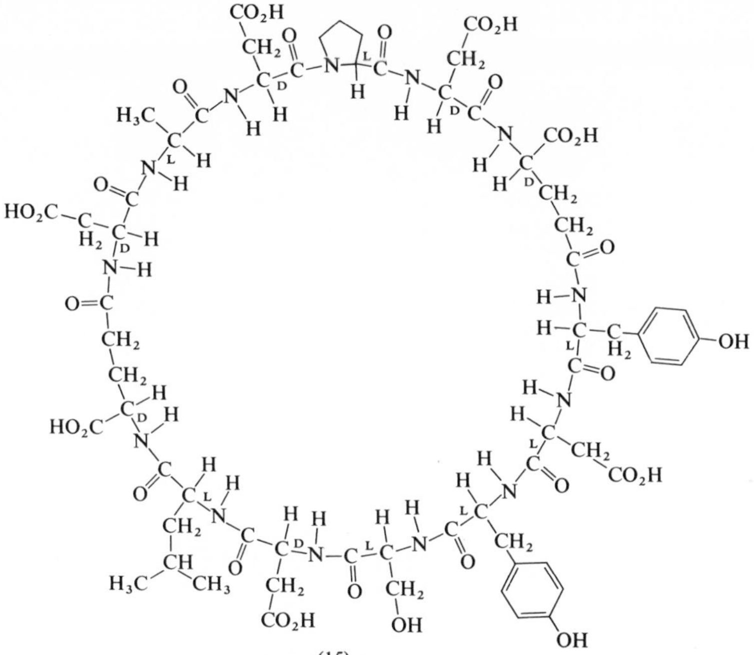

(15)

with the disappearance of aspartic acid served to identify those peptides containing L-aspartic acid. The composition of the two L-aspartate peptides enabled structure (15) to be established for mycobacillin[117].

3.3.3.3 *Alamethicin*

Of current interest also is alamethicin (16), an extracellular peptide from *Trichoderma viride*, which, like some macrocyclic depsipeptides, can transport cations through membranes. It is an octadecapeptide containing seven α-aminoisobutyric acid residues and resembles mycobacillin in having a γ-glutamyl linkage in the ring, although all its residues are of the L-configuration. The presence of α-aminoisobutyric acid as the major component considerably complicated the determination of the sequence. Not only does it render alamethicin and linear sequences derived from it inert towards enzymic digestion, but it has a very low colour yield on amino acid analysis and is frequently incompletely cleaved in the Edman degradation when *N*-terminal. Although partial acid hydrolysis under mild conditions yielded only three peptides, one of which was a linear octadecapeptide, Edman degradation of these could only be carried out satisfactorily for about four cycles and a host of smaller peptides produced by rather more severe acid hydrolysis had to be studied before enough information was obtained to be able to deduce the linear sequence of alamethicin.

A single free carboxyl group can be detected by titration of the intact peptide, indicating the presence of two glutamine or isoglutamine residues,

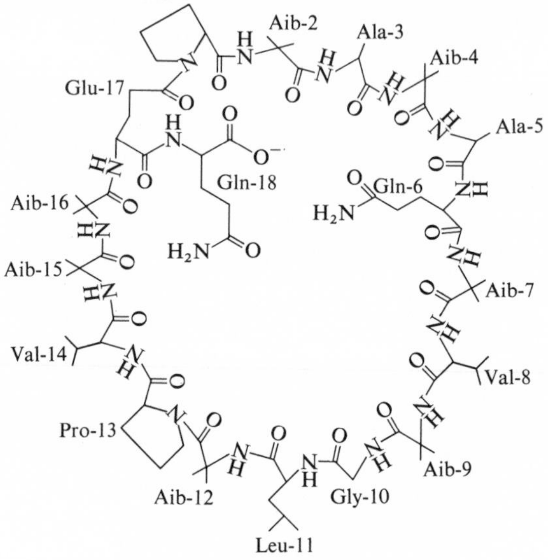

(16)

From Payne, Jakes and Hartley[118], reproduced by permission of Cambridge University Press.

and when it was established that residues 6 and 18 were γ-amidated it only remained to determine whether in alamethicin itself the imino-group of proline-1 was linked to the α-carboxyl of glutamine-18 or the γ-carboxyl of glutamic acid-17. Treatment with diborane reduced the free carboxyl group

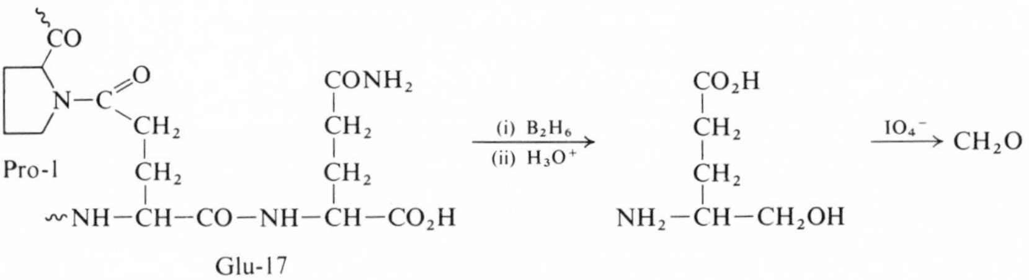

Figure 3.16

to a primary alcohol, and after acid hydrolysis reaction with periodate liberated formaldehyde. This could only have arisen from a free α-carboxyl group via γ-amino-δ-hydroxyvaleric acid (Figure 3.16), so the alamethicin macrocycle must have a pendant glutamine residue as depicted in (16)[118]. It was also concluded that the alamethicin sample studied was contaminated with similar material containing an extra α-aminoisobutyric acid residue at position 3 or 5 in place of alanine.

Confirmation of the sequence of alamethicin has come from a mass spectrometric study of *N*-decanoylpeptide methyl esters prepared from a partial acid hydrolysate. The spectra obtained were relatively simple and readily interpretable on the basis of amino acid (~NH—CHR—C≡O⁺) and aldimine (~$\overset{+}{N}$H=CHR) fragments. The α-linked nature of the glutamic

$$\sim NH-\underset{|}{\overset{|}{C}}H-\overset{+}{C}\equiv O \quad (CH_2CH_2CO_2Me \text{ side chain}) \qquad \sim\overset{+}{N}H=CH-CH_2-CO_2Me$$

$$\sim NH-CH(CH_2-CH{=}CO)-\overset{+}{C}\equiv O \qquad \sim\overset{+}{N}H=CH-CH_2-CH{=}CO$$

Figure 3.17

residues at 6–7 and 17–18 was indicated by the presence of the ions depicted in Figure 3.17 in the spectra of peptides containing these residues. However, the ready fission of the glutamic acid-17 γ-amide link to proline-1 again precluded to formation of peptides with this 1–17 link intact[119].

References

1. Wayne-Moss, C., Lambert, M. A. and Diaz, F. J. (1971). *J. Chromatog.*, **60,** 137
2. Gehrke, C. W., Kuo, K. and Zumwalt, R. W. (1971). *J. Chromatog.*, **57,** 209
3. Samejima, K., Dairman, W. and Udenfriend, S. (1971). *Analyt. Biochem.*, **42,** 222
4. Ellis, J. P. and Prescott, J. M. (1971). *J. Chromatog.*, **61,** 152
5. McKerrow, J. H. and Robinson, A. B. (1970). *Analyt. Biochem.*, **41,** 565
6. Mercier, J. C., Grosclaude, F. and Ribadeau-Dumas, B. (1971). *Eur. J. Biochem.*, **23,** 41
7. Valyulis, R. A. and Stepanov, V. M. (1971). *Biokhimiya,* **36,** 866
8. Sprössler, B., Heilmann, H. D., Grampp, E. and Uhlig, H. (1971). *Hoppe-Seyler's Z. Physiol. Chem.*, **352,** 1524
9. Zuber, H. (1968). *Hoppe-Seyler's Z. Physiol. Chem.*, **349,** 1337
10. Gibbons, I. and Perham, R. N. (1970). *Biochem. J.*, **116,** 843
11. Riley, M. and Perham, R. N. (1970). *Biochem. J.*, **118,** 733
12. Spande, T. F., Witkop, B., Degain, Y. and Patchornik, A. (1970). *Advan. Protein. Chem.*, **24,** 97
13. Inglis, A. S. and Edman, P. (1970). *Analyt. Biochem.*, **37,** 73
14. Sakiyama, S. and Sakai, S. (1971). *Bull. Chem. Soc. Japan,* **44,** 1661
15. Inui, T. (1971). *Bull. Chem. Soc. Japan,* **44,** 2515
16. Niall, H. D. (1971). *Nature New Biol.*, **230,** 90
17. Li, C. H. and Dixon, J. S. (1971). *Arch. Biochem. Biophys.*, **146,** 233
18. Tanaka, M., Haniu, M., Matsueda, G., Yasunobu, K. T., Mayhew, S. and Massey, V. (1971). *Biochemistry,* **10,** 3041
19. Brewer, H. and Ronan, R. (1970). *Proc. Nat. Acad. Sci. USA,* **67,** 1862
20. Smithies, O., Gibson, D., Fanning, E. M., Goodfliesh, R. M., Gilman, J. G. and Ballantyne, D. L. (1971). *Biochemistry,* **10,** 4912
21. von Wilme, M. (1970). *Angew. Chem. Int. Edn.*, **9,** 267

22. Laursen, R. A. (1971). *Eur. J. Biochem.*, **20,** 89
23. Boigne, J. M., Boigne, N. and Rosa, J. (1970). *J. Chromatog.*, **47,** 238
24. Birr, C., Just, C. and Wieland, T. (1970). *Annalen,* **736,** 88
25. Kulbe, K. D. (1971). *Analyt. Biochem.*, **44,** 548
26. Waterfield, M. and Haber, E. (1970). *Biochemistry,* **9,** 832
27. Vance, D. E. and Feingold, D. S. (1970). *Analyt. Biochem.*, **36,** 30
28. Deyl, Z. (1970). *J. Chromatog.*, **48,** 231
29. Birr, C., Reitel, C. and Wieland, T. (1970). *Angew. Chem. Int. Edn.*, **9,** 731
30. Braunitzer, G., Schrank, B. and Ruhfus, A. (1970). *Hoppe-Seyler's Z. Physiol. Chem.*, **351,** 1589
31. Braunitzer, G., Schrank, B., Ruhfus, A., Peterson, S. and Peterson, U. (1971). *Hoppe-Seyler's Z. Physiol. Chem.*, **352,** 1730
32. Rochat, C., Rochat, H. and Edman, P. (1970). *Analyt. Biochem.*, **37,** 259
33. Tschesche, H. and Wachter, E. (1970). *Eur. J. Biochem.*, **16,** 187
34. Weygand, F. and Obermeier, R. (1971). *Eur. J. Biochem.*, **20,** 72
35. Fairwell, T., Barnes, W. T., Richards, F. F. and Lovins, R. E. (1970). *Biochemistry,* **9,** 2260
36. Seiler, N., Schneider, H. H. and Sonnenberg, K-D. (1971). *Analyt. Biochem.*, **44,** 451
37. Stark, G. R. (1968). *Biochemistry,* **7,** 1796
38. Kubo, H., Nakajima, T. and Tamura, Z. (1971). *Chem. Pharm. Bull.*, **19,** 210
39. Yamashita, S. (1971). *Biochem. Biophys. Acta,* **229,** 301
40. Matsuo, H., Fujimoto, Y., Kobayashi, H., Tatsuno, T. and Matsubara, H. (1970). *Chem. Pharm. Bull.*, **18,** 890
41. Barber, M., Jolles, P., Vilkas, E. and Lederer, E. (1965). *Biochem. Biophys. Res. Commun.*, **18,** 469
42. Reviewed by Van Lear, G. E. and McLafferty, F. W. (1969). *Ann. Review of Biochemistry,* **38,** 299
43. Vilkas, E. and Lederer, E. (1968). *Tetrahedron Letters,* p. 3089; Thomas, D. W. (1968). *Biochem. Biophys. Res. Commun.*, **33,** 483
44. White, P. A. and Desiderio, D. M. (1971). *Analyt. Letters,* **4,** 141
45. LeClerc, P. A. and Desiderio, D. M. (1971). *Analyt. Letters,* **4,** 305
46. Vliegenthart, J. F. G. and Dorland, L. (1970). *Biochem. J.* **117,** 318
47. Shemyakin, M. M., Ovchinnikov, Yu.A., Vinogradova, E. A., Kiryushin, A. A., Feigina, M.Yu., Aldanova, N. A., Alakhov, Yu.B., Lipkin, V. M., Miroshnikov, A. I., Rozynov, B. V. and Kazaryan, S. A. (1970). *F. E. B. S. Letters,* **7,** 8
48. Toubiana, R., Bennett, J. E. G., Sach, E., Das, B. C. and Lederer, E. (1970). *F. E. B. S. Letters,* **8,** 207
49. Ovchinnikov, Yu.A., Kiryushin, A. A., Gorlenko, V. A. and Rozynov, B. V. (1971). *Zhur. Obschei, Khim.*, **41,** 660
50. Paz, M. A., Bernath, A., Henson, E., Blumenfeld, O. O. and Gallop, P. M. (1970). *Analyt. Biochem.*, **36,** 527
51. Roepstorff, P., Norris, K., Severinson, S. and Brunfeldt, K. (1970). *F. E. B. S. Letters,* **9,** 235
52. Polan, M. L., McMurray, W. J., Lipsky, S. R. and Lande, S. (1970). *Biochem. Biophys. Res. Commun.*, **38,** 1127
53. LeClercq, P. A., Smith, L. C. and Desiderio, D. M. (1971). *Biochem. Biophys. Res. Commun.*, **45,** 937
54. Morris, H. R., Williams, D. H. and Ambler, R. P. (1971). *Biochem. J.*, **125,** 189
55. Gray, W. R., Wajcik, L. H. and Futrell, J. H. (1970). *Biochem. Biophys. Res. Commun.*, **41,** 1111
56. Milne, G. W. A., Kiryushin, A. A., Alakhov, Yu.A., Lipkin, V. M. and Ovchinnikov, Yu.A. (1970). *Tetrahedron,* **26,** 299
57. Gray, W. R. and del Valle, U. E. (1970). *Biochemistry,* **9,** 2134
58. Geddes, A. J., Graham, G. N., Morris, H. R., Lucas, F., Barber, M. and Wolstenholme, W. A. (1969). *Biochem. J.*, **114,** 695
59. Roepstorff, P., Spear, R. K. and Brunfeldt, K. (1971). *F. E. B. S. Letters,* **15,** 237
60. McLafferty, F. W., Venkataraghavan, R. and Irving, P. (1970). *Biochem. Biophys. Res. Commun.*, **39,** 274
61. Lowry, P. J. and Chadwick, A. (1970). *Nature (London),* **226,** 219
62. Lande, S., Witter, A. and de Wied, D. (1971). *J. Biol. Chem.*, **246,** 2058

63. Schally, A. V., Baba, Y., Nair, R. M. G. and Bennett, C. D. (1971). *J. Biol. Chem.*, **246**, 6647
64. Baba, Y., Matsuo, H. and Schally, A. V. (1971). *Biochem. Biophys. Res. Commun.*, **44**, 459
65. Nair, R. M. G., Barrett, J. F., Bowes, C. Y. and Schally, A. V. (1970). *Biochemistry*, **9**, 1103
66. Bøler, J., Enzmann, F., Folkers, K., Bowers, C. Y. and Schally, A. V. (1969). *Biochem. Biophys. Res. Commun.*, **37**, 705
67. Burgus, R., Dunn, T. F., Ward, D. N., Vale, W., Amoss, M. and Guillemin, R. (1969). *Compt. Rend. Acad. Sci. Paris*, **268**, 2116
68. van Euler, U. S. and Gaddum, J. H. (1931). *J. Physiol.*, **72**, 74
69. Chang, M. M., Leeman, S. E. and Niall, H. D. (1971). *Nature New Biol.*, **232**, 86
70. Kato, H. and Suzuki, T. (1970). *Biochemistry*, **10**, 972
71. Kato, H. and Suzuki, T. (1970). *Experientia*, **26**, 1205
72. Ferriera, S. H., Bartelt, D. C. and Greene, L. J. (1970). *Biochemistry*, **9**, 2583
73. Dunn, R. S. and Perks, A. M. (1970). *Experientia*, **26**, 1200
74. Azegami, M., Ishii, S. and Ando, T. (1970). *J. Biochem. (Japan)*, **67**, 523
75. Iwai, K., Nakahara, C. and Ando, T. (1971). *J. Biochem. (Japan)*, **69**, 493
76. Chance, R. E., Ellis, R. M. and Bromer, W. W. (1968). *Science*, **161**, 165
77. Steiner, D. F., Clark, J. L., Nolan, C., Rubenstein, A. M., Margoliash, E., Aten, B. and Oyer, P. (1968). *Recent Progr. Hormone Res.*, **25**, 207
78. Solokangas, A., Smyth, D. G., Markussen, J. and Sundby, F. (1971). *Eur. J. Biochem.*, **20**, 183
79. Clark, J. L., Cho, S., Rubenstein, A. H. and Steiner, D. F. (1969). *Biochem. Biophys. Res. Commun.*, **35**, 456
80. Oyer, P. E., Cho, S. and Steiner, D. F. (1970). *Fed. Proc.*, **29**, 533
81. Frank, B. H. and Veros, A. H. (1968). *Biochem. Biophys. Res. Commun.*, **32**, 155
82. Neher, R., Riniker, B., Zuber, H., Rittel, W. and Kahnt, F. W. (1968). *Helv. Chim. Acta*, **51**, 917
83. Barg, W. F., Englert, M. E., Davies, M. C., Colucci, D. F., Snedeker, E. H., Dziobowski, C, and Bell, P. M. (1970). *Biochemistry*, **9**, 1671
84. Beesley, T. E., Harman, R. E., Jacob, T. A., Hammick, C. F., Vitali, R. A., Veber, D. F., Wolf, F. J., Hirschmann, R. and Denkewalter, R. G. (1968). *J. Amer. Chem. Soc.*, **90**, 3255
85. Potts, J. T., Niall, H. D., Keutmann, H. T., Brewer, H. B. and Deftos, L. J. (1968). *Proc. Nat. Acad. Sci.*, **59**, 1321
86. Csordás, A. and Michl, H. (1970). *Monatsh.*, **101**, 182
87. Mutt, V., Jorpes, J. E. and Magnussen, S. (1970). *Eur. J. Biochem.*, **15**, 513
88. Brown, J. C. and Dryburgh, J. R. (1971). *Canad. J. Biochem.*, **49**, 867
89. Heathcote, J. G. and Washington, R. J. (1970). *Internat. J. Protein. Res.*, **2**, 117
90. Tschesche, H. and Wachter, E. (1970). *Eur. J. Biochem.*, **16**, 187
91. Greene, L. J. and Bartelt, D. C. (1969). *J. Biol. Chem.*, **244**, 2646
92. Tschesche, H. and Wachter, E. (1970). *Hoppe-Seylers Z. Physiol. Chem.*, **351**, 1449
93. Hochstrasser, K., Illchmann, K. and Werle, E. (1970). *Hoppe-Seyler's Z. Physiol. Chem.*, **351**, 1503
94. Sato, S. and Tamiya, N. (1971). *Biochem. J.*, **122**, 453
95. Eaker, D. L. and Porath, J. (1967). *Abstr. 7th Int. Cong. Biochem. Tokyo*, **Vol. 3**, p. 499
96. Yang, C. C., Yang, H. J. and Huang, J. S. (1969). *Biochem. Biophys. Acta*, **188**, 65
97. Botes, D. P. and Strydom, D. J. (1969). *J. Biol. Chem.*, **244**, 4147
98. Strydom, A. J. C. and Botes, D. P. (1971). *J. Biol. Chem.*, **246**, 1341
99. Botes, D. P. (1971). *J. Biol. Chem.*, **246**, 7383
100. Hayashi, K., Takechi, M. and Sasaki, T. (1971). *Biochem. Biophys. Res. Commun.*, **45**, 135
101. Botes, D. P., Strydom, D. J., Anderson, C. G. and Christensen, P. A. (1971). *J. Biol. Chem.*, **246**, 3131
102. Chang, C. C. and Hayashi, K. (1969). *Biochem. Biophys. Res. Commun.*, **37**, 841
103. Chang, C. C. and Yang, C. C. (1971). *Biochem. Biophys. Res. Commun.*, **43**, 429
104. Sato, A., Sato, S. and Tamiya, N. (1970). *Biochim. Biophys. Acta*, **214**, 483
105. Narita, K. and Lee, C. Y. (1970). *Biochem. Biophys. Res. Commun.*, **41**, 339
106. Takechi, M., Sasaki, T. and Hayashi, K. (1971). *Naturwissenschaften*, **58**, 323
107. Nakai, K., Sasaki, T. and Hayashi, K. (1971). *Biochem. Biophys. Res. Commun.*, **44**, 893
108. Mebs, D., Narita, K., Iwanaga, S. and Samejima, Y. (1971). *Biochem. Biophys. Res. Commun.*, **44**, 711

109. Yang, C. C., Yang, H. J. and Chiu, R. C. (1970). *Biochem. Biophys. Acta,* **214**, 355
110. Endo, Y., Sato, S., Ishii, S. and Tamiya, N. (1971). *Biochem. J.,* **122,** 463
111. Samuelsson, G., Segar, L. and Olson, T. (1968). *Acta Chem. Scand.,* **22,** 2624
112. Samuelsson, G. and Pettersson, B. (1971). *Acta Chem. Scand.,* **25,** 2048
113. Samuelsson, G. and Pettersson, B. (1971). *Eur. J. Biochem.,* **21,** 86
114. Studer, R. O. (1969). *Experientia,* **25,** 899
115. Meyer, W. E., Templeton, G. E., Grable, C. I., Sigel, C. W., Jones, R., Woodhead, S. H. and Sauer, C. (1971). *Tetrahedron Letters,* 2357
116. Bohmann, G. (1970). *Tetrahedron Letters,* 3065
117. Sengupta, S., Banerjee, A. B. and Bose, S. K. (1971). *Biochem. J.,* **121,** 839
118. Payne, J. W., Jakes, R. and Hartley, B. S. (1970). *Biochem. J.,* **117,** 757
119. Ovchinnikov, Yu.A., Kiryushkin, A. A. and Kozhevnikova, I. V. (1971). *Zhur. Obshchei Khim.,* **41,** 2085
120. Ko, A. S. C., Smyth, D. G., Markussen, J. and Sundby, F. (1971). *Eur. J. Biochem.,* **20,** 190

4
Spectroscopic, Solution and Theoretical Studies Relating to the Conformations of Peptides

G. C. BARRETT
Oxford Polytechnic

4.1 INTRODUCTION

4.1.1 General

Any general description of a polypeptide, from the point of view of molecular structure, must emphasise the variety of functional groups which it may contain, in a typical case. In addition to the regular spacing of the peptide functional group along the poly(α-aminoacyl) backbone of the molecule, side-chains of amino acid residues represented in proteins contain a variety of aromatic and simple aliphatic functional groups, which may interact with each other, with backbone amide groups, and with solvent, to determine the solution conformation of a particular primary structure.

In principle, the conformation of a peptide may be deduced through a global assessment of the interactions of its constituent groupings, as for any other acyclic compound, and detailed computations for polypeptides have been a major feature of recent work. The structures of several globular proteins, determined in recent years through x-ray crystallographic analysis, reveal the conformational relationships adopted within each of their constituent amino acid residues and provide a factual basis against which predictions for peptide conformations may be checked. However successful the relatively naive conformational-analysis approach, or the more sophisticated molecular-orbital calculations may be in establishing local conformational relationships in proteins, long-range interactions stabilising the total polypeptide tertiary structure may distort the behaviour of local regions in the structure, compared with their behaviour as isolated oligopeptides; but nevertheless, the theoretical approach has been sufficiently promising to provide a justification for continuing studies on the same lines. Since reliable calculations depend upon reliable estimates of interactions, which in their turn depend upon accurate experimental studies, practical aspects remain an area in which further refinements in technique and interpretation are necessary for continuing progress in determining polypeptide solution conformations. Spectroscopic methods, in particular, complement the ultimate detail provided by x-ray crystallography through contributing information on the dynamic properties of the predominant static condition of a particular primary structure.

4.1.2 Conventions of nomenclature

The run of the polypeptide backbone through a molecule determines its secondary structure, and the general disposition of its side chains. The secondary structure is defined in terms of the torsion angles ψ_i and ϕ_i, adopted by linked sequences of atoms along the polypeptide backbone. By convention[1], ψ_i is the torsion angle for the C^α—C' bond of the *i*th residue in a polypeptide, and is the angle between the planes containing the neighbouring atoms N_i—C_i^α—C_i' and C_i^α—C_i'—N_{i+1} (Figure 4.1), and ϕ_i and ω_i are the corresponding torsion angles for the N_i—C_i^α and C_i'—N_{i+1} bonds, respectively. These angles are positive when the atoms defining them take up a right-handed helicity when viewed (from either direction) in Newman projection (Figure 4.2) and lie between the limits -180 degrees $<$ torsion angle $\leqslant +180$ degrees. Torsion angles are all 180 degrees for the extended polypeptide conformation (Figure 4.1), and for the right-handed α-helix for poly(L-alanine), implicating *trans*-peptide bonds ($\omega = 180$ degrees), $\phi = -57$ degrees, and $\psi = -47$ degrees.

Rotation about single bonds in the polypeptide backbone brings about

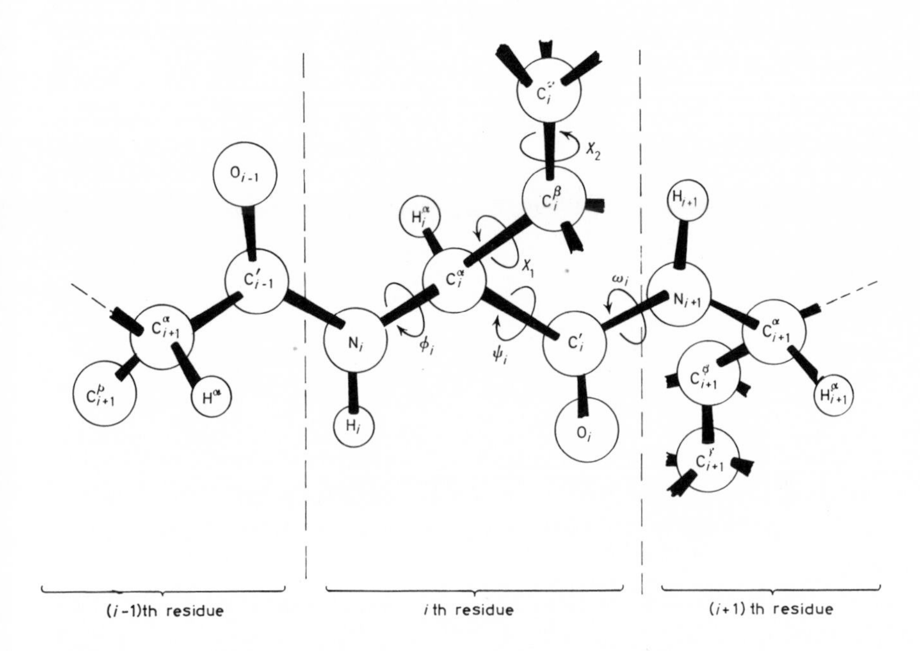

Figure 4.1 A section of a polypeptide in the fully-extended conformation (L-amino acid residues)

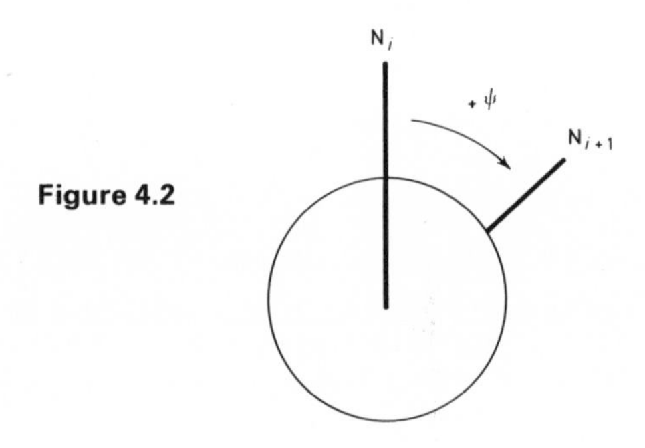

Figure 4.2

potential energy changes arising from distance-dependent non-bonded interactions; conformational energy maps are useful representations of variations in potential energy of the molecule accompanying variations in backbone torsion angles ϕ and ψ, for each amino acid residue (Figure 4.3).

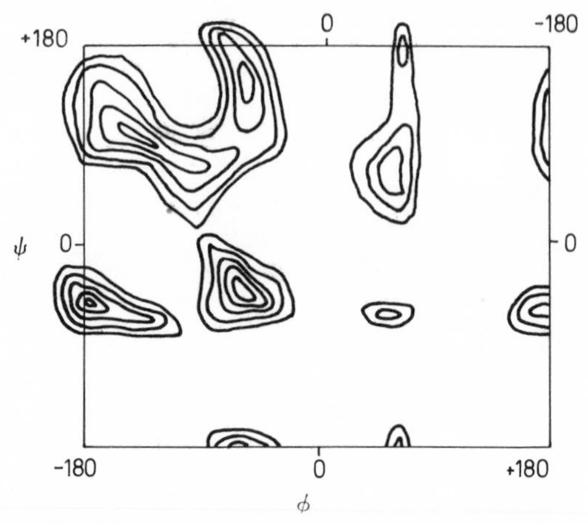

Figure 4.3 Conformational energy map for *C*-terminal residue in *N*-acetyl-L-Ala-L-Ala-L-Ala-L-Ala-NHMe, with preceding residues in 'α-helix' conformation; energy wells are shown, calculated for 20 degree increments of ψ_4 and ϕ_4 (reproduced, with permission of authors and publishers, from Ref. 196)

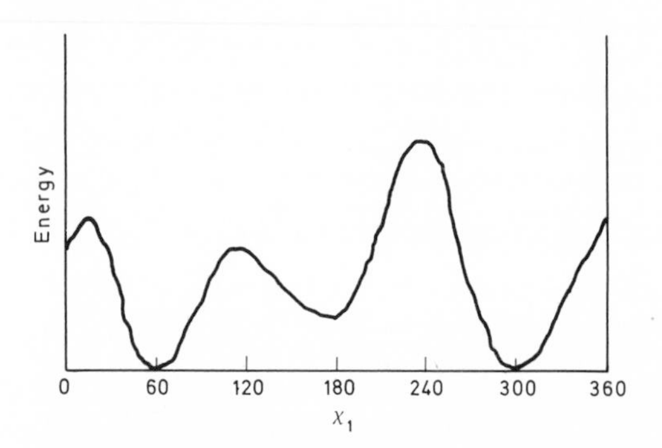

Figure 4.4 Energy–torsion angle relationship for the C^{α}—C^{β} bond of the valyl residue in cyclo(Gly-Val) (reproduced, with the permission of authors and publishers, from Ref. 202)

Sub-maps show the energy changes accompanying rotations about side-chain single bonds, χ_1, χ_2, etc. (Figure 4.4) for an amino acid residue contained in a fixed secondary structure. Since van der Waals' radii are of limited accuracy, freely-allowed values of ϕ and ψ may be represented by two areas on the ϕ–ψ map, one enclosed by an unbroken line (———), and a larger area (------) using minimum values of van der Waals' radii assessed partly intuitively.

4.1.3 Relationship of current conventions to older usage

The expression 'dihedral angle' is perhaps more common than the terms 'internal rotation angle' or 'torsion angle' in the literature of recent years relating to conformations about single bonds, but the latter expression, used as defined above, is recommended[1] for polypeptides. Earlier conformational energy maps place $\phi = \psi = 0$ degrees at the lower left-hand corner, but current recommendations[1] set this concordance at the centre of the map, requiring torsion angles reported previously to have 180 degrees subtracted from them to conform to the new recommendation*. Positive or negative helicity of the backbone single bonds is shown more clearly in the new convention.

4.2 ORDERED STRUCTURAL FEATURES OF POLYPEPTIDES

The regular spacing of the amide group along the polypeptide backbone, and the tendency towards inter- and intra-molecular hydrogen bond formation involving the amide proton and carbonyl oxygen atom, account for the tendency towards adoption by polypeptides of ordered conformations. These are favoured by solvents of low polarity, by regularities in primary structure, and by the absence of destabilising side-chain interactions. The

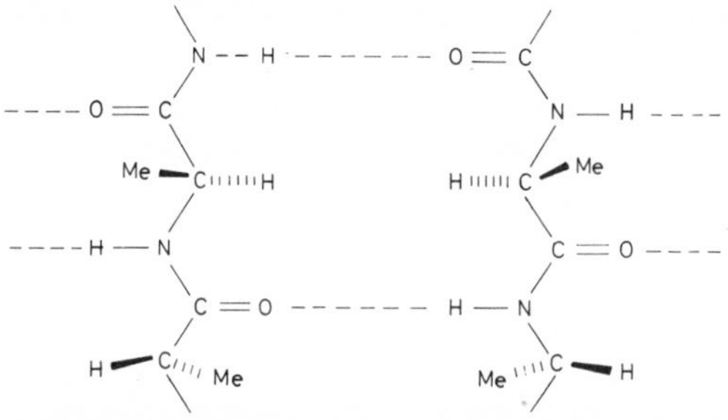

Figure 4.5 Two poly(L-alanine) chains in β-conformation of antiparallel pleated sheet

α-helix, now illustrated for polypeptides in every advanced textbook in organic chemistry and biochemistry and arguably the most dramatically appealing of all macromolecular conformations, and the β-conformation (Figure 4.5)[2], are the most important periodic conformations for polypeptides, involving intra- and inter-molecular or intra-chain hydrogen-bonding, respectively. Cyclic oligopeptides acquire a degree of ordered structure partly through transannular hydrogen bonding of alternate

*Torsion angles given in this chapter all conform to the new recommendation.

amide bonds, giving a ten-membered hydrogen-bonded ring, which is now recognised as a feature of the secondary structure of globular proteins, and referred to[3] as a 3_{10}-bend. It can be visualised as part of a '3_{10}-helix' involving every amide group in intramolecular hydrogen-bond formation[3]. Alternative nomenclature, the 'β-bend'[4], is more evocative of the effect of this unit in changing the direction of the polypeptide backbone in a globular protein, and two types are recognised[5], type I (illustrated in Figure 4.6) being the most general for L-amino acid residues, while type II (in which the central four atoms adopt the opposite helicity to type I) is permissible with a glycyl

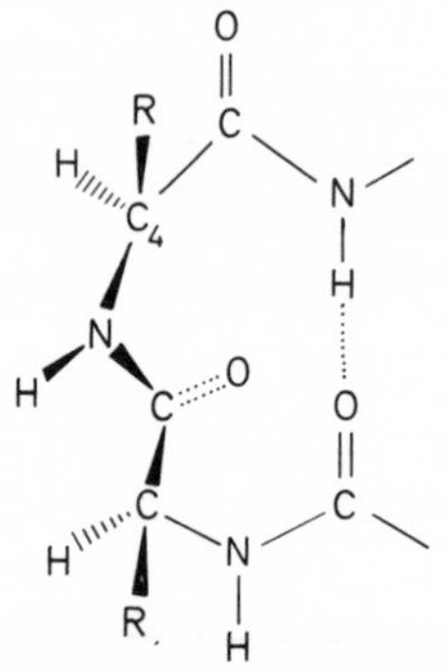

Figure 4.6 Type I 3_{10}-bend

residue (R = H at C-4) and can be discerned at three localities in, for example, α-chymotrypsin[6]. The carboxyl-terminal half of the insulin A-chain has been described[7] as 'casually helical', since it contains a number of β-bends; the expression 'gently helical' is also used[3] in relation to regions of horse heart ferricytochrome c.

Generally, globular proteins in their native state retain some periodic structure, though this can be small in some cases (e.g. elastase[8]). The term 'random' is not appropriate for irregular secondary structure[8, 9], the terms unordered, or aperiodic, or 'without long-range order'[9], or 'remainder'[9] being used for polypeptides and proteins in this state.

4.3 PHYSICAL METHODS APPLIED TO THE STUDY OF PEPTIDE CONFORMATIONS

4.3.1 General

Optical rotation and circular dichroism measurements* remain the most widely-used of the techniques which can supply information about the solution conformations of peptides. The routine applications of the methods are based largely on empirical comparisons, but although there is a valid theoretical basis to this approach, the models used for comparisons have recently been realised to be less than satisfactory for quantitative studies. However, o.r.d. and c.d. methods give reliable indications of changes in

*Optical rotatory dispersion (o.r.d.) and circular dichroism (c.d.) may be referred to as chiroptical or chirospectroscopic methods in the organic chemical literature, but these terms have not yet been taken up in the biochemical literature.

peptide conformation accompanying physical changes in the environment of the peptide. The methods are less successful for obtaining detailed information on local conformations and this is only obtainable, in principle, if the optical rotatory power developed in regions of the tertiary structure is assignable to chromophores in these regions. Techniques for conformational probing based upon fluorescence spectroscopy or on electron spin resonance spectroscopy (e.s.r.), have shown greater potential than studies of local conformations in polypeptides by o.r.d. or c.d. Instrumentation for nuclear magnetic resonance (n.m.r.) spectroscopy has developed now to the point where spectra can be obtained showing every proton resonance in a relatively complex, metal-free, peptide (e.g. antamanide), and furthermore, absorption peaks can be unambiguously assigned to their respective nuclei; spin–spin coupling constants for adjacent nuclei provide conformational relationships in favourable cases.

Brief reference to other physical methods in this chapter, in contrast to the prominence given to the techniques detailed in the preceding paragraph, illustrates the relative importance attached to the various methods in current conformational studies. Pioneering studies demonstrating the value in this field of each spectroscopic method date from some five years or more in the past; conformational assignments based upon the use of only one physical method are becoming less common as a result of better dissemination of information on available methods, for which collective reviews are becoming available[10–12].

4.3.2 Optical rotatory dispersion and circular dichroism

The c.d. features[13] of unordered, β-structural, and α-helical conformations of poly(L-lysine) are summarised in Table 4.1. These are associated with $n \rightarrow \pi^*$ and $\pi \rightarrow \pi^*$ transitions of the amide group[14, 15], the features characteristic of the ordered structures being generated by coupling of electric transition moments of ordered arrays of amide groups. O.R.D. curves show less characteristic features for each macromolecular conformation[13], but a turn

Table 4.1 Circular dichroism of Poly(-lysine)

Conformation	*Positive c.d. features*	*Zero c.d.* (nm)	*Negative c.d. features*
		250	Very weak negative c.d. max. 238–240 nm
Unordered	Weak positive c.d. max. at 218 nm	234	
		211	Strong negative c.d. max. at 197 nm
		250	Medium intensity c.d. max. at 217 nm
β-Structured	Strong positive c.d. max. at 195 nm	207	
α-Helical		250	Strong negative c.d. maxima at 222, 208 nm
	Intensely strong positive c.d. max. at 191 nm	202	

point in the o.r.d. curve at 233 nm is characteristic of the α-helix, a trough being diagnostic of the left-handed form, while a peak at this wavelength indicates right-handed helicity; c.d. curves adopt practically mirror-image forms about the wavelength axis for the opposite helix senses (cf. Refs. 16 and 17).

The use of o.r.d. and c.d. to assess oligo- and polypeptide conformations in solution is based on empirical comparisons of data with those of poly(L-α-amino acids) in different conformationally pure states. The data for poly(L-lysine) have been used as 'standards' for the unordered, β-structural and α-helical forms of polypeptides, and c.d. curves have been computed from them[13,18] for assessing the proportions of these conformations in proteins from their c.d. spectra. Now that several protein structures are known in detail through x-ray crystal analysis, it has been possible to compare predictions with fact; there are severe limitations to the quantitative accuracy of the predictions based upon c.d. parameters of a protein at a number of wave-lengths[19], and there is a growing realisation[9,20–23] that limitations of the empirical method are inevitable since no poly(amino acid) is entirely appropriate for use as a general 'standard'. The α-helix content deduced from c.d. spectra shows good agreement with c.d. spectra calculated for the available x-ray structures, and estimates of β-structure show fair agreement; however, unordered content is assessed with only qualitative success[9]. There remains good justification for continuing theoretical studies of polypeptide c.d., and recent work includes assessment of the c.d. of the unordered state[24–26]; this is characterised by a weak positive c.d. maximum near 218 nm (which is not specifically diagnostic of an extended chain conformation with three-fold left-handed helical sequences[25], as earlier claimed) and a large negative c.d. maximum (Table 4.1), and is satisfactorily accounted for on a theoretical basis[26]. New c.d. spectrometers capable of penetrating to deeper than 160 nm confirm the presence of a 165 nm maximum, assigned to a $n \rightarrow \sigma^*$ transition[27] rather than to alternative transitions advocated by other workers, but it is unlikely that sufficient diagnostic use can be made of c.d. data from this region to justify the experimental difficulties.

First reports of the infra-red c.d.[28] or o.r.d.[29] of representative polypeptides have appeared. The spectroscopic features may be related to the rotational strengths of vibrational transitions of the macromolecule, and although i.r.c.d. is zero for the unordered state and small for the α-helix, it is larger for β-structures and may have some value in conformational assignments of the latter state[28,29]. Preliminary studies of o.r.d. measurements on solutions of poly(γ-benzyl-L-glutamate), subjected to short duration pulsed electric fields applied either parallel or perpendicular to the light beam, have been reported[30]. In principle, additional information on molecular geometry is available with this technique, through evaluation of a relaxation time, τ, permitting the sense and length of a helix to be determined.

The c.d. of poly(L-lysine) has been used to determine the degree of ordering of structures of various peptide growth hormones[20]. The magnitude of the 222 nm c.d. peak indicates *c.* 30% α-helix, with best fits of experimental data obtained by including 10–15% β-structure and quite good agreement with these assignments is obtained if c.d. at 208 nm (where the β and unordered forms are 'iso-dichroic')[20] is used instead. However, aromatic side-chain chromophore contributions must limit the accuracy of these assignments even further[20]; some consideration has been given more generally to side-chain contributions[31], and the limitations they impose upon the empirical use of o.r.d. and c.d. to determine backbone conformations[20, 31].

Part of the limitation to the use of poly(L-lysine), as a model for the unordered conformation adopted by regions of proteins, may be traced to the adoption of different types of β-conformation by different primary structures[21]. One extreme (I-β)[21] may be represented by β-structured poly(L-lysine), with c.d. maxima at 195, 217 nm, while another (II-β) may be represented by β-structured poly(S-carboxymethyl-L-cysteine), with c.d. maxima at 198, 227 nm. β-Structured poly(L-serine), with c.d. maxima at 197, 222 nm, shows either a third variant, or a mixed β-conformation[21]. Related studies[22] show that synthetic poly(L-amino acids) may be satisfactory models for the c.d. contributions of regions of protein tertiary structures, though with the c.d. of poly(L-serine) in solutions of high salt content as a better model for the unordered polypeptide c.d. contribution. With this modification, c.d. spectra calculated for myoglobin, lysozyme and ribonuclease on the basis of their known tertiary structures agree at all characteristic points with experimental c.d. spectra[22]. The c.d. behaviour of poly(L-glutamic acid) has been used for similar purposes[32], and computed c.d. spectra for this poly(amino acid) are reported for 10, 20, 30, 50, 70 and 100% α-helical content[32]. These curves show an isosbestic point at 210 nm, with increasing negative ellipticity at longer wavelengths due to increasing backbone α-helicity. The c.d. of charged poly(L-glutamic acid) is similar to that of poly(L-alanylglycylglycine) in dilute aqueous solution or in cast films, when the latter polypeptide is known (x-ray analysis) to adopt the polyglycine-II conformation[33].

As noted above, the 'depth' of the o.r.d. trough at 233 nm gives only a rough estimate of the α-helix content of a polypeptide but it is a valuable means of monitoring conformational changes. For example, optical rotation measurements at 233 nm have been used to show that lactoperoxidase undergoes no apparent change of α-helix content (estimated to be 17%) during binding of anions F^-, N_3^-, or CN^- [34], also that the stabilisation of the poly(L-glutamic acid) α-helix is brought about by binding acetic acid, from solutions of pH which would otherwise favour denaturation[35].

4.3.3 O.R.D. and C.D. of oligopeptides and synthetic poly(amino acids)

Many examples of the general use of o.r.d., and now more particularly of c.d., for conformational assignments to peptides and proteins, are to be found in the recent literature. Acyclic oligopeptides longer than six or so residues appear to show some propensity towards ordered conformations in solution. Interpretations of o.r.d. and c.d. data for these peptides are based broadly upon the principles outlined in the preceding paragraphs, often without explicit discussion of the limitations of the method in the original paper. Most of the work relates to the more reliable use of these methods for detecting changes in conformation brought about by physical changes in the environment of the protein.

The o.r.d. of short peptides may be interpreted, through arithmetical treatment of rotational contributions from each residue[36], to indicate differences in conformation between related primary structures, but in practice n.m.r. methods may be more direct. Terminal *N*-thiobenzoyl oligopeptides, PhC(:S)·NH·CHR^1·CO(NH·CHR^n·CO)$_n$OH, show c.d. features near 390 nm, due to perturbation of the thiobenzamide chromophore by the asymmetric centres in the *N*-terminal and adjacent amino acid residues, which may be interpreted to reveal conformational differences at the *N*-terminus through a series of oligopeptides[37]. This, essentially a conformational probe approach, illustrates an alternative use of o.r.d. or c.d. for assessing peptide conformation, though the principle has been little used for the purpose.

Relatively short acyclic molecules can adopt a coiled or otherwise ordered conformation, and the enhanced c.d. of a C_{13}-ketone, ascribed to part-helix formation[38], is an example outside the peptide field. A rather longer run of a peptide chain is required before conditions can be found for the display of c.d. characteristic of an ordered conformation; in isoleucine oligomers Me_3C·O·CO·Ile$_n$·OMe ($n = 2$–8[39, 40], or $n = 1$–6[41]), o.r.d., c.d. and u.v. data for 2,2,2-trifluoroethanol solutions show the adoption of an associated β-structural conformation for $n = 6$ [41], or at the heptamer level[39, 40]. Isoleucine, considered to be disruptive of the α-helical conformation when involved in an ordered region of a peptide[39], in fact forms a more stable ordered conformation as the protected heptamer than the corresponding alanine heptapeptide derivative[40], which, together with aspartic and glutamic acid derivatives, has been studied earlier in this way.

Sequential oligopeptides H(L-Tyr-L-Ala-L-Glu)$_n$OH are unordered in solution at pH 7.4 for $n = 1$–9, but α-helical for large values of n, the onset of ordering being seen in the c.d. of the $n = 13$ oligomer[42]. C.D. data for tyrosine-containing peptides include contributions from perturbation of the side-chain phenolic chromophore by the α-carbon asymmetric centres, which tend to obscure the optical rotatory power generated in ordered regions of the backbone, especially near 224 nm[43, 44], when only a low degree of order is maintained. Tyrosine oligopeptides H(L-Tyr)$_n$OH adopt an unordered conformation in hydroxylic polar solvents for $n =$ up to 12[44], but poly(L-tyrosine) is α-helical in propane-1,2-diol[44], in DMF[44, 45], and in water–ethanol mixtures containing more than 45% ethanol[43]; it is mixed α-helical–β-structural in ethanol–water mixtures containing between 20 and 40% ethanol[43], and largely β-structural in water at low pH[46].

The o.r.d. and c.d. behaviour of representative poly(L-α-amino acids) has been committed to the literature of the period preceding the coverage of this review, though, as indicated in the foregoing paragraphs, there is continuing scope for the collection of more data. Poly(L-aspartate esters) are known to be unusual in adopting either a right-handed or a left-handed sense to their α-helical conformations in solution, the helix sense being determined by solvent; the near mirror-image c.d. curves of poly(β-benzyl-L-aspartate) in trimethyl phosphate and in chloroform, respectively, reflect the opposite handedness of the backbone helix[16]. Poly(1-benzyl-L-histidine) adopts unordered conformations in solutions containing small amounts of common mineral acids, but a stoichiometric amount of perchloric acid provides a remarkable contrast, causing α-helical ordering[47]. An alternating sequential polypeptide, poly(γ-benzyl-L-glutamyl-γ-benzyl-D-glutamate), appears to adopt an α-helical (mainly left-handed) conformation, as judged by o.r.d.[48–50] and n.m.r.[49, 50] studies. Poly(L-lysine) forms a more stable α-helix than poly(L-ornithine) at pH 11.75, where both poly(amino acids) are uncharged[51]. At pH 11, poly(L-lysine) is completely helical at 15 °C, but 90 % α-helix, 5 % β-form, and 5 % unordered at 25 °C[52]; at pH 7, it is unordered[52].

The view that the tertiary structure of a protein is determined by its full amino acid sequence has been established in a number of papers (e.g. ribonuclease A[53] and staphylococcal nuclease[54]), and, considered at the level of individual amino acid residues, nearest-neighbour interactions are predominantly important in determining local conformations[55]. Long-range interactions are, however, likely to be crucial for determining the overall shape of the protein molecule. In support of this, sequences 1–36, 37–87, 1–87, 37–119, and 120–153 of sperm whale myoglobin, cleaved from the protein (at proline residues) by treatment with sodium in liquid ammonia, are shown by o.r.d. and c.d. data to be much less α-helical (27, 28, 24, 41 and 18 %, respectively) in isolation than they are when part of the intact protein (97, 67, 79, 74 and 71 %, respectively, from the X-ray structure)[56]. Proline is a helix-breaking residue in polypeptides, and these fragments therefore contain intact helical sequences in the native tertiary structure. Cyanogen bromide cleavage of lysozyme at its two methionine residues results in loss of α-helical structure, while cleavage of disulphide bonds of the enzyme results in only *c.* 50 % disruption of the native α-helical content, as judged from c.d. data[57]. Corresponding results are obtained with cyanogen bromide cleavage products of myoglobin and of reduced *S*-carboxymethylated α-lactalbumin[58] and of ribonuclease A[32]. These facts suggest[32, 57, 58] that the primary structure, or peptide bonds in a particular sequence, may be more effective than disulphide bonds in maintaining regionally-ordered secondary structure. Addition of methanol to the myoglobin fragments in solution promotes some re-coiling, a maximum effect being reached at *c.* 60 % methanol levels, suggesting that less-hydrophilic regions are an important source of long-range interactions[56].

Cyclic oligopeptides are conformationally mobile in spite of some rigidity imposed by transannular hydrogen bonding. While a 14-membered ring containing amide groupings is particularly stable[59, 60], cyclic oligopeptides containing only α-amino acid residues cannot attain optimum ring sizes from this point of view, and a number of conformations of similar energy, or mixed conformations of different types of ordering, may be identified. For Gramicidin S, a cyclic decapeptide, 20 rotational parameters for the molecule with *trans*-amide bonds must be considered, making a complete conformational analysis impossible[61]. Cyclic hexapeptides containing glycine, leucine, tyrosine and histidine residues are only partly rigidified by their two transannular hydrogen bonds, c.d. studies into the far-u.v. revealing some conformational mobility[62]. The cyclodepsipeptide beauvericin shows solvent-dependent c.d. behaviour, associated similarly with conformational mobility[63]. The weak c.d. band at 230 nm in cyclo(Gly$_5$-L-Leu) corresponds with that seen in c.d. spectra of unordered acyclic polypeptides and is ascribed to overlapping $n \rightarrow \pi^*$ bands of isolated amide groups[62]. Cyclic hexapeptides of alanine have been assigned the β-pleated sheet conformation through n.m.r. studies[64]; their c.d. in non-polar solvents or in ethanol consists of weak $n \rightarrow \pi^*$ features (210–230 nm) and strong Cotton effects of opposite sign in the 180–205 nm region (split $\pi \rightarrow \pi^*$ bands), providing a useful reference c.d. spectrum for this conformation[64].

The peptide antibiotic stendomycin contains 14 α-amino acid residues, with the carboxyl group of its *C*-terminal residue linked in an ester bond with the side-chain of the D-*allo*-threonine residue at position 8; five residues of stendomycin have the D-configuration. It shows the c.d. expected of an α-helix in 2,2,2-trifluoroethanol; in water at 5 °C, it shows a helix-like c.d. spectrum which reverts to a spectrum characteristic of an unordered oligopeptide as a result of warming to 40 °C[65]. Gramicidin A′, a terminal *N*-formyl acyclic pentadecapeptide with most residues in the L-configuration, shows positive c.d. maxima at 224 and 212 nm, and a negative maximum at 194 nm in 2,2,2-trifluoroethanol, indicative of a left-handed α-helical conformation. Since the NH—C^{α}H proton coupling constants for this peptide are larger than expected for the α-helix, a less-coiled variant (the $\pi_{L,D}$-helix) is indicated[66]. Thyrocalcitonin, a polypeptide hormone containing 32 amino acid residues, is α-helical to only a small extent in water, but to *c.* 50% in 2-chloroethanol[67]; cleavage of its *C*-terminal region disulphide loop does not affect the c.d. behaviour. The conformational equilibria of polypeptide hormones may determine their effectiveness at receptor sites of low hydrophilicity, at which coiling is suggested to take place[67].

4.3.4 O.R.D. and C.D. of natural polypeptides and proteins

The c.d. and o.r.d. of longer polypeptides and proteins carrying most of the 'natural' amino acids includes contributions from both backbone and side-chain chromophores. Lipid or carbohydrate residues, when present, do not normally contribute significantly but they influence c.d. and o.r.d. data indirectly through stabilising α-helical regions in lipo- and glyco-proteins[68]. A number of proteins have low α-helix content (α-chymotrypsinogen, trypsin, elastase, soyabean trypsin inhibitor, histones F1 and F2a and ribonuclease after disulphide bond cleavage), and, although at least 20–30% α-helix content is required so that the c.d. of a protein is dominated by its characteristic spectral features[68], smaller α-helix contents can be detected by c.d. The enzyme elastase shows weak negative c.d., $[\theta] = 9900$ deg cm^2 dmol^{-1}, representative of an unordered secondary structure, though the addition of sodium dodecyl sulphate causes a considerable degree of ordering as shown by the appearance of negative c.d. bands centred at 207 and 218 nm[8]; removal of the detergent by dialysis leaves the protein with enhanced β-structure. The effect of sodium dodecyl sulphate has been observed also by Jirgensons and Capetillo[69], for various proteins and histones. In addition these workers report that the 240–300 nm c.d. features of these polypeptides are less intense, showing that aromatic side-chains possess greater rotational freedom in the presence of the detergent. Protein isolation procedures employing additives of this type may therefore cause conformational changes, and thus provide a protein with tertiary structure unrepresentative of the native state.

Proteins studied recently include α-lactalbumin, whose c.d. is interpreted[70] to show contributions by tryptophan and cystine residues, and to contain 25% α-helix, 15% β-structure and 60% unordered regions. Earlier o.r.d.–c.d. data for insulin have been supplemented by measurements into the far-u.v. wavelength region[21]. Soluble α- and γ-crystallins from dogfish lens show little or no α-helix content but mainly β- and unordered-structure[21]. Different forms, four in number, of neurotoxin II exist in equilibrium dependent upon pH and temperature (o.r.d. and difference spectrophotometry)[72], T_1-Ribonuclease studied at pH 1–12, between 4 and 70 °C, and in 8 M-urea, is shown to have a low degree of ordered structure in the native state, the enhanced 240–320 nm c.d. illustrating coupling between buried tyrosine residues[73]. Allantoicase studied by o.r.d., c.d. and i.r. methods, is shown to contain α-helical, β-structured, and unordered regions; a Cotton effect at 418 nm is ascribed to asymmetric binding of Mn^{2+} within the protein[74]. Kappa-type Bence–Jones proteins cleaved by limited tryptic digestion give fragments showing some β-structure[75]; immunoglobulins have been cleaved similarly and studied by c.d.[76]. Immunoglobulins from nine related species all show c.d. features (at 217 nm) consistent with their possession of a common, evolution-preserved, β-structure; an exception occurs with the structurally-unique lamprey immunoglobulin, which shows evidence of some α-helix content[23].

The c.d. of β-lactamase-II resembles that of a β-structured polypeptide[77], but a careful study shows that these features arise from carbohydrate impurities. C.D. changes during the titration of ribonuclease A with a specific inhibitor, cytidine 3′-monophosphate, are attributed to a change in tertiary structure around an environment-sensitive tyrosyl side-chain[78]. 1:1-Combination of haptoglobin (an α_2-serum globulin) with haemoglobin causes structural alterations in the haem environment, since changes in the 400–600 nm c.d. spectra accompany complex formation[79]. Raising the pH of cytochrome c oxidase solutions causes conformational changes, as revealed by variations in c.d. in the Soret region (near 420 nm)[80]; the polypeptide backbone of this enzyme is *c.* 50% α-helical[81, 82] but its o.r.d. and c.d. behaviour are significantly influenced by its oxidation state[81]. The positive maximum at 197 nm, falling to zero c.d. at 204 nm, indicates the presence of other ordered structures, possibly β-structural, in addition to α-helix[82]. Horse heart ferricytochrome c has been subjected to similar studies, particularly in the Soret region[83]; ferritin has a less ordered structure than its apo-protein[84]. Bovine erythrocyte cupro-zinc protein and its apo-protein show few differences in c.d. spectra in the far-u.v. wavelength region[85], suggesting that the metal atoms do not play a dominant role in determining the secondary structure of the protein component (no α-helix, some β-structure), while reduction of the disulphide bonds followed by *S*-carboxymethylation causes reduction in the intensity of the c.d. bands, suggesting that disulphide linkages exert a dominant effect in the maintenance of secondary and tertiary structure in this protein[85].

Thermal denaturation of A and B variants of bovine α-lactoglobulin assessed by o.r.d. and sedimentation velocity measurements, occurs through two consecutive first-order unfolding processes, causing increased β-structural content and progressive aggregation[86]. Denaturation of sperm whale myoglobin, horse heart cytochrome c, and bovine α-chymotrypsinogen, using a homologous series of water-soluble amides, has been studied by o.r.d. and isotropic spectroscopic methods[87]; an amide R^1.CO.NR^2R^3 is a more effective denaturant where R^1 is a long chain hydrocarbon residue, and R^2 and R^3 are alkyl groups rather than hydrogen atoms[87].

C.D. studies of membrane polypeptides have progressed with some vigour. If the limitations due to light-scattering effects now placed upon the interpretation of c.d. data of membrane suspensions can be overcome (correction methods have been proposed[88–90]), then the role of membrane polypeptides in membrane function[91] will be considerably clarified at the molecular level. Recent studies include the 190–230 nm c.d. of red blood cell membranes[88, 89] whose protein component is *c.* 50% α-helical[89], mitochondria membranes[89], intact nuclei and associated fragments[92].

4.3.5 Nuclear magnetic resonance spectroscopy

Applications of n.m.r. techniques to polypeptide studies have been reported in a fast-growing body of literature, now well-served by reviews[10, 93–103]. These include the n.m.r. of biopolymers (including polypeptides and proteins)[93], of proteins[94–97], and particular discussion of lysozyme[95, 96], ribonuclease[95, 96], and cytochrome c[95].

Conformational study of short peptides and cyclic oligopeptides is simplified by the establishment of a 'Karplus-like' relationship connecting the vicinal ^{1}H magnetic resonance coupling constant J(N—C$^\alpha$) with the torsion angle ϕ for the NH—C$^\alpha$H bond in a peptide residue[104]. Torsion angles for L,L- and L,D-diastereoisomers of alanylalanine, determined in this way, are in satisfactory agreement with those calculated on the basis of intramolecular interactions[105]; results for unordered polypeptide torsion angles determined in this way are also reasonable[104]. The involvement of the peptide NH proton in hydrogen-bonding or, conversely, its shielding from solvent, may be investigated by exchange studies with deuterated solvents and by analysis of temperature coefficients of NH proton resonances in non-hydroxylic solvents[106]. These principles have been illustrated in work on antamanide (1), a cyclic decapeptide which shows that a conformation is adopted presenting all carbonyl oxygen atoms at one side of the

Pro-Phe-Phe-Val-Pro
↑ ↓
Pro-Phe-Phe-Ala-Pro

(1)

molecule, and side-chains on the opposite face[106]; this requires that all amide bonds are in the *trans*-configuration, supported by c.d. measurements (large negative ellipticity in the 190–205 nm wavelength region).

An n.m.r. study has been made of the conformation of actinomycin, containing two pentapeptide lactone rings →L-Thr-D-Val-L-Pro-Sar-L-MeVal⌉. One of these lactone rings, studied in isolation, reveals a stretched transannular hydrogen bond between the D-valyl-NH proton and the sarcosyl carbonyl oxygen atoms[107], a torsion angle $\phi = 120$ degrees for the valyl NH—C$^{\alpha}$H bond, and a threonine C_{α}—C_{β} side-chain torsion angle $\chi_1 =$ 90 degrees[107], with all amide groupings in the *trans*-configuration. When interactions between the two lactone rings in actinomycin D itself are considered[108], the existence of hydrogen bonds between the D-valine NH proton in one peptide lactone ring and the L-*N*-methylvalyl carbonyl oxygen atom in the other ring are revealed (n.m.r. magnetic non-equivalence of threonine and valine–NH protons). The 220 MHz n.m.r. spectrum of actinomycin D[109] is only marginally better-resolved (sarcosine and threonine C$^{\alpha}$H protons at τ 5.2–5.5, better proline C_{α} and C_{γ} peaks) than that obtainable at 100 MHz[107], for which a complete assignment of resonances is possible. The conformation proposed by Lackner[107] for one of the pentapeptide lactone rings of actinomycin D is not entirely satisfactory[110], since it allows hindrance between the pyrrolidine ring of the proline residue and the sarcosine *N*-methyl group; possibly the assignment of all-*trans* amide groupings[107] will need revision. The more recent study[110] gives estimated torsion angles for side-chain C$^{\alpha}$—C$^{\beta}$ bonds of the valyl and *N*-methylvalyl residues, derived from n.m.r. coupling constants.

Related studies of cyclic oligopeptides include cyclo(tri-L-prolyl)[111], serratamolide and related cyclotetradepsipeptides[59], and a 220 MHz study of evolidine, a cyclic heptapeptide[112]. Cyclic hexapeptides are conformationally-mobile, a study of cyclo(Pro-Ser-Gly-Pro-Ser-Gly) at 220 MHz in water and in DMSO suggesting[113] a rapid interconversion between two similar β-conformations (β_D and β_L). A similar interconversion of β_{LT} and Ω_C conformations for cyclo(Ser-Pro-Gly-Ser-Pro-Gly)[114] has been observed through a study of NH-C$^{\alpha}$H vicinal coupling constants, NH-exchange rates, and the temperature-dependence of chemical shifts of NH protons (to show, for example, the all-*trans* configuration of amide bonds in the β_{LT}-conformation)[114]. Related n.m.r. studies of cyclic hexapeptides in methanol and in DMSO have been reported[115].

At a simpler structural level, the n.m.r. of amino acids and small peptides, and their terminal amino- and carboxyl-derivatives, has been thoroughly studied[116, 117]. Rotational isomerism in amino acid side-chains of alanine, valine and phenylalanine has been studied qualitatively, through solvent-dependent n.m.r. behaviour of the *N*-acetyl amino acid alkyl esters[117], with the conclusion that intramolecular interactions have a larger effect than dielectric constant of the medium in determining the relative energies of the different rotamers[117, 118]. Incidentally, the more concentrated solutions used for n.m.r. measurements compared with other absorption spectroscopic methods may introduce a source of non-comparability of interpretation of peptide conformations[119]. N.M.R. data and conformational calculations on a series of compounds of general structure $R^1NH{\cdot}CHR^2R^3$ lead to a relationship between torsion angle and NH—C$^{\alpha}$H coupling constant of the form $J(\phi) = 7.9 \cos^2 \phi - 1.55 \cos \phi + 1.35 \sin^2 \phi$, which gives values for torsion angles for amino acid residues in peptides in agreement with a proportion of available experimental data[120]. At present, this approach takes no account of side-chain interactions and while the results are consistent for known torsion angles for leucine, methionine and tryptophan residues in peptides, this is not so for valine, isoleucine, phenylalanine and tyrosine residues[120];

no doubt this somewhat illogical grouping of conforming and non-conforming amino acids will be rationalised by further calculations, bringing into consideration intramolecular interactions which have hitherto been assumed to be insignificant[120]. N.M.R. spectroscopic data for di-[121, 122] and tripeptides[122] have been reported, the latter (for peptides of glycine and valine) being discussed in relation to the mutual interactions of their amino acid and dipeptide moieties and as models for polypeptides. Oligopeptides containing aromatic amino acid residues have been studied by n.m.r., and where these are capable of some ordered secondary structure, the chemical shifts of the aromatic protons vary slightly in response to changes in the backbone conformation[123].

N.M.R. spectra of synthetic poly(L-α-amino acids) can be interpreted to reveal conformational transitions between ordered and unordered states. The peptide $C^{\alpha}H$ resonances are upfield in an α-helical conformation relative to their positions in the spectrum of the unordered polypeptide, for solutions in polar non-aqueous solvents[124]. This observation may be useful in conformational diagnosis of natural oligopeptides (e.g. to show that the backbone conformation of adrenocorticotrophic hormone, ACTH, is disposed so as to permit no interaction between its 1–10 and 11–24 amino acid sequences[124], and to show [125] that Gramicidin A′ adopts an ordered structure different from the classical α-helix, since its NH—$C^{\alpha}H$ coupling constants are significantly larger than those required for the α-helix). An interesting example, though unrepresentative of the general poly(amino acid) case, is the use of 220 MHz n.m.r. to study the stages of the conformational transition which occurs on dissolving poly(L-proline)-I in water[126]. This conformation is determined by *cis*-amide bonds; progressive isomerisation from the carboxyl terminus to the all-*trans* isomer, poly(L-proline)-II, is accompanied by changes in pyrrolidine ring geometry which are in their turn characterised by changes in vicinal coupling constants of the methylene protons[127].

The observation of double $C^{\alpha}H$-proton resonances during the transition from the α-helix to the unordered state by poly(γ-benzyl-L-glutamate) has been discussed[128] in terms of an exchange phenomenon which is slow on the n.m.r. time-scale ($\tau = 10^{-2}$ s). Peak area ratios of the $C^{\alpha}H$ and side-chain proton peaks in this poly(amino acid) may be used to determine intermediate unfolding stages; proton-decoupled ^{13}C n.m.r. is proving of great value in polypeptide studies[129], exemplified with poly(γ-benzyl-L-glutamate) in studies of the dissociation of the α-helical conformation in chloroform – trifluoroacetic acid mixtures, which is accompanied by an upfield shift in ^{13}C resonances (3.0 p.p.m. for the α-carbon and 2.7 p.p.m. for the amide carbon atom)[129]. Denaturation of proteins may be followed analogously by n.m.r., determination of the areas of various resonances allowing the identification of a multi-stage denaturation process if several peaks in the n.m.r. spectrum are found to vary in area at different rates[130]. The inhomogeneity of primary structure of proteins, of course, generates more complexity in n.m.r. spectra compared with the spectra of poly(amino acids) but at the same time the spectra may provide limited conformational information as a result of their complexity. Lysozyme, for example, at pH 2.8 tends to unfold, but three clusters of protonated arginine side-chains, comprising seven arginine residues in all, are shown to unfold before the bulk of the molecule, since the NH resonances of these side-chains start to sharpen before many other proton resonances[130]. α-Lactalbumin, studied in this way, is shown to be more easily denatured than lysozyme[130]. Calf thymus histones F1, F2a$_1$, and F2b, of known primary structure, have been subjected to variations of ionic strength and pH in solution, conformational changes being followed by 100 and 220 MHz n.m.r.[131, 132]. The F1 fraction (MW 21 000) suffers reduction in mobility in the second quarter of its sequence (residues 51 to 100), with increased ionic strength as a result of a small amount of β-structuring[132], both inter- and intra-molecular. Fraction F2a (MW 11 300) shows increased tendency to α-helix formation in its carboxyl half (residues 55 to 72) on similar treatment[132], with aggregation taking place on ageing. The regions susceptible to ordering in this way are neither the regions rich in basic residues, nor the regions rich in glycine and proline (which are unlikely to favour helix formation because of the helix-breaking character of these amino acid residues). These interpretations are based on observed line-broadening and reduction in area of CH_3 and CH_2 proton resonances in hydrophobic residue side-chains, of threonine, aspartic acid, and glutamic acid residues, while peaks associated with glycine and basic side-chains are unaffected[131, 132]. Fraction F2b similarly shows an increase in structuring in the middle run of its primary structure as a result of increases in pH and in ionic strength[131].

Side-chain resonances may be interpreted to reveal aspects of tertiary structure. An example from enzyme active site studies involves ribonuclease T[133]: C2-proton resonances of three histidine residues may be discerned in the n.m.r. spectrum of the enzyme, and their behaviour as a function of pD in D_2O allows calculation of the pK values of their imidazole substituents. With other evidence, two of the three histidine side-chains are implicated at the active site, and since one of these shows an upfield shift of the C2-proton resonance when the glutamic acid side-chain at position 58 is carboxymethylated, the glutamyl side-chain is also shown to be involved in the active site[133]. Poly(β-benzyl-L-aspartate), in comparison with the methyl ester analogue, shows side-chain n.m.r. features[134] which may be interpreted to show the absence of interactions between the benzyl groups when the polypeptide backbone is in the left-handed α-helical conformation (in chloroform solution). Consequently, there is no α-helix-stabilising contribution from these groups, as confirmed by denaturation studies, whereas in contrast, the right-handed α-helix adopted by poly(γ-benzyl-L-glutamate) depends upon stabilisation of this type[134].

Helix to unordered conformational changes may be followed by studying the n.m.r. changes of solvent molecules, e.g. poly(γ-benzyl-L-glutamate) in chloroform–trifluoroacetic acid[135]. The proton shift of the small proportion of trifluoroacetic acid needed in this case to break up the α-helix of the poly(amino acid), relative to comparison solvent mixtures, plotted against trifluoroacetic acid content, gives discontinuous plots around the conformational transition point. These arise from differences in solvation behaviour of the ordered and unordered states. Similar studies with $Cl_2CH{\cdot}CO_2H$–$CHCl_3$ or CH_2Cl_2 solvent mixtures for poly(γ-methyl-L-glutamate) have been reported[136], together with spin–lattice relaxation time data for this system. Water bound to macromolecular polypeptides shows different line-width from polypeptide to polypeptide[137]. However, a correlation does not at present appear likely to arise between linewidth and conformation, though for ionic poly(amino acids), the degree of hydration of side-chains is reduced by aggregation, suggesting a possible application.

The conformation of poly(β-alanine) in aqueous solution is shown by ^{1}H n.m.r. and deuterium exchange studies to be unordered[138].

Lanthanide cations have been proposed as n.m.r. probes for enzymes[139]; Gd^{III} and Eu^{III} are excellent probes for tightly-bound ligands and association of a substrate analogue, β-methyl-N-acetylglycosamine, with the Eu^{III}–lysozyme complex gives an intermediate in which the resonances of the acetamido and glycosidic methyl groups of the substrate are shifted by +1.6 and −2.8 p.p.m. respectively. The lanthanides substitute other metals in metallo-enzymes (e.g. Ca^{2+} in staphylococcal nuclease)[139], and appear to show considerable promise in studying local conformations.

N.M.R. studies of nuclei, other than those discussed earlier in this section, include ^{35}Cl line-width studies designed to reveal the completion of dithiothreitol reduction of disulphide bonds in enzymes in the presence of $Hg^{2+}HgCl_4^{2-}$; the disulphide bonds of α-lactalbumin are shown by this technique to be more accessible than those of lysozyme[140]. The molecular Zeeman effect for $HCO^{15}NH_2$ has been exploited[141] to calculate the α-proton chemical shift due to the re-orientation of amide geometry taking place in poly(L-alanine) while it undergoes the α-helix to unordered state transition; the observed shift is opposite to that calculated on the basis of anisotropic shielding of the amide group and it is therefore a solvent effect[141].

Proton relaxation-enhancement n.m.r. studies provide an alternative means of conformational probing of localities in proteins[142]. Measurement of the spin–lattice relaxation times of water protons of hydrated paramagnetic metal ions, e.g. Mn^{2+}, by pulsed n.m.r. spectroscopy when bound to a protein, and their enhancement in relation to conformational changes in the environment of the ion, provides a means of following changes in conformation near enzyme active sites, for example those resulting from binding of substrates or inhibitors[143].

4.3.6 Fluorescence measurements

Fluorescence spectroscopic studies have provided a useful body of information on conformational relationships between constituent functional groups in polypeptides and proteins. Variations in the intrinsic fluorescence of tryptophan and tyrosine aromatic side-chains, when present in polypeptides and proteins, can be interpreted in conformational terms, but the alternative approach, using reagents which introduce fluorescent groupings (fluorophores) into polypeptides and proteins is now being studied more intensively[97, 142, 144]. There is a risk that the native conformation is disturbed by the introduction of the fluorescent probe, but complementary physical and chemical reactivity studies are usually applied to assess this possibility.

Early studies of the fluorescence of tryptophan and tyrosine side-chains in α-chymotrypsin revealed the interaction of the side-chains of these residues with carbonyl or amide NH groups in the backbone of the surrounding polypeptide[145,146]. The work of intervening years on native proteins has not promised better focusing than this of conformational information at the molecular level. However, the effect of quenching agents on tryptophan emission, can for example, show the accessibility of the indole grouping of this amino acid, to reactants or to solvent; the two tryptophan residues in sperm whale myoglobin are shown to be in non-polar environments by the quenching effects of various inhibitors and substrates[147], iodate quenching being more efficient as the protein is denatured, indicating progressive exposure of the tryptophan residues[147]. Variation of the quantum yield with pH in this case suggests quenching of a tryptophan side-chain fluorescence by a neighbouring protonated imidazole of a histidine residue[147]. Tyrosine residues are less intensely fluorescent than tryptophan residues, but they provide natural probing points in many proteins, illustrated by work on ribonuclease[148]; three exposed tyrosine residues in this enzyme lose their fluorescence through *O*-acetylation, implying that the other three tyrosine residue side-chains are buried in the native conformation[148], a conclusion supported by the development of further fluorescence when the tris(*O*-acetyl) enzyme is denatured.

The theoretical basis of fluorescence spectroscopy applied to biological systems has been reviewed[142,149,150]. Excitation of fluorophore electrons as a result of light absorption produces changes in charge distribution, and also possibly changes in the polarisation of solvent molecules in the vicinity of the fluorophore. Re-emission of light occurs within 1–10 ns, during which time a variety of structural changes may occur within the excited state, to be reflected in the measured spectra[142]. Where calculated and measured intrinsic lifetimes, τ_0, of the excited state do not agree, a change in geometry is implicated. Conversely, a constrained environment to the fluorophore is revealed by agreement between calculated and measured values of τ_0, illustrated by the non-covalent binding of *N*-1-naphthyl-*N*-phenylamine to bovine serum albumin or to erythrocyte membranes[142].

4.3.6.1 *Nanosecond fluorescence polarisation spectroscopy*

Measurement of the polarisation of the fluorescent light emitted from a fluorophore, as an explicit function of time in the nanosecond range, can provide an estimate of the rotational motion of the group. The growing importance of nanosecond fluorescence polarisation spectroscopy for studying polypeptides and proteins in solution, and proteins in membranes, is now emerging, particularly in view of its potential value, together with other physical methods, for advancing present understanding of the molecular architecture of membranes[149,150]. An indication of the area in which the technique can be used, and its jargon, is provided in a study of the reversible binding of 1-anilino-8-naphthalenesulphonic acid and the covalent bonding of 5-dimethylamino-1-naphthalenesulphonyl chloride (dansyl chloride), to membrane fragments *in vitro*[150]. These probes are strongly immobilised on the membrane proteins, but a dramatic increase in mobility of the probe follows the solubilisation of *c*. 60% of the protein component of the membrane by the action of a detergent (Triton X–100)[150]; studies of this type bring to an end the intuitive speculation of the liquid-like structure of membranes, better regarded as rigid or solid[150].

In this technique, a vertically-polarised light beam pulse is passed through a sample, and intensities of the parallel and perpendicular components, respectively, of the emitted light are measured as a function of time, t, $I_{\parallel}(t)$ and $I_{\perp}(t)$ respectively, of the emitted light are measured as a function of time, t. The rate of decay of the total emitted light as a function of time phores and these are determined by the constraints of their immediate environments. The anisotropy $r(t) = [I_{\parallel}(t) - I_{\perp}(t)]/[I_{\parallel}(t) + 2I_{\perp}(t)]$ is directly related to the motion of the fluorophore in its excited state, since tumbling of the fluorophore about its point of attachment causes rapid depolarisation, with $r(t)$ tending to zero; $r(t)$ expresses the change in orientation of the fluorophore between the times of absorption and re-emission of light. Exponential decay of the energy of the excited state involves a time factor, τ, the lifetime of the excited state; $S(t)$ = energy absorbed from light pulse = $a_0 \exp(-t/\tau)$, where a_0 is the incident light intensity, applies for the simplest case where no energy transfer from the fluorophore to neighbouring groups occurs. Then $r(t) = r_0 \Sigma[\alpha_i \exp(-t/\theta_i)]$, where θ_i is the rotational relaxation time of the unconstrained probe, r_0 is the anisotropy at zero time, immediately following the light pulse. Comparison of relaxation times of the fluorophore in different environments gives a direct estimate of constraints in each environment on the motion of the fluorophore[150].

The implication of this theory—that where energy transfer from the excited state of the fluorophore occurs, it is capable of being assessed—can be exploited to give conformational information for peptides[151]. An elegant example[152] illustrates the proximity of the lysine residue at position 21 in the ACTH analogue Synacthen (a tetracosapeptide with full ACTH activity) to the tryptophan residue at position 9. The dansyl fluorophore, introduced at the ε-amino-group of Lys^{21} may be shown, through fluorescence lifetime measurements, to be tumbling more rapidly than the rest of the molecule, with the aromatic moiety 24.5 ± 1.5 Å distant from the indole grouping. The dansylated peptide still possesses full ACTH activity, and the separation of the groupings is unaffected by urea or guanidine denaturants[153]. Progress in determining the mode of action of polypeptide hormones, such as ACTH, depends largely on a knowledge of the disposition of side-chain functional groups. These groups may be close together in the primary structure, or may be brought together by folding of the peptide chain (Schwyzer proposes the terms 'sychnologic' and 'rhegnylogic', respectively, to define these two situations, the latter probably exemplified by insulin[153]). It is clear that energy-transfer studies will provide some of this knowledge.

Brief mention may be made of the use of 1-anilinonaphthalene-8-sulphonic acid and related compounds to bind at hydrophobic sites in antibodies[154], and of the use of 2-*p*-toluidinonaphthalene-6-sulphonate to detect subtle conformational changes occurring in the conversion of chymotrypsinogen into chymotrypsin, and of pepsinogen into pepsin[155]. Glucagon, a polypeptide hormone with 29 amino acid residues, acquires a larger degree of secondary structure as a result of combination with aliphatic quaternary amines, as revealed by o.r.d. studies (trough at 233 nm to reveal α-helical content), and by fluorescence and polarisation-of-fluorescence studies of the tryptophan residue at position 25; u.v. difference spectroscopy also helps to identify conformational changes in this case[156]. Attempts to identify the conformational states of phosphorylase b by o.r.d. and c.d. studies are not successful[143], but conformational probes successfully reveal minor changes. 7-Chloro-4-nitrobenzo-2-oxa-1,3-diazole gives fluorescent substitution products (at reactive thiol groups) with proteins, giving a fully-active enzyme with phosphorylase b which undergoes local conformational changes in the presence of AMP[143]. Proton relaxation enhancement studies of the enzyme to which Mn^{2+} is bound, show a concentration dependence similar to that for fluorescence quenching by AMP, suggesting that the Mn^{2+} ion is bound at a similar site to the point of attachment of the fluorescent label[143].

4.3.7 Electron spin resonance spectroscopy

An alternative technique for conformational studies of regions of tertiary structure makes use of e.s.r. measurements of polypeptides and proteins carrying 'spin labels' – groups containing a localised unpaired electron. These groups are introduced in much the same way as the fluorescent probes discussed in the preceding section, and are usually amine *N*-oxides, R^1R^2N—O, carrying a functional group in R^1 or R^2 through which covalent bonding to side-chain functional groups in the protein, or binding to protein surfaces, may occur. 2,2,6,6-Tetramethyl-4-isocyanopiperidino-*N*-oxyl, which binds specifically to the haem group in cytochrome P-450 through its isocyano group, has been used in an attempt to relate the function of the haem group with conformational changes in the polypeptide environment[157]. Further examples are *N*-(2,2,6,6-tetramethylpiperidin-1-oxy)maleimide, a maleylating agent applied to allow an e.s.r. study of human serum high-density lipoproteins for comparison with i.r. and c.d. data[158], and *N*-(1-oxyl-2,2,6,6-tetramethyl-4-piperidinyl)-iodoacetamide, used to alkylate the cysteine residue at position 93 in the β-chain of haemoglobin, for e.s.r. study of its haptoglobin complex[79].

The shape of the e.s.r. spectrum of 'spin-labelled' polypeptides is an indication of the rotational mobility of the grouping carrying the unpaired electron; where neighbouring paramagnetic ions or nuclei are also present in the 'tagged' protein, the line shape of the e.s.r. signal is influenced by dipolar coupling, from which distances of separation may be calculated[159]. Most applications depend upon interpretation of the e.s.r. spectrum of the former type, to determine the rotational correlation time, τ_c, of the label[160, 161], from which the degree of constraint placed upon the label by its environment may be estimated. Generally, a broad e.s.r. signal corresponds to a strongly-constrained environment[158]. Where a spin label is incorporated into a microsome protein, a high value ($\tau_c = 3.3$ ns) is deduced from the e.s.r. signal, indicating considerable hindrance[157]; addition of aminopyrine increases the tag mobility ($\tau_c = 2.0$ ns) and the effect of 6M-urea or detergents is to increase rotational frequency of the label even further ($\tau_c = 1.2$ ns)[157]. Aspartate aminotransferase (pig heart) tagged with iodoacetamide and maleimide spin-label derivatives (up to four labels per dimer) retains *c.* 90%

of its enzymic activity and shows c.d. behaviour identical with that of the native enzyme[161]. E.S.R. changes following addition of substrates or analogues indicate increased mobility of the labels, which are more than 15 Å from the active site of the enzyme; this demonstrates the occurrence of substrate-induced conformational changes consistent with the 'induced-fit' theory of enzyme action.

There are few studies yet of synthetic poly(amino acids) using the spin-label technique, one example[162] being the demonstration of slow rotation of a nitroxide label at the end of α-helical poly(L-glutamate ester) chains.

4.3.8 Infrared and Raman spectroscopy

The amide absorption frequencies of polypeptides are sensitive to ordering of the polypeptide backbone and some simple correlations may be made (e.g. *cis*-amides: 1450, 1350 cm^{-1}; *trans*-amides: 1550, 1250 cm^{-1}; α-helix: 1650, 620 cm^{-1}, with left-handed helices showing higher amide-I frequencies than right-handed forms[16,163]; β-structure: 1685–1700, weak, but position dependent upon parallel or antiparallel arrangement, 1630, strong, 700 cm^{-1}; unordered: 1656, 650 cm^{-1}); a recent example demonstrates the conformational changes of protein constituents of human red blood cell membranes, from a mixture of α-helical and unordered conformations towards an antiparallel β-structure in the presence of ATP and Mg^{2+}, enhanced by added Na^+ and K^+ salts[164]. The ratio of areas of amide-I bands at 1655 and 1633 cm^{-1} indicates the ratio (α-helix + unordered)/(β-structure), and permits a deduction of an early stage in thermal denaturation of human serum albumin in solution, as a progressive increase in β-structuring[165].

Oligopeptides of γ-ethyl-L-glutamate tend to adopt β-structures in chloroform, beyond the pentamer level, while in trimethyl phosphate the chain length must be at least seven amino acid residues before any ordering occurs[166]; these conclusions are based on amide-I and -II stretching frequencies. Poly(L-histidine) at higher degrees of polymerisation adopts *trans*-amide bonds, as shown by the positions of the amide-I and -III bands[167]. I.R. spectra confirm the assignment, based on other spectroscopic methods, of the adoption of predominantly β-pleated sheet conformation by cyclohexapeptides of alanine[64].

Corresponding information from Raman scattering measurements may be less ambiguous than that from i.r. studies. The conformationally-sensitive amide-I, weaker amide-II and amide-III modes are seen in Raman spectra, too, and distinction between parallel and antiparallel β-forms for poly(amino acids) based on γ-benzyl-L-glutamic acid, L-leucine, L-valine, and L-serine is particularly clear using Raman spectroscopy[168]. Poly(hydroxy-L-proline) has been shown to adopt the same conformation in solid and aqueous solution states, through Raman studies[169]. Laser-excited Ar^+ 4879 Å Raman spectra of aqueous solutions of poly(L-lysine) and poly(L-glutamic acid) have been reported; for unordered poly(L-lysine), frequencies 1653, 1665 and 1683 cm^{-1} were observed, while in the α-helical state, little scattering occurs above *c*. 1660 cm^{-1}. The amide-I band appears at 1647 cm^{-1}, and the β-form amide-I band is at 1631 cm^{-1} in these spectra; similar results are obtained for the glutamic acid analogue[170].

4.3.9 Non-spectroscopic methods

Exact physical constants may be determined for various properties of polypeptides in solution, but the information is rarely of greater use than spectroscopic data in conformational analysis.

The α-helix to unordered conformation transition is accompanied by discontinuities in a variety of physical parameters, one being refractive index increments as a function of solvent composition, e.g., for solutions of poly(γ-benzyl-L-glutamate) in chloroform–dichloroacetic acid mixtures[171]. Reduced sedimentation velocity coefficients for these systems also show sudden changes in the region of solvent composition corresponding to macromolecular conformational change[86, 171].

Viscosity measurements give information on overall molecular shape, for which viscosity is one of the most sensitive of physical parameters. Glucagon is capable of undergoing large conformational changes and can exist in α-helical or β-structural forms; intrinsic viscosity measurements show that this polypeptide possesses a spherical shape in aqueous solution, aggregating on standing at pH 11 and gelling in concentrated solution to adopt the antiparallel β-conformation[172]. Similar studies of poly(proline)-II, correlate molecular weight with intrinsic viscosity for dilute solutions of a series of samples of molecular weights between 4400 and 99 000[173]; the technique can be used to determine the *cis–trans* isomerisation of poly(L-proline) in concentrated aqueous calcium chloride solutions[173]. Fibrous keratins adopt solvent-stabilised conformations which are modified in the presence of salts[174]; some ordering is brought about by F^-, SO_4^{2-}, HCO_3^-, $MeCO_2^-$, Li^+, Mg^{2+} and Ca^{2+}, while Cl^-, Br^-, I^-, SCN^-, and ClO_4^- ions produce the opposite effect. The viscosity of Li^+ salt solutions of the protein decreases in the order SO_4^{2-}, Cl^-, Br^- and I^-. These effects are related to modification of inter-chain hydration patterns, and are also reflected in variations in denaturation temperatures[174].

Thermodynamic parameters for poly(L-lysine) have been intensively studied. In addition to deductions from o.r.d. and viscometry data[175], potentiometric titration[52, 175, 176] has proved to be particularly informative and conveniently studied, and provides figures for the free energy and enthalpy changes accompanying the folding of the unordered form of poly(L-lysine) into the α-helix in 10% aqueous ethanol[176] (-120 and -120 cal mol^{-1}, respectively) and in water (-78 and -880 cal mol^{-1}, respectively), and from the unordered state to the β-form in 10% aqueous ethanol (-78 and -880 cal mol^{-1}, respectively) and in water (-140 and 870 cal mol^{-1}, respectively). These figures illustrate stabilisation by ethanol of the α-helix with respect to the unordered state for this poly(amino acid), consistent with the known free energy of transfer of the peptide group from water to 10% ethanol[176]. Non-bonded interactions of the polypeptide backbone are less favourable for the β-form than for the α-helix, while hydrophobic contacts between side-chains raise the entropy of the β-form relative to that of the α-helix[176]. Calorimetric measurements give a figure $-1200 \text{ cal mol}^{-1}$ for the conversion of an unordered lysine residue in poly(L-lysine) in 0.1 N KCl, into the conformation with α-helix torsion angles[52]. Potentiometric titration of poly(L-glutamic acid) is less helpful than c.d. for studying the effects of organic electrolytes (tetra-n-butylammonium chloride in concentrated aqueous solution) on disruption of the α-helical conformation[177].

Depolarised Rayleigh scattering measurements of dipeptides in aqueous solutions give molecular optical anisotropy data which identify principal optical polarisability planes, for the amide bond; these are conformation dependent and may be interpreted to assist conformational analysis[178].

4.4 THEORETICAL STUDIES OF CONFORMATIONS OF PEPTIDES

4.4.1 General

Approaches towards the prediction of the stable conformation of a peptide, through a summation of the interactions within the molecule as a whole and the free energy for all interactions involving the solvent, have progressed beyond the relatively qualitative level at which acyclic organic compounds in general are analysed[179, 180].

The total conformational energy of a polypeptide includes the contributions of intrinsic threefold torsional potentials, $V(\phi,\psi)$, about the N—C^{α} and C^{α}—C′ bonds in each amino acid residue, together with non-bonded steric repulsions, London dispersion energies, and electrostatic dipole interactions, on the assumption that the rigid amide bond between two amino acid residues maintains each residue independent of its neighbours[106, 179]. The objective of theoretical studies is not only to assess the minimum total energy of these interactions, but also to consider those conformations of near-minimum energy which are statistically significant in view of the motions of the molecule in solution. While computational methods for determining minimum energy conformations are now well-established, and discussed in Scheraga's review[179] and in later papers (e.g. Ref. 181), the multi-dimensional energy–conformation surface is extremely complex for a peptide, carrying a large number of local minima which are difficult to resolve.

Regularities have been noticed between primary structure and regional periodic content of proteins for which x-ray structures are available; some of the natural amino acids are recognised as helix-breaking (glycine, proline) or helix-distorting (valine, threonine, isoleucine), with most other amino acids being more or less inclined to participate in helix formation or formation of other ordered structures. These comparisons have encouraged further studies of the computation of polypeptide conformational energies; assistance from theoretical studies, particularly of preferred conformations of side-chains, for clarifying minor uncertainties in x-ray structures, is also a valid objective.

4.4.2 Molecular-orbital studies

The extended Hückel treatment and CNDO/2 approach applied to the calculation of conformational energies of residues in peptides, are complementary to some extent, the former giving reliable torsional potentials about the peptide bond, while CNDO/2 gives reasonable estimates for local partial charges, when applied to oligopeptides of glycine and alanine as representative examples[182]. Calculations have been presented showing coplanarity of terminal N, C^{α}, and C′ atoms with the peptide group, for an *N*-terminal glycine residue[183]; this plane is affected by β- and γ-atoms in side-chains of other residues. *C*-Terminal residue conformations are similar[184]. For a proline residue, C^{δ}, N, C^{α}, and C^{β} are nearly coplanar, with C^{γ} on the same side as C′[185]. This series of papers is completed by calculations on hydrogen bond energies[186], and by a discussion of the basis of the molecular-orbital calculations used[187]. The side-chains of serine, threonine, cysteine, valine, and α-aminobutyric acid[188], and aspartic acid[189] residues in polypeptides restrict backbone torsion angles through the steric demands of the γ-atom[188], while the δ-atoms and those beyond them have no influence[189]. Furthermore, the position of the γ-atom is restricted to three staggered conformations of the C^{α}—C^{β} bond and one of these positions ($\chi_1 = 60$ degrees) prevents the residue from participating in an α-helix, either left- or right-handed[188]. These deductions are borne out by the behaviour of some γ-atom residues in myoglobin[190], lysozyme and α-chymotrypsin[188]. Amino acids of this type include phenylalanine, tyrosine, tryptophan, histidine, leucine and norvaline[191], and these behave conformationally like alanine[192].

Alanine preceding proline in a peptide adopts the right-handed α-helix conformation (i.e. torsion angles $\phi = c.\ -60$ degrees, $\psi = -50$ degrees; the Ra conformation in the terminology proposed by Liquori[193]) but its torsion angles are sensitive to the geometry of the pyrrolidine ring of the proline residue[194]. Several workers[195, 196] have used *N*-acetyl oligopeptide *N*-methylamides, $CH_3{\cdot}CO{\cdot}(NH{\cdot}CHR''{\cdot}CO)_n{\cdot}NH{\cdot}CH_3$, as models of protein segments, mainly in attempting to show that near neighbour interactions have a dominating effect, on a local basis, in determining features of secondary structure seen in x-ray structures of myoglobin[196], lysozyme[195, 196], and α-chymotrypsin[197]. Conformational analysis of the peptide lactone rings of actinomycin D[110], has used the same general approach, based on conformational maps for dipeptide and tripeptide fragments of the antibiotic and including consideration of *cis*- and *trans*-configurations for its *N*-methyl amide groupings[110]. Gramicidin S, a cyclic decapeptide, has been the subject of conformational speculation for some time, a mixed α–β-conformation being favoured now, in solution[60, 198]; calculations have been reported[61] for some of the range of near-minimum energy conformations which this compound may adopt.

Calculations are usually initiated for peptides for which experimental data are available, and since reliable calculations depend upon reliable experimental estimates of interaction factors, further progress is to some extent determined by improvements in experimental techniques. Calculations of minimum conformational energies for amino acids themselves, illustrated for glycine, alanine, phenylalanine and proline[199], are also considered to be related to their behaviour as residues in peptides.

Side-chains of lysine and arginine (—$(CH_2)_4$—NH_2, and —$(CH_2)_3$—N=$C(NH_2)$—NH_2, respectively) can adopt a large number of different conformations, in principle, but few of these are exemplified in x-ray structures of proteins[200]. Molecular-orbital calculations for backbone torsion angles of these residues give very satisfactory results when compared with their conformations in the proteins, but variations in the side-chain torsion angles χ_1 and to a lesser extent χ_2 can destabilise these minimum energy forms[200]. The method developed by Pullman and co-workers[215], an all-valence-electrons molecular-orbital method (PCILO theory – perturbative configuration interaction using localised orbitals theory) has been applied to dipeptides containing glycine and alanine residues[192], with the conformational energy map for alanine being typical of all amino acids with β-carbon substituents. Continuations of these studies[201, 202] for valine, isoleucine and leucine residues show these to be very similar to each other (and to alanine, contrary to more qualitative computations), and show that they should adopt stable α-helical conformations in peptides (cf. Ref. 203). For phenylalanine, tyrosine, histidine and tryptophan the torsion angles corresponding to the most stable conformations ($\phi = 80$ degrees, $\psi = -40$ degrees) of these residues in dipeptides permit a seven-membered intramolecular hydrogen bond to form[204]; this point on the conformational energy map is not usually considered to be in a stable zone, but independent i.r. studies tend to support the formation of seven-membered hydrogen-bond loops in peptides[205], and further results of this approach (which has also involved aspartic acid and glutamic acid residues[206]) and comparisons with experimental evidence, will be of interest. Poly(L-aspartate esters) are predicted to be capable of adopting conformations in which the side-chain is transverse ('wrap-round') to the axis of an α-helical backbone, for long side-chains, or longitudinal (parallel to the α-helix axis) for short side-chains. Polarised i.r. spectra can give experimental support for the predictions[163], but the 'wrap-round' possibility has been excluded for poly(β-benzyl-L-aspartate) by n.m.r. studies[134]. Dipole–dipole interactions between side-chain groupings and the nearest backbone amide dipole, influence the backbone torsion angles considerably when geometrical factors favour coupling. Predictions on this basis, that poly(β-*o*- and *m*-chlorobenzyl-L-aspartates) should form left-handed α-helices while the *p*-chlorobenzyl analogue should form the right-handed α-helix in helicogenic solvents[207] has been dramatically confirmed by o.r.d. and c.d. results; these polymers give almost mirror-image o.r.d. and c.d. curves about the wavelength axis in dioxan[16, 17, 179, 207].

Calculations have been presented[208] to account for the higher energies of peptides with amide bonds in the *cis*-configuration, relative to their *trans*-isomers.

A helical region in a polypeptide or proteins can tolerate the presence of one helix-breaking residue (typically glycine) without serious disruption, but two or more tend to terminate a helical sequence[209]. The 'helix probability profile' of a protein is related to the α-helix-forming tendency of each of the amino acids present in the protein, as deduced from the α-helix-forming tendency of the corresponding poly(amino acids)[179, 210, 211]. Some difficulty in obtaining a quantitative expression of this tendency remains[179] but with deductions based on available x-ray evidence for proteins, some progress has been made towards predicting this aspect of conformation of proteins for which x-ray structures are not yet available. Thus, the helix probability profiles of lysozyme and α-lactalbumin are very similar to each other[210], despite many differences in primary structure. This supports the view that these two proteins have very similar three-dimensional structures (cf Refs. 130, 140). The minimum energy conformations of collagen, calculated on the basis of potential energies of non-bonded interactions of constituent fragments, are in satisfactory agreement with experimental data[212].

Detailed information on the role of solvent in stabilising polypeptide conformations is still out of reach, but experimental generalisations relating solvent polarity to propensity for ordering of polypeptides are well-known and exemplified elsewhere in this chapter. The low stability of α-helical polypeptide chains in water, and in aqueous solutions of high ionic strength, has been discussed from a thermodynamic standpoint[213]. Conversely, helix formation in organic solvents is more favoured and therefore must be accompanied by a greater relative decrease in conformational free energy[213]. The extrusion of water, bound to an unordered polypeptide, which takes place as the macromolecule folds into a partly- or fully-ordered conformation brings polar side-chains more into contact with the solvent and is therefore energetically favoured[214].

References

1. IUPAC–IUB Commission on Biochemical Nomenclature, 1969. (1970). *Biochemistry,* **9,** 3471; (1970). *Eur. J. Biochem.,* **17,** 193; (1971). *Biochim. Biophys. Acta,* **229,** 1
2. McGuire, R. F., Vanderkooi, G., Momany, F. A., Ingwall, R. T., Crippen, G. M., Lotan, N., Tuttle, R. W., Kashuba, K. L. and Scheraga, H. A. (1971). *Macromolecules,* **4,** 112
3. Dickerson, R. E., Takano, T., Eisenberg, D., Kallai, O. B., Samson, L., Cooper, A. and Margoliash, E. (1971). *J. Biol. Chem.,* **246,** 1511
4. Lewis, P. N., Momany, F. A. and Scheraga, H. A. (1971). *Proc. Natl. Acad. Sci. U.S.A.,* **68,** 2293
5. Venkatachalam, C. M. (1968). *Biopolymers,* **6,** 1425
6. Birktoft, J. J., Blow, D. M., Henderson, R. and Steitz, T. A. (1970). *Phil. Trans. Roy. Soc. B,* **257,** 67
7. Hodgkin, D. M. *Bakerian Lecture, Royal Society, London,* June, 1972
8. Visser, L. and Blout, E. R. (1971). *Biochemistry,* **10,** 743
9. Saxena, V. P. and Wetlaufer, D. B. (1971). *Proc. Natl. Acad. Sci. U.S.A.,* **68,** 969
10. Leach, S. J. (editor). (1970). *Physical Principles and Techniques of Protein Chemistry.* (New York: Academic Press)
11. Wetzel, R. (1968). *Stud. Biophys.,* **8,** 127
12. Young, G. T. (ed.). (1969). *Specialist Periodical Reports: Amino-acids, Peptides and Proteins,* Vol. 1 (London: Chemical Society); (1970), Vol. 2; (1971), Vol. 3
13. Greenfield, N. and Fasman, G. D. (1969). *Biochemistry,* **8,** 4108
14. Woody, R. W. (1968). *J. Chem. Phys.,* **49,** 4797
15. Pysh, E. S. (1970). *J. Chem. Phys.,* **53,** 2156
16. Giancotti, V., Quadrifoglio, F. and Crescenzi, V. (1972). *J. Amer. Chem. Soc.,* **94,** 297
17. Erenrich, E. H., Andreatta, R. H. and Scheraga, H. A. (1970). *J. Amer. Chem. Soc.,* **92,** 1116
18. Balasubramanian, D. (1971). *Indian J. Chem.,* **9,** 1164
19. Jirgensons, B. (1970). *Biochim. Biophys. Acta,* **200,** 9
20. Sonenberg, M. and Beychok, S. (1971). *Biochim. Biophys. Acta,* **229,** 88
21. Ettinger, M. J. and Timasheff, S. N. (1971). *Biochemistry,* **10,** 824, 831
22. Rosenkranz, H. and Scholtan, W. (1971). *Z. Physiol. Chem.,* **352,** 896
23. Litman, G. W., Frommel, D., Rosenberg, A. and Good, R. A. (1971). *Biochim. Biophys. Acta,* **236,** 647
24. Zubkov, V. A., Birshtein, T. M., Milevskaya, I. S. and Volkenshtein, M. V. (1971). *Biopolymers,* **10,** 2051

25. Balasubramanian, D. and Roche, R. S. (1970). *Chem. Commun.*, 862
26. Aebersold, D. R. and Pysh, E. S. (1970). *J. Chem. Phys.*, **53,** 2156
27. Tinoco, I. *Lecture at NATO Advanced Study Institute on Fundamental Aspects and Recent Developments in Optical Rotatory Dispersion and Circular Dichroism,* Pisa, Italy, September 1971 (to be published)
28. Venyaminov, S. Y. and Chirgadze, Y. N. (1970). *Dokl. Akad. Nauk S.S.S.R.,* **195,** 722
29. Chirgadze, Y. N., Venyaminov, S. Y. and Lobachev, V. M. (1971). *Biopolymers,* **10,** 809
30. Jennings, B. R. and Baily, E. D. (1970). *Nature (London),* **228,** 1309
31. Magar, M. E. (1971). *J. Theoret. Biol.,* **33,** 105
32. Brown, J. E. and Klee, W. A. (1971). *Biochemistry,* **10,** 470
33. Rippon, W. B. and Walton, A. G. (1971). *Biopolymers,* **10,** 1207
34. Maguire, R. J. and Dunford, H. B. (1971). *Canad. J. Biochem.,* **49,** 666
35. Cann, J. R. (1971). *Biochemistry,* **10,** 3707
36. Dunstan, D. R. and Scopes, P. M. (1968). *J. Chem. Soc. C,* 1585
37. Barrett, G. C. (1969). *J. Chem. Soc. C,* 1123
38. Snatzke, G. (1967). *Proc. Roy. Soc. A,* **297,** 43
39. Toniolo, C. (1971). *Biopolymers,* **10,** 1707
40. Goodman, M., Naider, F. and Toniolo, C. (1971). *Biopolymers,* **10,** 1719
41. Widmer, H. and Lorenzi, G. P. (1971). *Chimia,* **25,** 236
42. Schechter, B., Schechter, I., Ramachandran, J., Conway-Jacobs, A. and Sela, M. (1971). *Eur. J. Biochem.,* **20,** 301
43. Conio, G., Patrone, E. and Salaris, F. (1971). *Macromolecules,* **4,** 283
44. Engel, J., Liehl, E. and Sorg, C. (1971). *Eur. J. Biochem.,* **21,** 22
45. Damle, V. N. (1970). *Biopolymers,* **9,** 937
46. Patrone, E., Conio, G. and Brighetti, S. (1970). *Biopolymers,* **9,** 897
47. Cosani, A., Peggion, E., Terbojevich, M. and Acampora, M. (1971). *Macromolecules,* **4,** 390
48. Heitz, F. and Spach, G. (1971). *Macromolecules,* **4,** 429
49. Bovey, F. A., Ryan, J. J., Spach, G. and Heitz, F. (1971). *Macromolecules,* **4,** 433
50. Rydon, H. N., Hardy, P. M., Haylock, J. C., Marlborough, D. I., Storey, H. T. and Thompson, R. C. (1971). *Macromolecules,* **4,** 435
51. Grourke, M. J. and Gibbs, J. H. (1971). *Biopolymers,* **10,** 795
52. Chou, P. Y. and Scheraga, H. A. (1971). *Biopolymers,* **10,** 657
53. Gutte, B. and Merrifield, R. B. (1971). *J. Biol. Chem.,* **246,** 1922
54. Taniuchi, H. and Anfinsen, C. B. (1971). *J. Biol. Chem.,* **246,** 2291
55. Epand, R. M. and Scheraga, H. A. (1968). *Biochemistry,* **7,** 2864
56. Atassi, M. Z. and Singhal, R. P. (1970). *Biochemistry,* **9,** 4252
57. Ohta, Y., Hibino, Y., Asaba, K., Sugiura, K. and Samejima, T. (1971). *Biochim. Biophys. Acta,* **236,** 802
58. Hermans, J. and Puett, D. (1971). *Biopolymers,* **10,** 895
59. Hassall, C. H., Moschidis, M. C. and Thomas, W. A. (1971). *J. Chem. Soc. B,* 1757
60. Hassall, C. H. and Thomas, W. A. (1971). *Chem. in Brit.,* **7,** 145
61. De Santis, P. and Liquori, A. M. (1971). *Biopolymers,* **10,** 699
62. Bush, C. A. and Ziegler, S. M. (1971). *Biochemistry,* **10,** 1330
63. Ovchinnikov, Y. A., Ivanov, V. T. and Mikhaleva, I. I. (1971). *Tetrahedron Letters,* 159
64. Ivanov, V. T., Shilin, V. V., Kogan, G. A., Meshcheryakova, E. N., Senyavina, L. B., Efremov, E. S. and Ovchinnikov, Y. A. (1971). *Tetrahedron Letters,* 2841
65. Urry, D. W. and Ruiter, A. (1970). *Biochem. Biophys. Res. Commun.,* **38,** 800
66. Urry, D. W., Glickson, J. D., Mayers, D. F. and Haider, J. (1972). *Biochemistry,* **11,** 487
67. Brewer, H. B. and Edelhoch, H. (1970). *J. Biol. Chem.,* **245,** 2402
68. Jirgensons, B. (1970). *Latv. PSR Zinat. Akad. Vestis, Kim. Ser.,* 688; (1971). *Chem. Abstr.,* **74,** 107 327
69. Jirgensons, B. and Capetillo, S. (1970). *Biochim. Biophys. Acta,* **205,** 355
70. Robbins, F. M. and Holmes, L. G. (1970). *Biochim. Biophys. Acta,* **221,** 234
71. Jones, H. A. and Lerman, S. (1971). *Canad. J. Biochem.,* **49,** 426
72. Chicheportiche, R. and Lazdunski, M. (1970). *Eur. J. Biochem.,* **14,** 549
73. Sander, C. and Ts'o, P. O. P. (1971). *Biochemistry,* **10,** 1953
74. 'S-Gravenmade, E. J., Van der Drift, C. and Vogels, G. D. (1971). *Biochim. Biophys. Acta,* **251,** 393
75. Ghose, A. C. and Jirgensons, B. (1971). *Biochim. Biophys. Acta,* **251,** 14

76. Ghose, A. C. (1971). *Biochem. Biophys. Res. Commun.*, **45**, 1144
77. Dalgleish, D. G. and Peacocke, A. R. (1971). *Biochem. J.*, **125**, 155
78. Simons, E. R. (1971). *Biochim. Biophys. Acta*, **251**, 126
79. Makinen, M. W. and Kon, H. (1971). *Biochemistry*, **10**, 43
80. Yong, F. C. and King, T. E. (1971). *Eur. J. Biochem.*, **20**, 111
81. King, T. E., Bayley, P. M. and Yong, F. C. (1971). *Eur. J. Biochem.*, **20**, 103
82. Myer, Y. P. (1971). *J. Biol. Chem.*, **246**, 1241
83. Skov, K. and Williams, G. R. (1971). *Canad. J. Biochem.*, **49**, 441
84. Wood, G. C. and Crichton, R. R. (1971). *Biochim. Biophys. Acta*, **229**, 83
85. Wood, E., Dalgleish, D. G. and Bannister, W. (1971). *Eur. J. Biochem.*, **18**, 187
86. Sawyer, W. H., Norton, R. S., Nichol, L. W. and McKenzie, G. H. (1971). *Biochim. Biophys. Acta*, **243**, 19
87. Herskovits, T. T., Jaillet, H. and De Sena, A. T. (1970). *J. Biol. Chem.*, **245**, 6511
88. Singer, S. J. and Glaser, M. (1971). *Biochemistry*, **10**, 1780
89. Urry, D. W., Masotti, L. and Krivacic, J. R. (1971). *Biochim. Biophys. Acta*, **241**, 600
90. Gordon, D. J. and Holzwarth, G. (1971). *Arch. Biochem. Biophys.*, **142**, 481
91. Cf. *Current Topics in Membranes and Transport* (1970). Vol. I
92. Wagner, T. and Spelsberg, T. C. (1971). *Biochemistry*, **10**, 2599
93. Rowe, J. J. M., Hinton, J. and Rowe, K. L. (1970). *Chem. Rev.*, **70**, 1
94. Hill, H. A. O. (1971). *NMR*, **4**, 167 (P. Diehl, editor) (Berlin: Springer Verlag)
95. Wuethrich, K. (1970). *Chimia*, **24**, 409
96. Phillips, W. D. (1971). *Probes of Structure and Function of Macromolecules and Membranes, Proc. Colloq. Johnson Res. Foundation*, 1969, **1**, 75 (B. Chance, editor). (New York: Academic Press)
97. Schwyzer, R. (1969). *Proc. Int. Congr. Pharmacol.*, 196 (R. Eigenmann, editor) (Schwabe, 1970: Basel)
98. Roberts, G. C. K. and Jardetzky O. (1970). *Advan. Protein Chem.*, **24**, 447
99. Sheard, B. and Bradbury, E. M. (1970). *Progr. Biophys. Molec. Biol.*, **20**, 187
100. Robin, M. B., Bovey, F. A. and Basch, H. (1970). *Chemistry of Amides*, 1, (J. Zabicky, editor) (New York: Interscience)
101. Cohn, M. (1970). *Quart. Rev. Biophys.*, **3**, 61
102. Mildvan, A. S. and Cohn, M. (1970). *Advan. Enzymol.*, **33**, 1
103. Allerhand, A. and Trull, E. A. (1970). *Ann. Rev. Phys. Chem.*, **21**, 317
104. Tonelli, A. E. and Bovey, F. A. (1970). *Macromolecules*, **3**, 410
105. Tonelli, A. E., Brewster, A. I. and Bovey, F. A. (1970). *Macromolecules*, **3**, 412
106. Tonelli, A. E., Patel, D. J., Goodman, M., Naider, F., Faulstich, H. and Wieland, T. (1971). *Biochemistry*, **10**, 3211
107. Lackner, H. (1970). *Tetrahedron Letters*, 2807, 3189; *Chem. Ber.*, **103**, 2476; (1971). *Tetrahedron Letters*, 2221
108. Conti, F. and De Santis, P. (1970). *Nature (London)*, **227**, 1239
109. Arison, B. H. and Hoogsteen, K. (1970). *Biochemistry*, **9**, 3976
110. De Santis, P., Rizzo, R. and Ughetto, G. (1971). *Tetrahedron Letters*, 4309
111. Deber, C. M., Torchia, D. A. and Blout, E. R. (1971). *J. Amer. Chem. Soc.*, **93**, 4893
112. Kopple, K. D. (1971). *Biopolymers*, **10**, 1139
113. Torchia, D. A., Di Corato, A., Wong, S. C. K., Deber, C. M. and Blout, E. R. (1972). *J. Amer. Chem. Soc.*, **94**, 609
114. Torchia, D. A., Wong, S. C. K., Deber, C. M. and Blout, E. R. (1972). *J. Amer. Chem. Soc.*, **94**, 616
115. Kopple, K. D., Go, A., Logan, R. H. and Savrda, J. (1972). *J. Amer. Chem. Soc.*, **94**, 973
116. Marshall, G. R. (1970). *Proc. First Amer. Peptide Symp. (1968)*, 151, (B. Weinstein, editor). (New York: Marcel Dekker)
117. Newmark, R. A. and Miller, M. A. (1971). *J. Phys. Chem.*, **75**, 505
118. Cavanaugh, J. R. (1970). *J. Amer. Chem. Soc.*, **92**, 1488
119. Temussi, P. A. and Goodman, M. (1971). *Proc. Natl. Acad. Sci. U.S.A.*, **68**, 1767
120. Ramachandran, G. N., Chandrasekaran, R. and Kopple, K. D. (1971). *Biopolymers*, **10**, 2113
121. Conti, F., Pietronero, C. and Viglino, P. (1970). *Org. Mag. Resonance*, **2**, 131
122. Nagai, M., Nishioka, A. and Yoshimura, J. (1970). *Bull. Chem. Soc. Jap.*, **43**, 1323
123. Cohen, J. S. (1971). *Biochim. Biophys. Acta*, **229**, 603
124. Patel, D. J. (1971). *Macromolecules*, **4**, 251

125. Glickson, J. D., Mayers, D. F., Settine, J. M. and Urry, D. W. (1972). *Biochemistry,* **11,** 477
126. Torchia, D. A. and Bovey, F. A. (1971). *Macromolecules,* **4,** 246
127. Torchia, D. A. (1971). *Macromolecules,* **4,** 440
128. Bradbury, E. M., Cary, P., Crane-Robinson, C., Paolillo, L., Tancredi, T. and Temussi, P. A. (1971). *J. Amer. Chem. Soc.,* **93,** 5916
129. Paolillo, L., Tancredi, T., Temussi, P. A., Trivellone, E., Bradbury, E. M. and Crane-Robinson, C. (1972). *J. C. S. Chem. Commun.,* 335
130. Bradbury, J. H. and King, N. L. R. (1971). *Aust. J. Chem.,* **24,** 1703
131. Boublick, M., Bradbury, E. M., Crane-Robinson, C. and Johns, E. W. (1970). *Eur. J. Biochem.,* **17,** 151
132. Boublick, M., Bradbury, E. M. and Crane-Robinson, C. (1970). *Eur. J. Biochem.,* **14,** 486
133. Rüterjans, H. and Pongs, O. (1971). *Eur. J. Biochem.,* **18,** 313
134. Silverman, D. N., Taylor, G. T. and Scheraga, H. A. (1971). *Arch. Biochem. Biophys.,* **146,** 587
135. Liu, K.-J. and Lignowski, S. J. (1970). *Biopolymers,* **9,** 739
136. Sato, K. and Nishioka, A. (1971). *Polymer J.,* **2,** 379
137. Kuntz, I. D. (1971). *J. Amer. Chem. Soc.,* **93,** 514, 516
138. Glickson, J. D. and Applequist, J. (1971). *J. Amer. Chem. Soc.,* **93,** 3276
139. Morallee, E. G., Nieboer, E., Rossotti, F. J. C., Williams, R. J. P., Xavier, A. V. and Dwek, R. A. (1970). *Chem. Commun.,* 1132; Dwek, R. A., Richards, R. E., Morallee, K. G., Nieboer, N., Williams, R. J. P. and Xavier, A. V. (1971). *Eur. J. Biochem.,* **21,** 204
140. Magnuson, J. A. and Magnuson, N. S. (1971). *Biochem. Biophys. Res. Commun.,* **45,** 1513
141. Tigelaar, H. and Flygare, W. H. (1972). *J. Amer. Chem. Soc.,* **94,** 343
142. Radda, G. K. (1971). *Biochem. J.,* **122,** 385
143. Birkett, D. J., Dwek, R. A., Radda, G. K., Richards, R. E. and Salmon, A. G. (1971). *Eur. J. Biochem.,* **20,** 494
144. Stryer, L. (1970). CIBA Foundation Symp. on *Molecular Properties of Drug Receptors,* 133 (R. Porter, editor), (London: Churchill)
145. Burshtein, E. A. (1968). *Stud. Biophys.,* **9,** 137
146. Cowgill, R. W. (1970). *Biochim. Biophys. Acta,* **200,** 18
147. Kirby, E. P. and Steiner, R. F. (1970). *J. Biol. Chem.,* **245,** 6300
148. Cowgill, R. W. and Lang, N. K. (1970). *Biochim. Biophys. Acta,* **214,** 228
149. Radda, G. K. (1971). *Current Topics in Bio-energetics,* **4**
150. Wahl, P., Kasai, M. and Changeaux, J.-P. (1971). *Eur. J. Biochem.,* **18,** 332
151. Radda, G. K. in Ref. 96
152. Schwyzer, R. and Schiller, P. W. (1971). *Helv. Chim. Acta,* **54,** 897
153. Schwyzer, R. (1972). *Lecture at Chemical Society Protein Group Meeting, Swansea,* April 1972
154. Parker, C. W. and Osterland, C. K. (1970). *Biochemistry,* **9,** 1074
155. Wang, J. L. and Edelman, G. M. (1971). *J. Biol. Chem.,* **246,** 1185
156. Bornet, H. and Edelhoch, H. (1971). *J. Biol. Chem.,* **246,** 1785
157. Raikhman, L. M., Annaev, B. and Rozantsev, E. G. (1971). *Dokl. Akad. Nauk SSSR,* **200,** 387
158. Gotto, A. M. and Kon, H. (1970). *Biochemistry,* **9,** 4276
159. Leigh, J. S. (1970). *J. Chem. Phys.,* **52,** 2608
160. McConnell, H. M. and McFarland, B. G. (1970). *Quart. Rev. Biophys.,* **3,** 91
161. Timofeev, V. P., Polianovsky, O. L., Volkenstein, V. M. and Lichtenstein, G. I. (1970). *Biochim. Biophys. Acta,* **220,** 357
162. Sanson, A. and Ptak, M. (1970). *Compt. Rend. Acad. Sci., Ser. D,* **271,** 1319
163. Bradbury, E. M., Carpenter, B. G. and Stephens, R. M. (1972). *Macromolecules,* **5,** 8
164. Graham, J. M. and Wallach, D. F. H. (1971). *Biochim. Biophys. Acta,* **241,** 180
165. Palm, V. (1970). *Z. Chem.,* **10,** 31
166. Goodman, M., Masuda, Y. and Verdini, A. S. (1971). *Biopolymers,* **10,** 1031
167. Muehlinghaus, J. and Zundel, G. (1971). *Biopolymers,* **10,** 711
168. Koenig, J. L. and Sutton, P. L. (1971). *Biopolymers,* **10,** 89
169. Deveney, M. J., Walton, A. G. and Koenig, J. L. (1971). *Biopolymers,* **10,** 615
170. Wallach, D. F. H., Graham, J. M. and Oseroff, A. R. (1970). *FEBS Letters,* **7,** 330
171. De Groot, K., Feyen, J., De Visser, A. C., Van de Ridder, G. and Bantjes, A. (1971). *Kolloid-Z., Z. Polym.,* **246,** 578

172. Epand, R. M. (1971). *Canad. J. Biochem.*, **49,** 166
173. Mandelkern, L. and Mattice, W. L. (1971). *J. Amer. Chem. Soc.*, **93,** 1769
174. Ebert, G. and Wendorff, J. (1971). *Fortschrittsber., Kolloide Polym.*, **55,** 58
175. Barskaya, T. V. and Ptitsyn, O. B. (1971). *Biopolymers,* **10,** 2181
176. Pederson, D., Gabriel, D. and Hermans, J. (1971). *Biopolymers,* **10,** 2133
177. Steigman, J. and Cosani, A. (1971). *Biopolymers,* **10,** 357
178. Avignon, M. and Bothorel, P. (1971). *Compt. Rend. Acad. Sci. Ser. C,* **272,** 355
179. Scheraga, H. A. (1971). *Chem. Rev.*, **71,** 195
180. Goodman, M., Verdini, A. S., Choi, N. S. and Masuda, Y. (1970). *Topics in Stereochemistry,* **5,** 69
181. Crippen, G. M. and Scheraga, H. A. (1971). *Arch. Biochem. Biophys.*, **144,** 453, 462
182. Scheraga, H. A., Momany, F. A., McGuire, R. F. and Yan, J. F. (1971). *J. Phys. Chem.*, **75,** 2286
183. Ponnuswamy, P. K. and Sasisekharan, V. (1970). *Biochim. Biophys. Acta,* **221,** 153
184. Ponnuswamy, P. K. and Sasisekharan, V. (1970). *Biochim. Biophys. Acta,* **221,** 159
185. Ramachandran, G. N., Lakshminarayanam, A. V. and Balasubramanian, R. (1970). *Biochim. Biophys. Acta,* **221,** 165
186. Chidambaram, R., Balasubramanian, R. and Ramachandran, G. N. (1970). *Biochim. Biophys. Acta,* **221,** 182
187. Balasubramanian, R., Chidambaram, R. and Ramachandran, G. N. (1970). *Biochim. Biophys. Acta,* **221,** 196
188. Ponnuswamy, P. K. and Sasisekharan, V. (1971). *Biopolymers,* **10,** 565
189. Ponnuswamy, P. K. and Sasisekharan, V. (1971). *Biopolymers,* **10,** 583
190. Watson, H. C. (1969). *Progr. Stereochem.*, **4,** 299
191. Chandrasekaran, R. and Ramachandran, G. N. (1970). *Int. J. Protein Res.*, **2,** 223
192. Pullman, B., Maigret, B. and Perahia, D. (1970). *Theoret. Chim. Acta,* **18,** 44
193. Liquori, A. M. (1969). *Quart. Rev. Biophys.*, **2,** 65
194. Damiani, A., De Santis, P. and Pizzi, A. (1970). *Nature (London),* **226,** 542
195. Popov, E. M. and Lipkind, G. M. (1971). *Molec. Biol.*, **5,** 624
196. Jeronimidis, G. and Damiani, A. (1971). *Nature, New Biol.*, **229,** 150
197. Lipkind, G. M. and Popov, E. M. (1971). *Molec. Biol.*, **5,** 667
198. Schwyzer, R. and Ludescher, U. (1969). *Helv. Chim. Acta,* **52,** 2033
199. George, J. M. and Kier, L. B. (1970). *Experientia,* **26,** 952
200. Pullman, B., Coubeils, J. L., Courriere, P. and Perahia, D. (1971). *Theoret. Chim. Acta,* **22,** 11
201. Maigret, B., Perahia, D. and Pullman, B. (1971). *Biopolymers,* **10,** 491
202. Caillet, J., Pullman, B. and Maigret, B. (1971). *Biopolymers,* **10,** 221
203. Ostroy, S. E., Lotan, N., Ingwall, R. T. and Scheraga, H. A. (1970). *Biopolymers,* **9,** 749
204. Maigret, B., Pullman, B. and Perahia, D. (1971). *Biopolymers,* **10,** 107
205. Marraud, M., Neel, J. and Avignon, M. (1970). *J. Chim. Phys. Physicochim. Biol.*, **67,** 959
206. Maigret, B., Perahia, D. and Pullman, B. (1971). *Biopolymers,* **10,** 1649
207. Yan, J. F., Momany, F. A. and Scheraga, H. A. (1970). *J. Amer. Chem. Soc.*, **92,** 1109
208. Tonelli, A. E. (1971). *J. Amer. Chem. Soc.*, **93,** 7153; (1972). **94,** 346
209. Scheraga, H. A., Ananthanarayanan, V. S., Andreatta, R. H. and Poland, D. (1971). *Macromolecules,* **4,** 417
210. Lewis, P. N. and Scheraga, H. A. (1971). *Arch. Biochem. Biophys.*, **144,** 584
211. Scheraga, H. A., Von Dreele, P. H., Lotan, N., Ananthanarayanan, V. S., Andreatta, R. H. and Poland, D. (1971). *Macromolecules,* **4,** 408
212. Tumanyan, V. (1971). *Molec. Biol.*, **5,** 499
213. Ptitsyn, O. B. (1970). *Discus. Faraday Soc.*, **49,** 70
214. Go, M., Go, N. and Scheraga, H. A. (1971). *J. Chem. Phys.*, **54,** 4489
215. Maigret, B., Pullman, B. and Dreyfus, M. (1970). *J. Theoret. Biol.*, **26,** 321

5
Biological Activity of Structural Variants of Insulin

D. G. SMYTH
National Institute for Medical Research, Mill Hill, London

5.1 INTRODUCTION

Structure–activity studies in proteins and peptides have started with the assumption that certain amino acid residues are essential for function and that modification of these will lead to total loss of activity. In the study of enzymes the expectation has been fulfilled. With ribonuclease, for example, it was found that modification of either of two of the histidine residues or one of the lysine residues resulted in a product devoid of activity and this led to recognition of the 'active site' of the enzyme[1]. With peptide hormones, residues with abnormally high reactivity have been sought but none has been found. It appears that hormones do not possess an 'active centre' comparable to that found in enzymes.

The mechanism of action of a peptide hormone is considered to involve interaction of the hormone with specific biological receptors, an interaction which causes activation of molecules at the receptor site and leads to the expression of physiological activity. The *S*-peptide *S*-protein system of ribonuclease[2] has provided a useful analogy: the separate fragments, 1 to 20 and 21 to 124, exhibit no enzyme activity but in a mixture the two fragments interact to generate the full activity of native ribonuclease. Structure–function studies in this system have involved determination of the ability of modified *S*-peptides to interact with *S*-protein and produce enzyme activity; these experiments bear a formal resemblance to the determination of the ability of modified peptide hormones to interact with physiological receptors and exhibit hormonal activity. It is interesting that a comparable system has now been found at the opposite end of the ribonuclease molecule. The fragments containing residues 1 to 119 and 120 to 124 are totally inactive but a mixture of the fragments 1 to 119 and 111 to 124 can generate the full activity[3]. The production of activated complexes by the interaction of inert polypeptides that are complementary in structure may well constitute a general pattern of molecular events in protein and peptide biology.

In the context of structure–activity relationships in insulin, the relevant structure is the one taken up when the hormone interacts with the biological receptor. Elucidation of the 3-dimensional structure of crystalline zinc insulin[4–6] and the accumulation of evidence that a similar or identical structure persists in solution under physiological conditions of temperature and pH has literally added a new dimension to the study of structure and function. The results of the chemical studies of structures necessary for activity, presented in this review, are based on the amino acid sequence of insulin (illustrated in Figure 5.1 as part of the proinsulin molecule) and interpretations are then made within the steric framework of the 3-dimensional model.

5.2 DERIVATIVES OBTAINED BY MODIFICATION OF FUNCTIONAL GROUPS

5.2.1 Amino

Evidence that the three NH_2 groups of insulin are not essential for function was presented in 1950 when it was reported that a heavily acetylated insulin retained much of the hypoglycaemic activity of the natural hormone[10]. Since that time, homogeneous derivatives have been isolated with substituents attached to the NH_2 group of glycine, phenylalanine or lysine (Table 5.1). In the main, the original conclusion has been confirmed.

Lindsay and Shall, employing the 8 M urea chromatography system of Bromer[11], were able to isolate three *N*-monoacetyl, two diacetyl derivatives and a triacetyl derivative by the reaction of insulin with hydroxysuccinimide acetate[12]. Under the mild conditions used, competitive reaction at the OH group of tyrosine would be slight. Each of the *N*-acetyl derivatives gave essentially the full activity of insulin in the mouse-convulsion assay.

Brandenburg and his colleagues[13] obtained acetyl derivatives by the reaction of insulin with *p*-nitrophenyl acetate in dimethyl sulphoxide. In

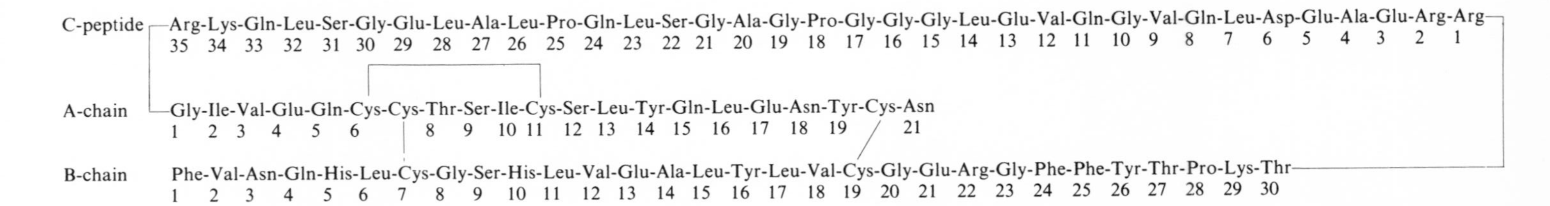

Figure 5.1 Primary structure of human proinsulin. The sequence of the A- and B-chains is based on the experiments of Nicol and Smith[7] and the sequence of the *C*-peptide on the experiments of Oyer *et al.*[8] and Ko *et al.*[9]; the basic residues at positions 1, 2, 34 and 35 of the *C*-peptide are assumed by analogy with porcine and bovine proinsulins

Table 5.1 Hypoglycaemic activities of *N*-acyl derivatives of insulin

Insulin derivatives	*Substituent*	*Mouse convulsion assay mean potency* (U/mg)	*Reference*
Crystalline bovine insulin (International Standard)		24.0	Bangham and Mussett[106]
Gly^{A1}-monosubstituted insulins			
acetyl	$CH_3CO—$	23.6	Lindsay and Shall[12]
carbamyl	$NH_2CO—$	20.5	Massey, Smyth, Sabey, Stewart and Webb, unpublished
acetoacetyl	$CH_3COCH_2CO—$	19.8	Lindsay and Shall[18]
thiazolidine-4-carbonyl	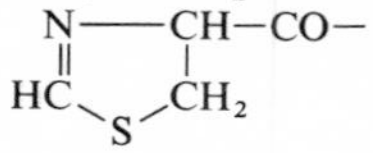 N=CH–S–CH₂–CH(–CO–)–N ring (HC=N—CH(CO—)—CH₂—S—)	12.8	Lindsay and Shall[19]
maleyl	$HO_2CCH{=}CHCO—$	14.5	Salokangas, Smyth, Sabey, Stewart and Webb, unpublished
arginyl	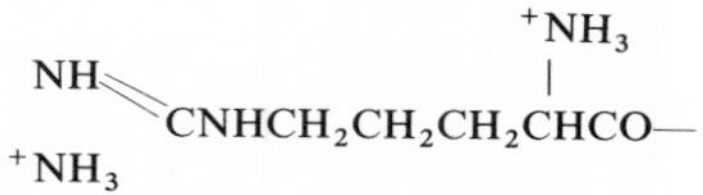 $(NH{=})(^{+}NH_3)CNHCH_2CH_2CH_2CH(^{+}NH_3)CO—$	14.2	Weinert, Kircher, Brandenburg and Zahn[107]
phthaloyl	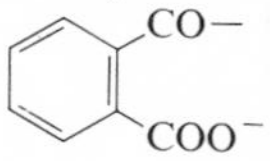 $C_6H_4(CO—)(COO^-)$ (benzene ring)	9.9	Salokangas, Smyth, Sabey, Stewart and Webb, unpublished
Phe^{B1}-monosubstituted insulins			
acetyl		22.4	Lindsay and Shall[12]
carbamyl		23.4	Massey, Smyth, Sabey, Stewart and Webb, unpublished
acetoacetyl		24.5	Lindsay and Shall[12]
thiazolidine		23.2	Lindsay and Shall[19]
maleyl		18.4	Salokangas, Smyth, Sabey, Stewart and Webb, unpublished
phthaloyl	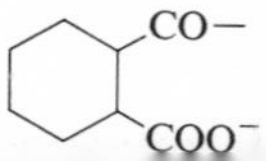 $C_6H_{10}(CO—)(COO^-)$ (cyclohexane ring)	15.6	Salokangas, Smyth, Sabey, Stewart and Webb, unpublished

	HO, OH, COO^-, C, NH—CS—		Arquilla[21]
p-iodophenylalanine (B1)	$I-C_6H_4-CH_2CH(NH_3^+)CO-$	16.0	Krail, Brandenburg and Zahn[24]
Lys^{B29}-substituted insulins			
acetyl		25.7	Lindsay and Shall[12]
thiazolidine		22.8	Lindsay and Shall[19]
Gly^{A1}, Phe^{B1}-disubstituted insulins			
acetyl		20.2	Lindsay and Shall[12]
acetoacetyl		17.0	Lindsay and Shall[12]
carbamyl		21.6	Massey, Smyth, Sabey, Stewart and Webb, unpublished
maleyl		13.5	Salokangas, Smyth, Sabey, Stewart and Webb, unpublished
phthaloyl		11.5	Salokangas, Smyth, Sabey, Stewart and Webb, unpublished
phenylthiocarbamyl	PhNHCS—	4.8	Africa and Carpenter[81]
fluoresceinthiocarbamyl		0.9	Bromer, Sheehan, Berns and Arquilla[21]
Gly^{A1}, Lys^{B29}-disubstituted insulins			
acetyl		21.6	Lindsay and Shall[12]
t-butyloxycarbonyl	$(CH_3)_3C-O-C(=O)-$	19.0	Geiger, Schone and Pfaff[23]

Table 5.1 *continued*

Insulin derivatives	*Substituent*	*Mouse convulsion assay mean potency* (U/mg)	*Reference*
Gly^{A1}, Phe^{B1}, Lys^{B29}-trisubstituted insulins			
acetyl		21.6	Lindsay and Shall[12]
carbamyl		21.2	Massey, Smyth, Sabey, Stewart and Webb, unpublished
acetoacetyl		10.4	Lindsay and Shall[18]
glutaryl		8.3	Ko, Smyth, Sabey, Stewart and Webb, unpublished
maleyl		11.0	Salokangas, Smyth, Sabey, Stewart and Webb, unpublished
succinyl		11.2	Ko, Smyth, Sabey, Stewart and Webb, unpublished
fluoresceinthiocarbamyl		0.1	Bromer, Sheehan, Berns and Arquilla[21]
phthaloyl		10.1	Salokangas, Smyth, Sabey, Stewart and Webb, unpublished
alanyl		10.0	Levy and Carpenter[22]
lysyl		9.9	Levy and Carpenter[22]
t-butyloxycarbonyl		5.7	Levy and Carpenter[22]

this solvent, acetylation takes place most rapidly at the A1 glycine and the B29 lysine. Monoacetylglycine (A1) insulin and monoacetyllysine (B29) insulin were obtained in crystalline form and a diacetyl (A1, B29) insulin and triacetyl (A1, B1, B29) insulin were also isolated. The di- and tri-acetylated derivatives, however, had additional acetyl groups attached to some of the tyrosine residues. When the activities were measured *in vitro* in the fat-pad assay, a surprising result was obtained. Monoacetyl-lysine (B29) insulin was almost as active as insulin but monoacetylglycine (A1) insulin exhibited only 40% of the activity. The di- and tri-acetyl derivatives were less active than the A1 glycine derivative. It was suggested that modification of the A1 glycine NH_2 group leads to a small change in conformation, the resulting structure combining less favourably with the biological receptor. A fair indication of this was obtained with the triacetylated derivative by measurement of circular dichroism and dispersion; but only minor differences from insulin were seen with the monoacetylglycine derivative.

The monoacetylglycine insulin was prepared also by a different route[14]. The NH_2 group of isolated A-chain was acetylated and the modified A-chain was combined with natural B-chain according to a procedure for the reconstitution of insulin from the A- and B-chain *S*-sulphonates[15]. The product appeared identical with the derivative obtained directly by acetylation of insulin.

Modification of the NH_2 group of the A1 glycine leads to a reduced tendency to aggregate[16]. This has been demonstrated by gel filtration of a series of carbamyl derivatives of insulin. Furthermore, monocarbamylglycine (A1) insulin is susceptible to digestion by trypsin at neutral pH whereas insulin and the monocarbamylphenylalanine derivative are relatively resistant[17]. The susceptibility of the glycine derivative to enzymic digestion indicates that it possesses a more open structure than insulin but it remains to be seen whether this has involved a change in the disposition of the A- and B-chains within the insulin monomer or whether it is the intermolecular association that has been affected.

The attachment of substituents larger than acetyl to the A1 glycine results in a more significant loss of activity. Thus, coupling of the acetoacetyl[18] or the dimethylthiazolidine carbonyl group[19] to the A1 glycine causes up to 50% inactivation *in vivo*. A similar decrease in activity results from succinylation of the A1 glycine, which involves the attachment of an anionic substituent[17]. It is notable that N^{A1}, N^{B1}, N^{B29}-trisuccinylinsulin, which has six more negative charges than insulin, still retains much of the activity of insulin. This would suggest that overall charge is not a critical factor in the interaction between insulin and its biological receptors.

The NH_2 group of the B1 phenylalanine residue seems to be remote from the region involved in activity. Substitution of this group by carbamylation[20] or acetylation[12] provides fully active molecules and substitution even by fluorescein thiocarbamyl, a large polycyclic group, leads only to partial inactivation[21]. The attachment of this radical to both the A1 glycine and the B1 phenylalanine leads to almost total inactivation.

In the fluorescein thiocarbamylation reaction the B1 phenylalanine was much more reactive than the A1 glycine. This contrasts with the reactivities exhibited during the modification of insulin by t-butyloxycarbonyl azide[22, 23], which takes place mainly at the A1 glycine and B29 lysine. The reversible protection of glycine and lysine with the t-BOC group has been exploited for selectively removing the B1 residue by Edman degradation; the resulting desphenylalanine insulin was extended by the attachment of iodophenylalanine to provide a homogeneous insulin with a radioactive substituent. The iodine was attached at a site that is relatively unimportant for hormonal activity; but the site may be involved in antigenic activity[12]. The iodine-labelled derivative exhibited 60% of the activity of insulin[24].

The isolation of lysine B29 substituted insulins with full activity precludes an important role for this amino acid in insulin function. The conclusion has been confirmed[25] by the preparation of a highly active insulin fragment lacking the last three residues of the B-chain (Section 5.4). The isolation of fragments lacking the A1 glycine or the B1 phenylalanine has also supported the conclusions drawn on these residues. The region of the molecule occupied by A1 glycine can be important for biological activity; the structure at the B1 phenylalanine is less critical.

5.2.2 Tyrosine

Iodination has been extensively used for the incorporation of a radioactive isotope into the insulin molecule. For use in immunoassay[26], the iodinated insulins frequently contain several atoms of iodine per molecule[27] and the iodine may be attached to both tyrosine and histidine. These derivatives include a proportion of degraded molecules resulting from the exposure to oxidising agents during iodination or from radiation damage[28].

Mild iodination of insulin in aqueous solution takes place predominantly at the tyrosyl residues of the A-chain[29] and a limited reaction occurs also at tyrosine B16[30]. By carrying out the reaction in the presence of organic solvent or 8 M urea, iodination can be made to take place at both tyrosine B16 and tyrosine B26. From the rates of iodination in different media, conclusions have been drawn on the availability of the four tyrosine residues[29]. Tyrosine A14 appears to be fully exposed to the solvent. Tyrosine A19 is sterically influenced by the presence of the neighbouring disulphide bridge, tyrosine B26 is embedded in a non-polar region which is virtually independent of the protein conformation, and tyrosine B16 is concealed in a hydrophobic environment which depends on the preservation of the native structure. These predictions are well supported by the disposition of the four tyrosine residues in the 3-dimensional structure of crystalline insulin[4].

The results obtained by bioassay of the iodinated insulins are of particular interest. On the rat epididymal fat pad assay, a preparation containing an average of 1.5 atoms of iodine per insulin molecule retained 92% of the activity of insulin. Incorporation of 2.5 atoms of iodine reduced the specific activity to 56% and 4 atoms of iodine reduced the activity to 22%[29]. It should be noted that the lightly iodinated insulin would be expected to contain some unmodified insulin as well as some di-iodinated insulin and therefore the value of 92% obtained for the specific activity cannot be firmly assigned to a monoiodo derivative.

Iodinated insulin containing an average of 1.0 atom of iodine per molecule has been separated from residual insulin[31]. Early reports indicated that the monoiodo derivative possessed only 1% of the activity of insulin but further work has shown that the derivative retains 70% of the activity[32]. As this preparation represents a mixture of A14 and A19 modified derivatives, each carrying a single iodine atom, it would seem that the two tyrosines can accept an iodine substituent without substantial loss of activity. The possibility remains, however, that one of the two tyrosines can be iodinated with complete retention of activity while iodination of the other causes a degree of inactivation. The question will remain unresolved until homogeneous, monoiodinated derivatives are isolated. It is, however, clear that the incorporation of more than one atom of iodine into insulin gives rise to derivatives with reduced specific activity.

The differential reactivities exhibited by the tyrosine residues to iodination are not maintained in the reaction of insulin with cyanuric fluoride[33]. In this case, the reactive residues are A19 and B16. The B26 tyrosine becomes reactive after tryptic digestion, which liberates the CO_2H-terminal octapeptide of the B-chain. The lack of reactivity of this residue in native insulin was used to indicate that it occupies a concealed position in the molecular structure. Intrachain hydrogen bonding was proposed between the side chains of B26 tyrosine and B22 arginine and interchain bonding between the side chains of A14 tyrosine and B13 glutamic acid. Examination of the 3-dimensional crystalline structure[4] reveals that the first of these postulated linkages is feasible but the second is highly unlikely.

The tyrosine residues of insulin have been modified by reaction with tetranitromethane[34, 35] at neutral pH. Two of the four tyrosines were reactive and these proved to be the residues at A14 and A19. The main mononitrated product was a derivative in which the nitro substituent was attached to the A14 tyrosine alone. It was fully active in a blood sugar depression assay performed in diabetic mice[34].

It is not possible to compare the reactivities of the tyrosines during nitration and iodination with the reactivities exhibited to cyanogen fluoride since the reaction conditions were not identical. The latter reaction was conducted at pH 9.7 in the presence of dioxan and these factors are known to affect the aggregation state[36] and conformation[37] of insulin. The low reactivity of tyrosine A14 to cyanogen fluoride is probably related to the anomalous pK value of the A14 hydroxyl group rather than to an inherent inaccessibility of the tyrosine ring. From inspection of the model of crystalline insulin it is possible that the lower reactivity of the OH group of tyrosine A14 could be explained by the presence of a hydrogen bond between tyrosine A14 and the carbonyl group of B2 valine or the carboxyl group of glutamic acid A17.

Modification of the tyrosine OH groups of *N*-tricarbamyl insulin by acetic anhydride gave a derivative in which essentially all the tyrosine residues were acylated[38]. The unfractionated reaction mixture retained 25% of the activity of insulin in the mouse-convulsion assay. Under the preparative conditions used, partial crosslinking may have taken place through formation of mixed anhydrides at the CO_2H groups and reaction of these groups with the tyrosine OH groups of adjacent molecules[39]. Nevertheless, the substantial activity of the tetra *O*-acetyl preparation indicates that none of the tyrosine

OH groups is essential for function; but esterase effects, which could cause removal of *O*-acetyl groups during the bioassay, should not be discounted.

5.2 Imidazole of histidine

The two histidine residues at B5 and B10 are a characteristic feature of almost all insulins (Figure 5.2) and insulins with the two histidines but with many variations in the rest of the molecule exhibit essentially the same specific activity in the mouse-convulsion assay[40]. Guinea-pig insulin, which lacks the histidine at B10, has a considerably lower specific activity in the mouse-convulsion assay. This indirect evidence suggested that both histidines, and especially that at B10, are important for function.

Insulin derivatives with both the histidine and the amino groups modified have been obtained by reaction of insulin with acrylonitrile[41]. The B5 histidine was more reactive than the B10 histidine. This could be due to zinc binding[42] or to steric hindrance, but the latter is unlikely since the reaction was not accelerated by the presence of denaturing solvents. Lysine B29, on the other hand, was much more reactive to acrylonitrile in the presence of 8 M urea. This observation is consistent with previous evidence that the B29 residue is not freely accessible in native insulin; tryptic cleavage at the Lys–Ala bond takes place only slowly at neutral pH[43]. The mixture of cyanoethylated insulins, in which all the NH_2 groups and 50% of the histidine residues had been alkylated, exhibited 9 U/mg in the mouse-convulsion assay.

It is difficult to relate the decreased activity of cyanoethylated insulin to the blocking of a specific functional group. However, the rate of disappearance of biological activity during the reaction appeared to correspond to the rate of modification of the B5 histidine rather than to the modification of the NH_2 groups. The decreased activity was therefore attributed to the modification of the B5 histidine[41]. Studies on the photo-oxidation of insulin, a reaction which affects mainly histidine residues, have also shown that the histidine residues of insulin have some importance for activity[44, 45]. Whether these reactions lead to conformational changes in the insulin molecule which adversely influence activity or whether the B5 histidine is directly involved in the biological action of insulin is not known.

No insulin derivative with an acyl substituent attached to histidine has been reported. Acyl imidazoles, unlike acyl amino compounds, are unstable at physiological pH[46].

5.2.4 Guanidino of arginine

All insulins with known amino acid sequence, except guinea-pig and coypu, have arginine at position 22 on the B-chain (Figure 5.2). As this is the sole arginine in the molecule, attention has focused on this site as one that might be important for function.

```
      1        3             6        8             11  12        14  15       17                                       26  27  28           31
Man:  Glu-Ala-Glu-Asp-Leu-Gln-Val-Gly-Gln-Val-Glu-Leu-Gly-Gly-Gly-Pro-Gly-Ala-Gly-Ser-Leu-Gln-Pro-Leu-Ala-Leu-Glu-Gly-Ser-Leu-Gln

Pig:  Glu-Ala-Glu-Asn-Pro-Gln-Ala-Gly-Ala-Val-Glu-Leu-Gly-Gly-Gly-Leu-Gly-Gly  –   –  Leu-Gln-Ala-Leu-Ala-Leu-Glu-Gly-Pro-Pro-Gln

Ox:   Glu-Val-Glu-Gly-Pro-Gln-Val-Gly-Ala-Leu-Glu-Leu-Ala-Gly-Gly-Pro-Gly-Ala-Gly  –   –   –   –   –  Gly-Leu-Glu-Gly-Pro-Pro-Gln
```

Figure 5.2 Species variation in the primary structure of human[8,9], porcine[90] and bovine *C*-peptides[95,96]

Modification of arginine residues in proteins has been reported with the use of α,β-dialdehydes[47]. The reaction is considered to take place at NH_2 and guanidino groups but the nature of the derivatives formed has not been studied. When insulin was allowed to react with glyoxal[48], 80% of the molecules underwent reaction at arginine and it was assumed that the NH_2 groups of glycine, phenylalanine and lysine had also reacted. The specificity of the reaction was assessed by acid hydrolysis and amino acid analysis. The glyoxal modified arginine was stable to hydrolysis but the *N*-modified glycine and phenylalanine appeared to revert to the free amino acids. As the glyoxal reaction took place more rapidly in alkali denatured insulin, it was suggested that arginine B22 is normally present in a folded region of the molecule. On the basis of this and other experiments[49], it was proposed that the guanidino group participates in the maintenance of 3-dimensional structure; a hydrogen bond was postulated between arginine B22 and tyrosine B26.

Bunzli and Bosshard re-examined the reaction of glyoxal with insulin under mild conditions[50] and were concerned with the effect on biological activity. The sites of reaction in the modified protein again were determined by amino acid analysis. The values obtained for glycine and phenylalanine showed a decrease of one residue, which contrasts with the results of Nakaya *et al.*[48]. No decrease occurred in lysine or arginine and on this evidence the reaction was considered to be confined to the α-NH_2 groups.

It seems most likely that glyoxal causes transamination at NH_2 groups with formation of the corresponding keto derivatives[51] and that the corresponding amino acids would not be released on hydrolysis. The biological activity of the glyoxal treated insulin on rat adipose tissue was reported to be 3.8 U/mg (16% of insulin).

Treatment of the glyoxal modified insulin, which had α-keto groups at the *N*-termini of the A- and B-chains, with phenyl glyoxal resulted in 86% reaction at arginine. The activity of the unfractionated mixture of products

B-chain——Arg—
N
C
CH_2-CH_2 HN NH
C——CO
CH_2-CH_2

Figure 5.3 Structure of the cyclohexanedione derivative formed at arginine B22 in insulin. The nature of the modification of the guanidino group is that proposed by Toi *et al.*[53]

on the fat pad was 1% that of insulin[50]. This led the authors to conclude that the guanidino group of arginine plays an important role in the biological function.

It should be noted that none of the arginine modified derivatives were isolated from the reaction mixtures. Furthermore, the specificity of the reaction of glyoxal with protein functional groups has not been studied and possible reactions with the imidazole group of histidine or the hydroxyl group of tyrosine should not be overlooked.

An alternative approach to studying the effect of arginine modification on the activity of insulin would be to carry out the dialdehyde reaction on a

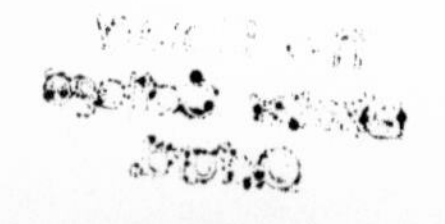

fully active derivative in which the NH_2 groups are blocked by small substituents. The change in biological activity accompanying the modification could then be attributed specifically to the modification at the guanidino group. The reaction of N^{A1}, N^{B1}, N^{B29}-tricarbamylinsulin (21 U/mg) with cyclohexanedione led to the isolation of a chromatographically pure product in which the arginine residue was completely blocked (Figure 5.3)[52]. The compound retained 30% of the activity of insulin. The diminished activity of the cyclohexanedione derivative of *N*-tricarbamylinsulin is not necessarily due to elimination of the guanidino function; it could be a consequence of the introduction of a bulky substituent. In either case, it appears that the guanidino group of arginine of insulin is not essential for function.

5.2.5 Carboxyl

The CO_2H-groups of insulin have been coupled to glycine ethyl ester with the aid of 1-ethyl-3-(3-dimethylaminopropyl) carbodi-imide, a water soluble di-imide[54]. All six of the CO_2H groups could be made to react but under mild conditions a fraction was obtained in which the modification was restricted to an average of one carboxyl group. Degradation of the product indicated that 80% of the reaction had occurred in the B-chain at one or both of the glutamic acid residues; the CO_2H-terminal alanine was unaffected. The remaining 20% of the reaction appeared to have taken place at the CO_2H group of asparagine A21.

The high biological activity of this mixture of monosubstituted derivatives (17.7 U/mg) confirms earlier reports[55] that at least one of the carboxyl groups of insulin is not essential for function. It was suggested that the reaction that had taken place at the CO_2H group of A21 asparagine did not lead to inactivation, which would indicate that the A21 CO_2H group is unnecessary for activity. However, since the mixture of monosubstituted derivatives exhibited 28% less activity than insulin that the activity loss could be associated specifically with A21 reaction. Judgement on the importance of this CO_2H group should be suspended.

When insulin was esterified in anhydrous hydrogen chloride–methanol, two products were formed in which the six CO_2H groups were present as methyl esters[56]. One of the products was formed by an $N \rightarrow O$ shift at the peptide bond linking tyrosine B26 to threonine B27. Such an ester linkage would revert to the normal peptide linkage under the conditions employed in bioassay. The hexamethyl ester of insulin was found to be devoid of biological activity. The absence of activity could be attributed to the chemical modification of one or more CO_2H groups needed for binding of the hormone to its receptor site. Optical rotatory dispersion studies, however, indicated a general change in conformation following the esterification reaction.

5.3 ANALOGUES OBTAINED BY CHEMICAL SYNTHESIS

The solid-phase method of peptide synthesis[57] has made possible the rapid preparation of a wide variety of A- and B-chains. Twenty-eight analogues of

Table 5.2 Relative potencies of hybridised insulins derived from synthetic analogues of ovine A-chain and natural B-chain [58—62]

	1	2	3	4	5	6	7	8	9	10	11	12	13	14	15	16	17	18	19	20	21	Biological activity* % Mouse-convulsion assay	Rat fat-pad assay
I	Gly	Ile	Val	Glu	Gln	Cys	Cys	Ala	Gly	Val	Cys	Ser	Leu	Tyr	Gln	Leu	Glu	Asn	Tyr	Cys	Asn	1.0–1.5	1.0–2.0
II												Ala										0.8–1.2	0.4–0.9
III												Ala							Phe			1.0	0.4
IV										Leu		Ala										0.6–1.0	
V		Leu	Leu							Leu		Ala										inactive	
VI												Ala										0.05	0.05
VII												Ala							Phe		Ala	1.0	2.7
VIII												Ala						Ala	Phe		Ala	1.0	4.1
IX			Leu									Ala										inactive	inactive
X		Ala										Ala										inactive	inactive
XI	Ala								Ser													0.2–0.4	0.3
XII		Gly							Ser													0.1–0.25	0.15
XIII	β-Ala											Ala										0.2	0.3
XIV			Ile									Ala										0.1	0.16
XV					Ala							Ala										0.2–0.4	0.33–0.37
XVI												Ala			Ala							0.1–0.4	0.33–0.37
XVII												Ala		Phe					Phe			1.3	1.0
XVIII												Ala		Ala								1.0–1.3	1.4
XIX						Ala					Ala											inactive	inactive
XX					Glu							Ala										1.5–2.0	2.1
XXI												Ala			Glu							2.5	4.5
XXII		Leu																				0.4	0.3
XXIII		Val																				0.6–1.0	0.3–0.7
XXIV					Glu							Ala			Glu			Ala	Phe		Ala	2.4	4.0
XXV					Glu							Ala		Ala	Glu			Ala	Phe		Ala	1.0–1.7	0.97
XXVI		Val			Glu							Ala		Ala	Glu			Ala	Phe		Ala	0.8	0.54
XXVII					Glu					Ala		Ala			Glu			Ala	Phe		Ala	1.8	
XXVIII					Glu					Ala		Ala		Ala	Glu			Ala	Phe		Ala	3.2	
XXIX					Glu							Ala	Ala	Ala	Glu			Ala	Phe		Ala	1.8	

*The activities are expressed as a percentage of the specific activity of natural insulin (24 U/mg). For comparison, the value obtained on hybridisation of natural A- and B-chains was 1.0 to 1.5%.

sheep A-chain[58–62] and 33 of B-chain[63–66] have been described. Each synthetic chain was allowed to hybridise with the complementary natural chain[67] and without fractionation the combination mixtures were tested for biological activity.

Hybrid insulins formed from synthetic A-chains and natural B-chain (Table 5.2) revealed that the replacement of A14 tyrosine or A19 tyrosine by phenylalanine was without effect of activity. Similarly A21 asparagine, A18 asparagine, A12 serine and A10 valine could be replaced by alanine without adverse effect; the side chains of these residues appear unimportant in the active molecule. Some residues could be replaced by similar residues but more radical changes led to decrease in activity. The glutamine residues at A5 and A15 were exchanged for glutamic acid and A2 isoleucine was replaced by valine without reduction in activity. When alanine was placed at A5 or A15, or A2 isoleucine was replaced by leucine, a less active product resulted. Positions sensitive to change were A1 glycine, A2 isoleucine and A3 valine. The positions where the 'natural' residues appeared to be irreplaceable were A6 cystine and A11 cystine.

In general, the residues that were resistant to change were the hydrophobic residues, many of which form the 'core' of the 3-dimensional crystal structure (Section 5.5.2). The conclusions drawn from the analogues with replacement at A1 glycine, A14 tyrosine, A19 tyrosine or A21 asparagine are in complete agreement with the data obtained from the study of derivatives and fragments (Sections 5.2 and 5.4). The essential requirement for the intrachain disulphide bridge at A6 to A11 is clearly demonstrated; additionally it is known that one of the sulphur atoms of this bridge can be replaced by a methylene group with some retention of activity[68]. Perhaps the most striking result of the hybridisation experiments was that an A-chain with eight replacements in its 21 residues was able to form a hybrid insulin with as much activity as the hybrid formed from natural A-chain.

Of the synthetic B-chains (Table 5.3), it is notable that a shortened chain lacking residues 1 to 3 and 28 to 30 gave rise to a hybridised insulin with as much activity as that formed from natural B-chain. This is in agreement with the results obtained with the fragments derived from intact insulin (Section 5.4). New findings were that changes at positions 4, 9, 10 and 27 were without effect on activity. The identification of B10 histidine as an unimportant residue is perhaps surprising in view of the role of this histidine in intermolecular aggregation[42]. The residues sensitive to change were B5 histidine, B22 arginine and B26 tyrosine, where replacement by alanine led to diminution of activity.

A major difficulty in these experiments is the low activities obtained on recombination of A- and B-chains, even when the chains were derived from natural insulin (0.5–1.2% of the activity of intact insulin, Ref. 69). The observed activity after hybridisation is a reflection not only of the intrinsic activity of the semi-synthetic analogue but also of the number of these molecules formed. In the case of a synthetic chain that gave little or no activity on hybridisation with its partner chain, it is possible that the combination was unfavourable; the low activity could be due to an unusually poor yield of the two-chain insulin molecule. On the other hand, when the activity obtained after hybridisation was equal to that obtained with natural

Table 5.3 Relative potencies of hybridised insulins derived from synthetic analogues of bovine B-chain and natural A-chain [63–66]

	1	2	3	4	5	6	7	8	9	10	22	23	24	25	26	27	28	29	30	Biological activity % Mouse-convulsion assay	Fat-pad assay
I	Phe	Val	Asn	Gln	His	Leu	Cys	Gly	Ser	His	Arg	Gly	Phe	Phe	Tyr	Thr	Pro	Lys	Ala	0.5–1.0	0.4–0.8
II			Ala																	0.6–1.1[a]	0.84
III				Ala																0.7[a]	0.59
IV									Ala											0.6[a]	0.45
V																				0.6[a]	0.46
VI				Ala																0.6[a]	0.51
VII																		Ala		0.5[a]	
VIII																	Ala			0.6[a]	
IX																Ala				1.0[a]	
X									Ala							Ala	Ala			1.5[b]	
XI									Ala							Ala				5.5[c]	
XII				Ala					Ala							Ala	Ala			1.1[b]	
XIII									Ala							Ala	Ala			0.2[b]	
XIV				Ala					Ala							Ala				5.0[c]	
XV				Ala					Ala						Ala					2.5[c]	
XVI									Ala							Ala				0.4[c]	
XVII										Ala							Ala			0.7[a]	
XVIII					Ala												Ala			0.5[a]	
XIX					Ala					Ala						Ala -	Ala			0.5[a]	
XX					Ala				Ala							Ala -	Ala			1.2[d]	
XXI					Ala				Ala	Ala						Ala -	Ala			1.3[d]	
XXII					Ala				Ala	Ala						Ala -	Ala			1.2[d]	
XXIII									Ala	Ala						Ala				5.4[e]	
XXIV					Ala				Ala	Ala						Ala				2.4[e]	
XXV				Ala					Ala	Ala						Ala				5.6[c]	
XXVI				Ala	Ala				Ala							Ala				2.6[c]	
XXVII				Ala	Ala				Ala	Ala						Ala				2.7[c]	
XXVII					Ala				Ala	Ala						Ala				2.8[c]	
XXIX									Ala		Ala					Ala -	Ala			0.1[a]	
XXX									Ala		Orn					Ala				1.0[a]	
XXXI									Ala		Ala					Ala				0.1[a]	
XXXII									Ala		His					Ala				0.1[a]	
XXXIII						Ala			Ala -	Ala		Ala								1.0[a]	
XXXIII						Ala			Ala -	Ala	Ala									<0.1[a]	

*The activity obtained on hybridisation of each synthetic B-chain with natural A-chain is compared with the activity obtained on hybridisation of natural B-chain with A-chain. In the different series of experiments, chains, it can be assumed that the analogue had a specific activity similar to that of insulin. Indeed when an A-chain with four modifications, glutamic acid at position 5 and alanine at positions 12, 18 and 21, was purified and combined with natural B-chain and the resulting insulin isolated, the analogue exhibited 75% of the activity of insulin[70]. This finding lends strength to conclusions drawn from the results of the hybridisation experiments.

5.4 FRAGMENTS OF INSULIN

The term 'fragment' is used to describe a portion of the insulin molecule which has the sequence of a natural insulin but lacks one or more amino acid residues from the NH_2 or CO_2H terminus. The discussion is limited mainly to fragments comprising sections of the A- and B-chains linked by disulphide bridges. Such fragments have been obtained by enzymic and chemical degradation of natural insulin or by chemical synthesis of an abbreviated chain and combination with its complementary chain. A fragment that exhibits even a small degree of biological activity would seem to possess the 'functional region' of the hormone molecule, formally equivalent to the active centre of an enzyme. A totally inactive fragment, on the other hand, can call attention to the 'missing' regions of the molecule that seem obligatory for function.

The CO_2H terminus of the B-chain is not necessary for the activity of insulin. Removal of B30 alanine by carboxypeptidase digestion led to the isolation of desalanine (B30) insulin, a fully active derivative[71]. This established, for the first time, that the full activity of insulin can be expressed by a fragment of the molecule. It is now known that the last three residues can be removed from the CO_2H terminus of the B-chain without adverse effect. A B-chain lacking these residues was obtained by reaction with sodium in liquid ammonia. Cleavage took place at the peptide bond linking B27 threonine and B28 proline to give a mixture of products; in one the CO_2H terminus was occupied by an amino alcohol and in another by an amino aldehyde. After purification, the shortened B-chain was combined with A-chain and the product was found to exhibit the full activity of insulin[72]. Thus, at least three residues at the CO_2H terminus of the B-chain are unnecessary for activity; but a limit is set by desoctapeptide (B23–B30) insulin which lacks the last eight residues. This fragment, obtained by the action of trypsin on insulin, is almost inactive[73, 74]. One or more of the residues between positions B23 and B27 would seem to be necessary for function, the most likely being the aromatic residues at B24, B25 and B26.

The CO_2H terminus of the A-chain is important for activity. Extensive digestion of desalanine (B30) insulin with carboxypeptidase resulted in removal of the A21 asparagine residue and the resulting derivative, desalaninedesasparagine insulin, was almost completely inactive[69]. It may be noted that desamido (A21) insulin, in which the A21 asparagine is replaced by aspartic acid, is a highly active analogue[75] and this suggests that the essential feature at the CO_2H terminus of the A-chain is an α-CO_2H group. The carboxyl group of A21 asparagine has been implicated in an electrostatic linkage with the guanidino group of arginine B22. If this were the case, insulin lacking the A21 asparagine might present a different conformation from the natural hormone. Desalaninedesasparagine insulin is, in fact, monomeric at physiological pH[71] and has been reported to exhibit different antigenic properties from insulin[76].

That the NH_2 terminus of the B-chain is not important for activity was suggested when zinc free insulin was digested with leucine aminopeptidase to release a number of amino acids without appreciable loss of activity[77]. More recently, residues have been removed from the NH_2 terminus of the B-chain one at a time. The reaction of t-butyloxycarbonyl azide with insulin was found to take place at the NH_2-groups of A1 glycine and B29 lysine; the B1 phenylalanine residue was unaffected[78, 79]. The NH_2 group of the phenylalanine was then allowed to react with phenylisothiocyanate and the entire B1 residue and protecting groups were removed by exposure to trifluoracetic acid. The resulting desphenylalanine insulin, which has been obtained crystalline[80], was fully active. Further cycles of the procedure have given rise to products lacking residues B2, B3 and B4 and the fragments are said to retain a considerable degree of activity. A full report is awaited.

Removal of both the A1 glycine and the B1 phenylalanine results in considerable inactivation. Reaction of the NH_2-terminal residues with phenylisothiocyanate provided N^{A1}, N^{B1}-diphenylthiocarbamyl insulin (Table 5.1) which on exposure to trifluoracetic acid gave rise to desglycine (A1) desphenylalanine (B1) insulin.

The specific activity of the isolated derivative was 10% that of insulin[81]. In view of the full activity exhibited by desphenylalanine insulin, it is clear that the low activity of desglycine desphenylalanine insulin can be explained by the absence of the A1 glycine residue.

The importance of the structure at the NH_2 terminus of the A-chain was clearly demonstrated by the synthesis of destetrapeptide (A1-A4) A-chain, which after combination with natural B-chain gave a totally inactive insulin fragment[82]. Furthermore, synthesis of deamino (A1) insulin by a similar route gave only a weakly active analogue[83]; the hypoglycaemic activity was 35% of the activity of insulin. The deamino derivative differs from insulin only by the replacement of the A1 amino group by hydrogen. Thus, the active insulin molecule has a definite requirement for the NH_2-group of the A1 glycine. A positive charge at this site, however, is not essential because *N*-carbamyl glycine (A1) and *N*-acetylglycine (A1) insulin exhibit a high *in-vivo* activity[12, 20].

The absence of activity in destetrapeptide (A1–A4) insulin sets a lower limit to the length of A-chain which must be present in an insulin fragment for the expression of activity. It was, therefore, to be expected that the two synthetic fragments comprising residues A16–A21 coupled through a disulphide bridge to B18–B21, and A20–A21 coupled to B12–B21, would be inactive[84]. The essential requirement for the NH_2-terminal region of the A-chain does not tally with a report that an active fragment of insulin, released during exposure to the enzymes of adipose cells, was constituted from disulphide-linked peptides derived from the CO_2H-terminal regions of the A- and B-chains[85].

The biological activity of the A- and B-chains of insulin has been investigated frequently but the results have varied. In view of the low activities involved, it is essential that chains isolated from insulin should not be contaminated by intact insulin or by each other and in this respect a synthetic chain should provide an unequivocal answer. A fully reduced synthetic A-chain was reported to have a potency approximately 3×10^{-5} times that of insulin on the evolution of $^{14}CO_2$ from ^{14}C-glucose in the isolated rat diaphragm[86]. It is worth mentioning that B-chain isolated from insulin has been reported to possess weak antagonistic activity[87] but another report has indicated weak insulin activity[88].

Insulin-like activity is exhibited in *in-vitro* test systems by certain small fragments of the B-chain. The tetrapeptide amide Arg-Gly-Phe-PheNH_2 caused a significant increase in glucose uptake and glycogen content in rat diaphragm and on glucose oxidation in fat cells[89]. The effect was not exhibited by Gly-Phe-PheNH_2 or by Arg-Gly-Phe-Phe, which suggests that the arginine residue is necessary for the observed activity. However, no other amino acid was tested for its ability to replace arginine and the specific activity of the tetrapeptide was less than 10^{-6} times the molar activity of insulin.

5.5 NATURALLY OCCURRING INSULINS

5.5.1 Proinsulin

Proinsulin consists of a single polypeptide chain commencing at the NH_2 terminus with the B-chain sequence of insulin, continuing through a connecting peptide (C-peptide), and terminating in the insulin A-chain (see Figure 5.1 and Ref. 8). It can be considered as a structural analogue of insulin in which amino acid residues are coupled at the CO_2H terminus of the B-chain and at the NH_2 terminus of the A-chain.

The degree of hypoglycaemic activity exhibited by proinsulin in the intact animal is somewhat indeterminate because its effect may vary with the levels of circulating proinsulins and converting enzymes. It is not clear whether the activities measured are those of the intact proinsulin molecule or whether the activity is due to release of insulin from proinsulin during the assay. Purified porcine proinsulin exhibited an activity of 3 U/mg in the mouse-convulsion assay[90]. Bovine proinsulin was found to give an activity of 10 U/mg when its effect was measured on blood sugar levels in fasted rats but it was less active in fed rats or fasted mice[91]. In view of the wide structural variations among the different proinsulins, it would be preferable that assay should be conducted in the homologous species.

It is interesting to note that proinsulin exerts a more protracted effect than insulin on blood sugar levels in animals and in man[91, 92]. This effect might be traced, in part, to structural features of proinsulin which are not present in insulin. First, proinsulin contains only intrachain disulphide bridges, which are more stable than interchain bridges[93], and secondly the A1 glycine of insulin, which is important for activity, is immune to attack by aminopeptidase when it is a component of the proinsulin molecule. Since the *in vivo* degradation of insulin takes place at the interchain disulphide bridges and at the NH_2 termini[94], the protracted action exerted by proinsulin could be a consequence of an improved *in vivo* stability. This calls attention to a general difficulty encountered in structure–activity studies. Certain structural modifications may lead to an alteration in the *in-vivo* stability of a hormone and the observed specific activity may differ from the activity at the target site.

The sequence of human[8, 9], bovine[95, 96] and porcine *C*-peptides[90], though differing far more than the corresponding insulins, do show some conservation (Figure 5.2). The NH_2-terminal and CO_2H-terminal residues of the *C*-peptides are identical and certain amino acids are conspicuous by their absence. These include trytophan, tyrosine, phenylalanine, histidine, cystine and methionine (methionine has recently been observed at position 14 in guinea-pig *C*-peptide[97]). The central region in all *C*-peptides is rich in glycine and non-polar residues.

No biological activity has been demonstrated for the isolated C-peptide. The role of the connecting segment is probably performed during the transformation of proinsulin to insulin. It appears to facilitate the formation of the native structure of insulin by ensuring the correct pairing of cysteine residues. Evidence for this hypothesis has been obtained by the *in-vitro* reduction and re-oxidation of bovine and rat proinsulins; approximately 70% of the immunological reactivity was recovered[98]. The insulin molecule itself refolds in poor yield when reduced and re-oxidised under the same conditions[99].

The regions of the C-peptide that show preservation of structure would seem to have a task to perform over and above the role of converting the bimolecular reaction of chain combination to a unimolecular reaction. In this connection, A1 glycine and B30 alanine are close in the 3-dimensional structure and it would not require a connecting peptide 30 residues long to join the two chains. An additional role of the C-peptide might be to exercise a protecting effect during the activation of proinsulin to insulin by preventing cleavage at arginine B22. Such protection could result if the C-peptide were to cover the B22–B23 region, which is on the outside of the insulin hexamer.

5.5.2 Species variation in structure

The sequence variations reported for naturally occurring insulins are illustrated in Figure 5.4. Clearly, there are certain stretches of the A- and B-chains that are strongly conserved and others where mutation has been frequent. The first seven residues and the last three residues of the A-chain, for example, are remarkably constant. The first five and the last four residues of the B-chain, on the other hand, show much variation. It is reasonable to assume that the constant amino acids perform an important or essential role in maintaining the conformation of the active molecule. The placement of the constant amino acids within a 3-dimensional structure and their possible roles in function will be discussed on the basis of the model obtained by x-ray crystallography[4].

One group of invariant residues comprise the three cystines, the leucine residues at A16, B6, B11 and B15, isoleucine at A2, glycine at B8 and B23, and valine at B18. These residues form a 'core' in the interior of the molecule.

A chain

1	2	3	4	5	6	7	8	9	10	11	12	13	14	15	16	17	18	19	20	21	22
Gly	Ile	Val	Glu	Gln	Cys	Cys	Ala	Ser	Val	Cys	Ser	Leu	Tyr	Gln	Leu	Glu	Asn	Tyr	Cys	Asn	
		Leu	Asp				Thr	Gly	Thr		Asp	Lys	Phe	Asp		Gln	Ser				Asp
							His	Asn	Ile		Asn	Ile	His			Met					
								His	Pro		Thr	Arg	Asn								
								Arg													
								Lys													

B chain

	1	2	3	4	5	6	7	8	9	10	11	12	13	14	15	16	17	18	19	20	21	22
	Phe	Val	Asn	Gln	His	Leu	Cys	Gly	Ser	His	Leu	Val	Glu	Ala	Leu	Tyr	Leu	Val	Cys	Gly	Glu	Arg
Met	Ala	Ala	Lys	Arg	Arg				Pro	Asn			Asp	Thr			Ser			Gln	Asp	Asp
Val	Tyr	Pro	Pro							Gln										Arg	His	
Ala			Ala																			
			Ser																			

23	24	25	26	27	28	29	30
Gly	Phe	Phe	Tyr	Thr	Pro	Lys	Thr
		Tyr	Arg	Asn	Ser	–	Ala
				Ile	Asn	Met	Ser
				Pro		Asn	Asp

Figure 5.4 Species variation in the primary structure of insulin. The data is assembled from the Atlas of Protein Sequence and Structure. (National Biomedical Research Foundation, Silver Springs, Maryland, 1970. M. O. Dayhoff and R. V. Eck, editors)

A second group of invariant residues, the B12 valine, B16 tyrosine and B24 phenylalanine, and possibly the aromatic residues at B25 and B26, are located at contact sites between the two monomers of the insulin dimer. The other invariant residues, A1 glycine, A5 glutamine and A19 tyrosine, are on the surface of the molecule in the aggregated form; it has been suggested that these residues may be involved in the maintenance of tertiary structure for they would certainly play a role in determining the surface topography. Being on the outside of the molecule, the A1, A5 and A19 residues would be able to participate in intermolecular linkages between insulin and receptor macromolecules and conservation of structure at this region may be mandatory for activity. Evidence supporting this has been obtained from the study of analogues lacking certain amino acids (Section 5.4) and from the study of derivatives with modifications at these sites (Section 5.2).

Most of the residues that differ among the species are found on the surface of the crystalline molecule. Because of their accessibility, these residues may determine the immunological properties. Thus, bovine insulin has alanine at A8 and valine at A9 whereas pig insulin bears threonine and isoleucine at these positions[100, 101]; the interchange of these amino acids must underlie the differences in antigenicity that exist between ox and pig insulins.

The residues on the surface of the insulin molecule may also influence the tendency to aggregate. Histidine at B10, for example, plays a key role in the binding of zinc in the crystalline hexamer and it is conserved in almost all insulins. It is notable that guinea-pig insulin, in which the B10 histidine is replaced by asparagine, does not appear to form a hexamer[102]. Furthermore, residues that normally participate in the contact site between dimers within the hexamer could understandably be different in guinea-pig insulin; and residues B14, B17 and B20 in the guinea-pig are, in fact, different from the corresponding residues in the insulins of other species.

5.6 CONCLUSIONS

No single amino acid residue in insulin, with the exception of cystine, has been shown to be essential for activity. It seems unlikely, therefore, that the action of insulin should involve the formation of a covalent linkage between the hormone and a receptor, comparable to the linkages formed transiently between an enzyme and its substrate.

The evidence is against a possible disulphide exchange reaction between the interchain disulphide bridge (A6–A11) and a thiol or disulphide group in a receptor: the intrachain disulphide can be replaced by the sterically similar CH_2—S linkage without serious effect on activity. Similarly it is probable that the interchain disulphides (A7–B7 and A20–B19) do not participate in the action of the hormone. It has been reported that insulin can be recovered intact after brief exposure to adipose cells, an exposure long enough to cause stimulation of glucose oxidation[103].

While no residue seems mandatory for function, there are certain positions in the molecule where change leads to marked decrease in activity. Many of these positions are occupied by hydrophobic residues, such as valine, isoleucine and leucine, which are located at the centre of the crystalline hexamer[4]. On the other hand, the hydrophilic residues which include glutamic and aspartic acids, serine, threonine and lysine and which are mostly on the outside of the molecule can in general be altered without much loss of activity. This implies that the characteristic shape of insulin and of its active analogues is impressed upon the molecule by the distribution of the hydrophobic residues; it also implies that the interaction of insulin with its receptors is not mediated by charged groups.

It is not possible, at present, to delineate clearly the components of the molecule that perform a specific role in the maintenance of conformation from others which may be involved directly in receptor interactions. Much evidence points to the importance of steric factors, the bulk of certain amino acid side chains and their disposition in the molecule, for the activity of the hormone. Information has accumulated, too, on the significance of the 'chemically reactive' components of the structure. These include the potentially reactive side chains such as the guanidino group at B22 and CO_2H group at A21, the polarity of certain groups and their ability to form hydrogen bonds, and the overall hydrophilic–hydrophobic properties of particular residues. The value of this information will appreciate when the receptor macromolecules have been isolated and made available for study.

Much progress has been made recently on the separation of insulin binding proteins from the surface of cells sensitive to insulin[104, 105]. This is particularly opportune because the interaction between insulin and 'insulin receptor protein' seems likely to hold the key to the manner in which the hormone triggers its biological effect. With the knowledge of the 3-dimensional structure of crystalline insulin, the ability to synthesise modified A- and B-chains and combine them to form active molecules, the ability to probe conformational change in solution with the rapidly advancing techniques of fluorescence quenching and n.m.r. spectroscopy, the stage is now set for investigating the first stage in the molecular action of insulin, the interaction of the hormone with its primary physiological receptor.

References

1. Crestfield, A. M., Stein, W. H. and Moore, S. (1963). *J. Biol. Chem.*, **238,** 2421
2. Richards, F. M. and Vithayathil, P. J., Protein structure and function, Brookhaven Symposia in Biology, No. 13, Brookhaven National Laboratory, Upton, New York, p. 115
3. Gutte, B., Lin, M. C., Caldi, D. G. and Merrifield, R. B. (1972). *J. Biol. Chem.*, **247,** 4763
4. Adams, M. J., Blundell, T. L., Dobson, E. J., Dodson, G. G., Vijayan, M., Baker, E. N., Harding, M. M., Hodgkin, D. C., Rimmer, B. and Sheat, S. (1969). *Nature (London)*, **224,** 491
5. Blundell, T. L., Cutfield, J. F., Cutfield, S. M., Dodson, E. J., Dodson, G. G., Hodgkin, D. C., Mercola, D. A. and Vijayan, M. (1971). *Nature (London)*, **231,** 506
6. Blundell, T. L., Dodson, G. G., Hodgkin, D. C. and Vijayan, M. (1971). *Rec. Prog. Hormone Res.*, **20,** 1. (New York: Academic Press)
7. Nicol, D. S. H. W. and Smith, L. F. (1960). *Nature (London)*, **187,** 483
8. Oyer, P. E., Cho, S., Peterson, J. D. and Steiner, D. F. (1971). *J. Biol. Chem.*, **246,** 1375
9. Ko, A. S. C., Smyth, D. G., Markussen, J. and Sundby, F. (1971). *Eur. J. Biochem.*, **20,** 190
10. Fraenkel-Conrat, J. and Fraenkel-Conrat, H. (1950). *Biochim. Biophys. Acta*, **5,** 89
11. Bromer, W. W. and Chance, R. E. (1967). *Biochim. Biophys. Acta*, **133,** 219
12. Lindsay, D. G. and Shall, S. (1971). *Biochem. J.*, **121,** 737
13. Brandenburg, D., Gattner, H. G. and Wollmer, A. (1972). *Z. Physiol. Chem.*, **353,** 599

14. Brandenburg, D. (1972). *Z. Physiol. Chem.*, **353,** 263
15. Katsoyannis, P. G. and Tometsko, A. (1966). *Proc. Nat. Acad. Sci. U.S.*, **55,** 1554
16. Massey, D. E. and Smyth, D. G. (1972). *Eur. J. Biochem.*, **31,** 470
17. Ko, A. S. C. and Smyth, D. G., unpublished results
18. Lindsay, D. G. and Shall, S. (1969). *Biochem. J.*, **115,** 587
19. Lindsay, D. G. and Shall, S. (1970). *Eur. J. Biochem.*, **15,** 547
20. Lindsay, D. G., Loge, O., Losert, W. and Shall, S. (1972). *Biochim. Biophys. Acta*, **263,** 658
21. Bromer, W. W., Sheehan, S. K., Berns, A. W. and Arquilla, E. R. (1967). *Biochemistry*, **6,** 2378
22. Levy, D. and Carpenter, F. H. (1967). *Biochemistry*, **6,** 3559
23. Geiger, R., Schone, H. H. and Pfaff, W. (1971). *Z. Physiol. Chem.*, **352,** 1487
24. Krail, G., Brandenburg, D. and Zahn, H. (1971). *Z. Physiol. Chem.*, **352,** 1595
25. Katsoyannis, P. G., Zalut, C., Harris, A. and Meyer, R. J. (1971). *Biochemistry*, **10,** 3884
26. Yalow, R. S. and Berson, S. A. (1960). *J. Clin. Invest.*, **39,** 1157
27. Bannerjee, R. N. and Gibson, K. (1962). *J. Endocrinol.*, **25,** 145
28. Glover, J. S., Salter, D. N. and Shepherd, B. P. (1967). *Biochem. J.*, **103,** 120
29. Massaglia, A., Rosa, V., Rialdi, G. and Rossi, C. A. (1969). *Biochem. J.*, **115,** 11
30. Garratt, C. J., Harrison, D. M. and Wicks, M. (1972). *Biochem. J.*, **126,** 123
31. Arquilla, E. R., Ooms, H. and Mercola, D. (1968). *J. Clin. Invest.*, **47,** 474
32. Arquilla, E. R., Miles, P. V. and Morris, J. W. S. (1971). *Proc. Biochem. Soc.*, 5
33. Aoyama, M., Kurihara, K. and Shibata, K. (1965). *Biochim. Biophys. Acta*, **107,** 257
34. Morris, J. W. S., Mercola, D. A. and Arquilla, E. R. (1970). *Biochemistry*, **9,** 3931
35. Gattner, H. G. (1971). *Z. Physiol. Chem.*, **352,** 7
36. Fredericq, E. (1956). *Arch. Biochem. Biophys.*, **65,** 218
37. Goldman, J. and Carpenter, F. H., 158th National Meeting of the American Chemical Society, New York, N.Y., Sept. BIOL 177
38. Massey, D. E., Smyth, D. G., Sabey, G. A., Stewart, G. A. and Webb, F. W., unpublished results
39. Brandenburg, D. (1971). *Z. Physiol. Chem.*, **352,** 8
40. Smith, L. F. (1966). *Amer. J. Med.*, **40,** 662
41. Bosshard, H. R., Jorgesen, K. H. and Humbel, R. E. (1969). *Eur. J. Biochem.*, **9,** 353
42. Tanford, C. and Epstein, J. (1954). *J. Amer. Chem. Soc.*, **76,** 2170
43. Young, J. D. and Carpenter, F. H. (1961). *J. Biol. Chem.*, **236,** 743
44. Weil, L., Seibles, T. S. and Herskovits, T. T. (1965). *Arch. Biochem. Biophys.*, **111,** 308
45. Weitzel, G., Schaeg, W., Boden, G. and Williams, B. (1965). *Liebigs Ann. Chem.*, **689,** 248
46. Stadtman, E. R. (1953). *J. Amer. Chem. Soc.*, **75,** 2022
47. Takahashi, K. (1968). *J. Biol. Chem.*, **243,** 6171
48. Nakaya, K., Horinishi, H. and Shibata, K. (1967). *J. Biochem. (Tokyo)*, **61,** 345
49. Aoyama, M., Kurihara, K. and Shibata, K. (1965). *Biochim. Biophys. Acta*, **107,** 257
50. Bunzli, H. F. and Bosshard, H. R. (1971). *Z. Physiol. Chem.*, **352,** 1180
51. Stark, G. R. (1970). *Advan. Protein Chem.*, **24,** 261
52. Ko, A. S. C. and Smyth, D. G., unpublished experiments
53. Toi, K., Bynum, E., Norris, E. and Itano, H. A. (1967). *J. Biol. Chem.*, **242,** 1036
54. Ozawa, H. (1970). *Biochemistry*, **9,** 2158
55. Mommaerts, W. F. H. M. and Neurath, H. (1950). *J. Biol. Chem.*, **185,** 909
56. Levy, D. and Carpenter, F. H. (1970). *Biochemistry*, **9,** 3215
57. Merrifield, R. B. (1963). *J. Amer. Chem. Soc.*, **85,** 2149
58. Weber, U., Schneider, F., Köhler, P. and Weitzel, G. (1967). *Z. Physiol. Chem.*, **348,** 947
59. Weber, U., Hörnle, S., Grierser, G., Herzog, K. H. and Weitzel, G. (1967). *Z. Physiol. Chem.*, **348,** 1715
60. Weber, U., Hörnle, S., Köhler, P., Nagelschneider, G., Eisele, K. and Weitzel, G. (1968). *Z. Physiol. Chem.*, **349,** 512
61. Hörnle, S., Weber, U. and Weitzel, G. (1968). *Z. Physiol. Chem.*, **349,** 1428
62. Weber, U., Herzog, K. H., Grossmann, H., Hörnle, S. and Weitzel, G. (1969). *Z. Physiol. Chem.*, **350,** 1425
63. Weber, U. and Weitzel, G. (1968). *Z. Physiol. Chem.*, **349,** 1431
64. Weitzel, G., Eisele, K., Zollner, H. and Weber, U. (1969). *Z. Physiol. Chem.*, **350,** 1480
65. Weitzel, G., Weber, U., Eisele, K., Zollner, H. and Martin, J. (1970). *Z. Physiol. Chem.*, **351,** 263

66. Weitzel, G., Weber, U., Martin, J. and Eisele, K. (1971). *Z. Physiol. Chem.*, **352,** 1005
67. Du, Y. C., Shang, Y. S., Lu, Z. X. and Tsou, C. L. (1961). *Scientica Sinica,* **10,** 84
68. Jöst, K., Rudinger, J., Klostermeyer, H. and Zahn, H. (1968). *Z. Naturforsch,* **23,** 1059
69. Zahn, H., Bremer, H. and Zabel, R. (1965). *Naturforsch,* **206,** 653
70. Weber, U., Herzog, K. H., Grossmann, H., Hartter, P. and Weitzel, G. (1971). *Z. Physiol. Chem.,* **352,** 419
71. Slobin, L. I. and Carpenter, F. H. (1963). *Biochemistry,* **2,** 16
72. Katsoyannis, P. G., Zalut, C., Harris, A. and Meyer, R. J. (1971). *Biochemistry,* **10,** 3884
73. Young, J. D. and Carpenter, F. H. (1961). *J. Biol. Chem.,* **236,** 743
74. Bromer, W. W. and Chance, R. E. (1967). *Biochim. Biophys. Acta,* **133,** 219
75. Carpenter, F. H. and Chrambach, A. (1962). *J. Biol. Chem.,* **237,** 404
76. Arquilla, E. R., Bromer, W. W. and Mercola, D. (1969). *Diabetes,* **18,** 193
77. Smith, E. L., Hill, R. L. and Borman, A. (1958). *Biochim. Biophys. Acta,* **29,** 207
78. Geiger, R., Schone, H. H. and Pfaff, W. (1971). *Z. Physiol. Chem.,* **352,** 1487
79. Geiger, R. (1971). *Z. Physiol. Chem.,* **352,** 7
80. Brandenburg, D. (1969). *Z. Physiol. Chem.,* **350,** 741
81. Africa, B. and Carpenter, F. H. (1970). *Biochemistry,* **9,** 3215
82. Katsoyannis, P. G. and Zalut, C. (1972). *Biochemistry,* **11,** 3065
83. Katsoyannis, P. G. and Zalut, C. (1972). *Biochemistry,* **11,** 1128
84. Kamber, B. (1971). *Helv. Chim. Acta,* **54,** 398,
85. Rudman, D., Garcia, L. A., Del Rio, A. and Akgun, S. (1968). *Biochemistry,* **7,** 1864
86. Marglin, A. and Cushman, S. W. (1967). *Biochem. Biophys. Res. Comm.,* **29,** 710
87. Ensinck, J. W., Mahler, R. J. and Vallance-Owen (1965). *Biochem. J.,* **94,** 150
88. Langdon, R. G. (1960). *J. Biol. Chem.,* **235,** PC15
89. Weitzel, G., Eisele, K., Gugliemi, H., Stock, W. and Renner, R. (1971). *Z. Physiol. Chem.,* **352,** 1735
90. Chance, R. E., Ellis, R. M. and Bromer, W. W. (1968). *Science,* **161,** 165
91. Puls, W. and Kroneberg, G. (1969). *Diabetologia,* **5,** 325
92. Galloway, J. A., Root, M. A., Chance, R. E., Rathmacher, R. P., Challoner, D. R. and Shaw, W. N. (1969). *Diabetes,* **18,** 341
93. Cecil, R. and Loening, V. E. (1960). *Biochem, J.,* **76,** 146
94. Varandani, P. T., Chroyer, L. A. and Nafy, M. A. (1972). *Proc. Nat. Acad. Sci. U.S.A.,* **69,** 1681
95. Steiner, D. F., Cho, S., Oyer, P. E., Terris, S., Peterson, J. D. and Rubinstein, A. H. (1971). *J. Biol. Chem.,* **246,** 1365
96. Salokangas, A., Smyth, D. G., Markussen, J. and Sundby, F. (1971). *Eur. J. Biochem.,* **20,** 183
97. Markussen, J., Smyth, D. G. and Sundby, F., unpublished experiments
98. Steiner, D. F. and Clark, J. L. (1968). *Proc. Nat. Acad. Sci. U.S.A.,* **60,** 622
99. Wilson, S., Dixon, G. H. and Wardlaw, A. C. (1962). *Biochim. Biophys. Acta,* **62,** 483
100. Sanger, F. and Thompson, E. O. P. (1953). *Biochem. J.,* **53,** 353
101. Brown, H., Sanger, F. and Kitai, R. (1955). *Biochem. J.,* **60,** 556
102. Yip, C., in Blundell, T., Cutfield, J. F., Dodson, G. G., Dodson, E., Hodgkin, D. C. and Mercola, D. A. (1971). *Proc. Biochem. Soc.,* September
103. Cutrecasas, P. (1971). *Proc. Nat. Acad. Sci. U.S.A.,* **68,** 1264
104. Cutrecasas, P. (1972). *Proc. Nat. Acad. Sci. U.S.A.,* **69,** 318
105. Freychet, P., Roth, J. and Neville, D. M. (1971). *Proc. Nat. Acad. Sci. U.S.A.,* **68,** 1833
106. Bangham, D. R. and Mussett, M. V. (1959). *Bull. World Hlth. Org.,* **20,** 1209
107. Weinert, M., Kircher, K., Brandenburg, D. and Zahn, H. (1971). *Z. Physiol,* **352,** 719

6
Procedures for Peptide Synthesis

CHARLES H. STAMMER
University of Georgia, U.S.A.

6.1 INTRODUCTION

The purpose of this review is to describe in some detail recent reports on the methods being used in peptide synthesis. It is now within the power of the specialist in peptide synthesis to prepare peptide chains containing 20–50 amino-acid residues in any desired sequence and having the full biological activity of a natural hormone or enzyme. This has come about through the tremendous ingenuity of workers in this field who have improved the classical procedures and investigated new protecting groups, coupling methods and completely new processes which accomplish syntheses using automated equipment and a minimum of man-hours. The past decade has seen these first steps toward the synthesis of proteins; the next will see the use of synthetic proteins in medicine.

The detailed strategy of peptide synthesis requires a tactical planning which allows various functional groups to remain blocked while others are freed for reaction in a subsequent step. This requires an arsenal of protecting groups with a complete spectrum of sensitivity to de-blocking agents. A great deal of effort continues to be made in this area, often by organic chemists not involved in peptide work. Protecting groups for amine and carboxyl functions are of crucial importance, but the side-chain functionality of poly-functional amino acids requires special treatment. Coupling methods are also often discovered by organic chemists since the synthesis of amides falls in their domain, but the determination of conditions which provide maximum yield with a minimum of side reactions often awaits the efforts of those heavily committed to peptide synthesis. Since the optical purity of the final product is mandatory, racemisation during the synthesis must be avoided at all costs. Special test procedures have been devised to determine the propensity of a given coupling method toward racemisation. The results of such tests indicate which procedures are best, but they do not allow determination of the amount of racemisation which has actually occurred at any given stage of a peptide synthesis. The reaction path by which racemisation actually occurs during coupling is still being studied in several laboratories and a considerable level of understanding has been reached. These studies have led to new coupling methods which are much more free from racemisation than any previously known. True understanding of a process invariably leads to practical results.

Probably the most important development of recent years is the use of resin supports in peptide synthesis. It is this 'solid-phase' method which makes the synthesis of proteins a possibility. The simplicity of the process allows its automation and eliminates the many man-hours of laboratory work previously required. The limitations of this process are just now becoming evident and the specialist is working to overcome them. Recent developments in this area will be thoroughly discussed. Several excellent reviews[1] have appeared recently describing various aspects of the field of peptide synthesis*.

6.2 BLOCKING GROUPS

Many new blocking groups have been designed in recent years. Some of these have been shown useful in practice and some have not yet been proved. As indicated previously, the peptide chemist requires a series of blocking functions having a progressively greater sensitivity to de-blocking reagents so that one function may be liberated in the presence of other fully protected ones. Work continues in this area with this as the primary goal.

---

*Abbreviations used in this chapter are in accord with those proposed by the IUPAC–WB commission on Biochemical Nomenclature (1972). *Biochem. J.*, **126**, 773

6.2.1 Amine blocking groups

6.2.1.1 Urethanes

The single most important contribution to amine protection was the introduction of the benzyloxycarbonyl (Z) group by Bergmann[2] in 1932. This protecting group incorporates the benzyl ester function which is readily cleaved by hydrogenolysis, sodium in liquid ammonia or hydrogen bromide in acetic acid. This group is introduced by reaction of the amino function

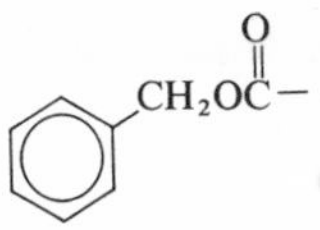

$$C_6H_5CH_2OC(=O)- \qquad CH_3-C(CH_3)_2-O-C(=O)-$$

Z 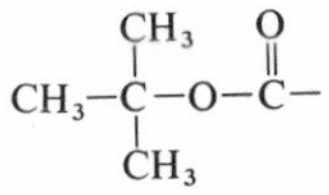 Boc

with the corresponding acid chloride in basic solution. Since 1932, many ring substituted Z-groups have been prepared and tested in peptide synthesis. Electron releasing groups increase while electronegative groups generally decrease the rate of cleavage by acids. A comparative study of *p*-substituted benzyloxycarbonyl groups has been made[3]. The Z-group is stable to the basic conditions used for the hydrolysis of esters, allowing amino functions to remain protected during carboxyl de-blocking. The side reactions accompanying use of the benzyloxycarbonyl group have been summarised elsewhere[1b]. The Z-group has proved itself extremely useful and is still extensively used today.

A relatively new urethane grouping is the t-butoxycarbonyl (Boc) group introduced by Carpino[4]. This group was originally incorporated using Boc–azide, but this reagent reacts sluggishly with amino acids. More recently, the acid chloride[5], pentachlorophenyl (Pcp) ester[6] and the acid fluoride[7] have been used. Several trifunctional amino acids were completely blocked when the acid fluoride was used[8]. The Boc group can be removed under very mildly acidic conditions, generally in the absence of water. Hydrogen chloride[4] in organic solvents such as nitromethane, ethyl acetate or methylene chloride, neat trifluoroacetic acid[9] or hydrogen fluoride[4] have been used. These acidic reagents often cause partial cleavage of triphenylmethyl (Trt), Z and benzyl ester protecting groups also. Independently, both Schnabel[10a] and Hiskey[10b] have reported the removal of the Boc group in the presence of all three of the above protecting groups by using a solution of boron trifluoride etherate in glacial acetic acid. A tetrapeptide in which the cysteine sulphhydryl was blocked by a Trt group was synthesised successfully by this method. It has also recently been shown[11] that β-mercaptoethanesulphonic acid in glacial acetic acid removes the Boc group without cleavage of benzyl ester or Z blocking groups. In 1970, Milne showed by mass spectrometry[12] that a tryptophan residue may be butylated during de-blocking of a Boc peptide with trifluoroacetic acid. Hydrogen chloride in ethyl acetate or acetic acid caused considerably less butylation, but did not eliminate this side reaction completely.

An extremely acid-labile new protecting group, *N*-2-biphenylyl-2-propoxycarbonyl, (Bpoc) group, has been recently introduced by Sieber and

Iselin[13] and used in the solid-phase synthesis of bovine parathyroid hormone[14]. The great usefulness of this group lies in the fact that it can be removed

Bpoc   Adoc

from an α-amino group some 2–9 × 10^3 times faster than the Boc group using 80% aqueous acetic acid or a 75% chloroacetic acid–methylene chloride mixture. Merrifield[14] found that the Bpoc group could be removed by a 0.5% trifluoroacetic acid–methylene chloride solution, without effecting side chain protecting groups such as Z and benzyl ethers and esters. Bpoc is introduced on to the amino function with the appropriate phenyl ester[13], azide[13] or acid fluoride[15]. In spite of the large size of the Bpoc group, it causes no steric diminution of coupling rates[13] as does the Trt group when present on the α-amino function.

The 1-adamantyloxycarbonyl (Adoc) group also appears to be a useful protecting group of the urethane type. It was first prepared as its acid chloride and its uses explored by Gerzon[16]. It has been used by Wuensch[17] in the synthesis of a hexapeptide fragment of secretin in which both the amino and imidazole functions of histidine were protected with the Adoc group. Removal is accomplished in anhydrous trifluoroacetic acid.

The t-amyloxycarbonyl[18] (Aoc) and the di-isopropylmethoxycarbonyl[19] (Dipmoc) groups have recently been described. The former seems to offer no advantage over Boc while the latter was shown[20] to withstand treatment

$CH_3CH_2C(CH_3)_2$–O–C(=O)– $[(CH_3)_2CH]_2CHOC(=O)$–

Aoc   Dipmoc

with hydrogen chloride in acetic acid better than the Z group, making it useful for side-chain protection during Boc removal. It can be removed rapidly with anhydrous hydrogen fluoride. A recent study[20] of substituted Z groups showed that both *m*- and *p*-chloro-Z were more stable to hydrogen bromide than Z but could be removed from the α-amino group of lysine in 2–3 h by treatment with hydrogen fluoride.

A new acetylenic urethane protecting group[21], dimethylpropynyloxycarbonyl (Dmpoc), has the advantage of undergoing ready hydrogenolysis in the presence of sulphur-containing amino acids. The sequences, Phe-Phe-Met-Gly and Cys(Bzl)-Gly-Phe-OMe were both synthesised using Dmpoc

HC≡C–$C(CH_3)_2$–O–C(=O)–   N–O–C(=O)– $CH_2OC(=O)$–

Dmpoc   Pipoc   Fmoc

at each stage with removal by hydrogenolysis. It is introduced via the trichlorophenyl ester, and the amino acid derivatives prepared were crystallized as dicyclohexylammonium salts.

Another blocking group which can be hydrogenolysed from sulphur-containing amino acids and peptides is the piperidinoxycarbonyl (Pipoc) function reported by Young[22]. It is stable to hydrogen bromide in acetic acid and trifluoroacetic acid at room temperature but is removed by hot aqueous acetic acid. These properties make it useful as a lysine ε-amino protecting group as shown by such use in a dipeptide synthesis. Other reducing conditions such as zinc–acetic acid, dithionite and electrolysis can also be used for cleavage of Pipoc.

The protecting groups most in use today are sensitive to removal under various conditions of acidity. In an attempt to develop a series of base sensitive groups, Carpino[23] has reported the fluorenylmethoxycarbonyl (Fmoc) function. Fmoc is removed by simple dissolution of the derivative in liquid ammonia, morpholine or ethanolamine giving cleavage products which are ether soluble. The group is stable to hydrogen chloride and bromide in organic solvents, trifluoroacetic acid and hydrogenolysis. It can be introduced using the crystalline acid chloride or azide. Fmoc has not been used as yet in a peptide synthesis but would appear promising.

6.2.1.2 *Miscellaneous blocking groups*

Very few other types of blocking groups have been used successfully in practice. The nitroarylsulphenyl group (Nps), introduced by treatment of an amino-acid derivative with a nitroarylsulphenyl chloride, has been used

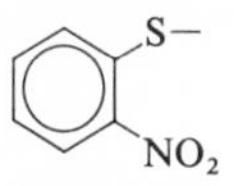

S–

NO_2

Nps

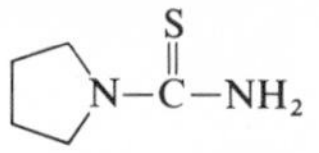

S

N–C–NH_2

(1)

extensively due to its extreme sensitivity to acid. The Nps amino acid derivatives are generally crystallised and stored as dicyclohexylamine salts. A new method[25] for the introduction of Nps uses the sulphenyl thiocyanate rather than the chloride in the presence of silver nitrate which binds the CNS ions formed. The yields of Nps derivatives were much improved. A reversal of this same equilibrium was used by Wuensch[26] for the removal of the Nps group. Treatment of a fully protected Nps pentapeptide and decapeptide with ammonium thiocyanate in the presence of 2-methylindole as scavenger gave good yields of peptide with only the terminal amino group liberated.

The simple chloroacetyl group has recently been used[27] in the synthesis of several peptides both as an *N*-terminal and as ε-amino protecting groups. It was useful because it was readily removed by *N*,*N*-pentamethylene thiourea (1). The by-product thiazolone can be removed with dilute acid after the liberated amino function has been acylated with the next chloroacetylamino acid. t-Butyl ester and trifluoroacetyl groups are not attacked by the thiourea reagent. An interesting report by D'Angeli[28] proposes the acetoacetyl as a useful *N*-protecting group. This group can be removed readily by hydroxyl-aminolysis and gives optically-pure coupling products. Several acetoacetyl amino acids (2) have been converted into optically stable oxazolones (3)

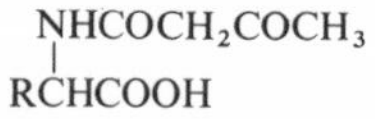

$NHCOCH_2COCH_3$
RCHCOOH
(2)

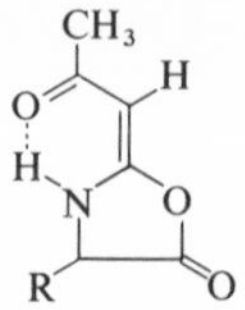

CH_3 H O H N O R O
(3)

with dicyclohexylcarbodiimide (Dcc) and the oxazolones were shown to react with amino-esters giving optically pure peptides. The special optical stability of (3) was given as the reason for obtaining optically pure peptides from (2) since it is well known that oxazolone formation is responsible for most racemisation during coupling.

A start has been made toward the synthesis of amine protecting groups which can be removed by photolysis. Patchornik[29] has recently prepared various *o*-nitrobenzyloxycarbonyl groups which photolyse quantitatively; however, the yields of free amine were only 30–90 %. This area shows excellent promise and should be investigated extensively.

6.2.2 Carboxyl blocking groups

In general, a carboxyl function must be blocked by a group having a sensitivity to de-blocking reagents different from those used to remove amine protecting groups. With the present emphasis on the stepwise construction of peptide chains from the *N*-terminal, the terminal carboxyl function must withstand repeated treatment with amine de-blocking reagents, usually acids of varying strength.

Recent work[30, 31] using anisyl (4) esters indicates how they may be used. They can be prepared in good yield by treatment of *N*-protected amino-acid

CH_3O—⟨benzene ring⟩—CH_2—
(4)

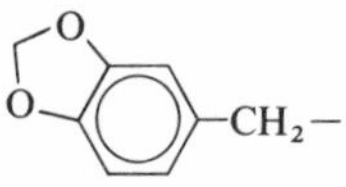

O O ⟨benzene ring⟩—CH_2—
(5)

salts with *p*-methoxybenzyl bromide. Removal occurs readily in trifluoroacetic acid, formic acid or a hydrogen chloride–nitromethane[30] solution. The Z-group is stable to the last two reagents. The amino protecting group Nps can be selectively removed[31] in the presence of an anisyl ester by treatment with *exactly* one equivalent of hydrogen chloride in methanol. A series of Z-dipeptide piperonyl esters (5) has been prepared[32] recently and is shown to be completely de-blocked to the free dipeptides in hydrogen bromide–acetic acid, but to the Z-dipeptides in trifluoroacetic acid. Only the Nps group was removed when Nps amino-acid piperonyl esters were treated with hydrogen chloride in dioxan.

Two ester groupings which are stable to the usual de-blocking agents are the phthalimidomethyl[33] (6) and *p*-bromophenacyl[34] (7) functions which are

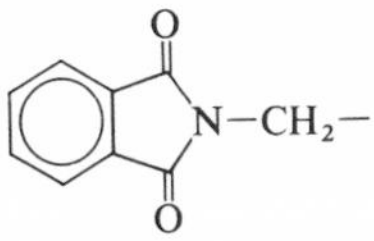

(6) 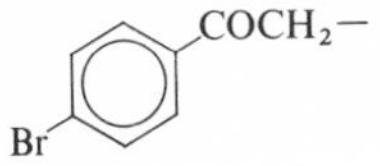 (7) 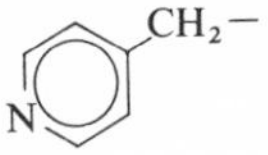 (8)

removed by treatment with zinc in acetic acid. They have been used only on amino acids so far, but should become more popular in the future.

A clever method for using the terminal carboxyl protecting group as a handle for purification of the intermediate products after each step of a peptide synthesis has been developed by Young and co-workers[35]. When the 4-picolyl group (8) was used to protect the carboxyl terminal of a growing peptide chain, the crude coupling product could be separated from the mixture at each stage due to the basicity of the picolyl group. Originally a sulphoethylsephadex column was used to retain the product while contaminants were removed. More recently[36a], a pentapeptide was prepared using an Amberlyst-15 resin column. Anhydrous ethyl acetate was used to put the ester on the column and a pyridine–dimethylformamide solution was used to elute the product. During the synthesis of Val^5-angiotensin, these workers[36b] were able to extract the desired product into 2M citric acid because the side chain carboxyl functions of both aspartic and glutamic acids were protected by the basic picolyl esters. The picolyl group is removed by alkali, catalytic hydrogenation or sodium-liquid ammonia, but is stable to strong acids.

Another type of ester blocking group is the *p*-methylthiophenyl (9) function which is stable to hydrogen chloride–methylene chloride (Boc removal) and hydrogen bromide–acetic acid (Z removal) has been reported[37].

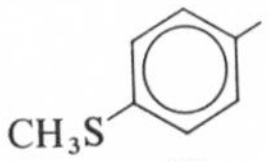

(9)

More importantly, this blocking group can be converted into an activated ester by peroxide oxidation to the sulphone. Recent uses of this ester will be discussed in Section 6.3.1.

6.2.3 Side-chain protecting groups

The mercapto group of cysteine, guanido function of arginine and the imidazole ring of histidine are side chain functionalities often requiring protection. Only the more recent work in this area will be discussed.

6.2.3.1 Cysteine mercapto group

The mercapto function is a very reactive nucleophile and is consequently easy to protect and easy to de-block. It can be protected with any of the *N*-blocking groups, but this is not generally useful in the synthesis of long peptides since selective de-blocking of the terminal amino group is not possible. Recently[38], *N*,*S*-bis-Boc-Cys has been prepared by treatment of cysteine with Boc-chloride. As with Boc, the commonly used trityl (Trt) and diphenylmethyl (Dpn) groups are removed under acidic conditions. A new mercapto protecting group reported by Wieland[39], dicarboethoxyethyl (Dce), is removed under basic conditions at room temperature. It was introduced on to the sulphur atom

$$\underset{\text{Dce}}{(C_2H_5O_2C)_2CHCH_2-} \qquad \underset{\text{Ibm}}{(CH_3)_2CHCH_2OCH_2-} \qquad \underset{\text{Tec}}{Cl_3CCH_2O\overset{O}{\overset{\|}{C}}-}$$

of cysteine with methylene malonic ester and used in the synthesis of glutathione. The use of *N*-trifluoroacetyl and a *C*-terminal ethyl ester allowed the removal of all the blocking groups in a last alkaline saponification step. Dce is stable to hydrogen bromide in acetic acid and trifluoroacetic acid. Another new *S*-protecting group is the t-butylthio function, which is put on the sulphur atom with t-butyl mercaptan under basic conditions and removed by reduction with sodium sulphite.

The groups described above are used primarily to protect the mercapto function during the building of a peptide chain. While Trt and Dpm have been used very frequently for this same purpose, Hiskey has introduced a clever method for the synthesis of disulphide linkages[41] which makes use of the differential reactivity of Trt- and Dpm-thio groups with thiocyanogen. Consequently, these protecting groups take on added importance when the synthesis of a disulphide containing peptide is attempted. Zervas[42] has recently reported a new method for the introduction of these groups using the corresponding carbinols in trifluoroacetic acid. In work related to selective disulphide syntheses, Hiskey[43] has described a new thiol protecting group which forms a carbon–sulphur linkage as reactive toward thiocyanogen as the Trt-thio function, but which is more stable to acid. The isobutoxymethyl (Ibm) group is introduced directly onto the thiol function using the corresponding alkyl chloride and is stable to hydrazine, dilute acetic acid and a 12M hydrochloric acid–acetone mixture. Hydrogen bromide–acetic acid, trifluoroacetic acid and boron trifluoride–acetic acid remove the Ibm group.

6.2.3.2 Miscellaneous functions

Previous mention has been made in this review of efforts to block the ε-amino group of lysine with groups which survive de-blocking of α-amino functions. The trichloroethoxycarbonyl group (Tec) has recently[44] been used for this purpose. This group survives hydrogenation and strong acids but can be removed using a zinc–acetic acid reagent. It is introduced via reaction of the required acid chloride with the lysine–copper complex, or Z-lysine.

Even though the guanido function has been protected with nitro, tosyl, Z and Boc groups, problems still arise when using any of these. The previously discussed Adoc group has most recently been used[45] to protect the guanidine function during the synthesis of some 32 arginine peptides without significant side reactions. Z-Arg was acylated with Adoc chloride to give the N^{δ}, N^{ω}-

bis(Adoc) derivative. The Adoc group increases the solubilities of arginine derivatives and can be removed by trifluoroacetic acid or hydrogen chloride. It resists hydrogenolysis which allows it to be retained during Z removal.

Most of the recent reports on imidazole protecting groups are directed toward use in the solid-phase method of peptide synthesis and will be discussed in that section. Fridkin[46] has used the 2,4-dinitrophenyl group (Dnp) for imidazole protection during the synthesis of a histidine polymer. Z-His(Dnp) was converted to its *N*-carbonic anhydride, polymerised and de-blocked using mercaptoethanol under very mild conditions.

Rearrangements of asparaginyl and glutaminyl peptides during various chemical interconversions causes the formation of mixtures which are difficult to separate. Weygand[47] introduced the 2,4-dimethoxybenzyl (Dmb) moiety as a protecting group for the amide functions of these two amino

CH_2- CH_3O OCH_3

Dmb

$[CH_3O-C_6H_4]_2-CH-$

Mbh

acids. These workers used two Dmb groups for each amide function. Marshall[48] now reports that a single Dmb group gives more crystalline derivatives of Asp(NH_2) and Glu(NH_2) and that rearrangements can still be minimised. The Dmb group is removed by hydrogen bromide–acetic acid or trifluoroacetic acid but is stable to hydrogenolysis. The bis(*p*-methoxy)-benzhydryl group (Mbh) has been used similarly[49] to protect the amide function of asparagine and glutamine. The Mbh group differs from Dmb in that it is introduced directly on to the amide nitrogen by reaction with the substituted benzhydrol in acetic acid solution. Like Dmb, it is removed by strong acids and is stable to hydrogenolysis, base and hydrazine.

6.3 COUPLING METHODS

The 'activation' of an amino-acid carboxyl group such that it will react with the amino function of another amino-acid molecule to give a new peptide linkage has been accomplished in many different ways. This structural modification of a carboxyl function must raise its reactivity to exactly the 'right' level so that coupling will occur rapidly and completely, without the intervention of inter- or intra-molecular side reactions. Reactivity of too high a level leads to racemisation by at least two mechanisms, while low activation leads to low coupling rates and a predominance of side reactions. Coupling methods can be characterised as those which proceed through readily isolable activated intermediates and those which do not. The active ester and *N*-carbonic anhydride methods fall into the first, while all other methods which use 'coupling' agents comprise the latter category. We shall discuss the recent work in both of these areas.

6.3.1 Active esters

The use of active esters has gained greatly in popularity in the past few years. A great deal of work using polyhalophenyl esters has been reported recently. Polychloro-, bromo- and fluoro-phenyl esters have all been used. Kapoor[50] has synthesised a pentapeptide esterase model using pentachlorophenyl (Pcp) active esters at each step. Kovacs[51] has recently shown that pentafluorophenyl (Pfp) esters couple very rapidly with amino-acid derivatives essentially without racemisation. Some 20 Pfp esters were prepared by dicyclohexylcarbodi-imide (Dcc) coupling of Z-amino acids with pentafluorophenol. These esters were difficult to purify and extremely sensitive to hydrolysis, but could be coupled in the crude state to give good yields of peptides. In a definitive study[52] of the coupling and racemisation rates of various active esters of Z-Lys(Bzl) with VolOMe, Kovacs established a sequence of active esters having decreasing ratios of coupling to racemisation rates. The 'best' ester has the largest ratio; i.e. it undergoes the most rapid coupling with the least racemisation. The order found was: Pfp (46) [53] > Dnp−2,4 (23) > Pcp (16) > Tbp−2,4,6 (5) > Su (4) > Tcp−2,4,6 (3) > Tcp−2,4,5 (2) > Dnp−2,6 (2) > Np (1). It is interesting to note that the *N*-hydroxysuccinimide ester racemised only 1.5 times faster than the Pfp ester, but the latter coupled some 7.5 times faster than the former. It is the extremely high coupling rate of the Pfp ester which makes it attractive. Using a radio-active tracer these workers also established that racemisation of Z-Cys(Bzl)-OPcp does not occur by the elimination-addition of benzyl mercaptan but rather by direct abstraction of the α-proton. In a later study, Kovacs[54] found that racemisation of Z-Cys(Bzl)-OPcp with triethylamine was occurring via the 'isoracemisation' mechanism in which the α-proton is removed to allow epimerisation without exchange with the surrounding medium. This is the first report that compounds of this structural type undergo isoracemisation.

A new series of active esters which undergo very rapid coupling with amino-acid derivatives all have a similar structural feature which allows intramolecular general-base catalysis of the reaction. Young[55] has prepared some 15 2-pyridylthio esters (10) of Boc amino acids and has shown that they react exothermically with amino esters. No racemisation occurred during peptide bond formation and it was found that these esters also reacted readily with

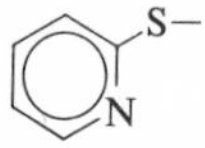

(10) 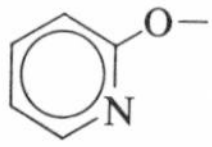 (11)

alcohols. The 2-pyridylthio esters were prepared by direct coupling of the thiol and Boc amino acid with Dcc in ethyl acetate. Concurrently, Morley[56] reported the synthesis of several 2-pyridyloxy (11) esters which also coupled exothermically with amino esters. Peculiarly, these esters could be obtained by allowing a Boc amino-acid to react with 2-pyridinol and Dcc in *pyridine* solution only. None of the other common solvents used for Dcc couplings gave the desired activated esters. As in the case of the pyridylthio esters, the pyridyloxy esters also allowed racemisation-free peptide bond formation. It was also shown that the coupling rates of these esters was strongly effected by the solvent: the order found was dioxan > ethyl acetate ≫ dimethylformamide > dimethyl sulphoxide. *p*-Nitrophenyl esters respond to solvent composition in exactly the opposite order and these facts were used to support the proposed difference in mechanism by which these esters undergo aminolysis. The pyridyloxy ester method was used in the solid-phase synthesis of oxytocin giving the pure product in 70% yield. Consistent with the high reactivity of the 2-pyridyloxy and 2-pyridylthio esters is the report[57] that the anhydrous sodium salt of 2-pyridinol causes tremendous rate increases when added to *p*-nitrophenyl ester solutions during coupling. 2-Pyridinol also acts as a catalyst, but its salt is *ca.* 500 times better. Some evidence for the formation of a reactive intermediate, perhaps the 2-pyridyl ester, was presented.

A third group of esters which undergo anchimerically assisted coupling are those of catechol (12; R = H). A new twist has been added to their use in peptide synthesis by Jones[58] who has prepared several Z-di- and tripeptides having *C*-terminal *O*-monobenzylcatechol ester groups (12, R = Bzl). When the *O*-benzyl group was removed with hydrogen bromide in acetic acid, with simultaneous Z removal from nitrogen, the ester became an activated ester. The peptides so prepared were allowed to polymerise giving sequential polypeptides.

The *p*-methylthiophenyl ester grouping (13) is also a 'convertible' carboxyl protecting group. Johnson[59] has described both of its uses in the synthesis of

O— OR (12) CH_3S O— (13) CH_3SO_2 O— (14)

a heptapeptide and depsipeptide. Methylthiophenyl esters are prepared by direct Dcc coupling of the *N*-protected amino acid with the phenol. Conversion of Mtb into its 'activated' form (14) was accomplished with a hydrogen peroxide–acetic acid reagent. Recently[60], the use of *m*-chloroperbenzoic acid as oxidising agent has been investigated and is recommended even in the presence of tryptophan, nitroguanidyl and amide functions. Of course, this method cannot be used when methionine, cysteine or cystine residues are present.

Several new ways to make activated esters have been reported. Cyanomethyl esters are prepared in good yield and without racemisation by reaction of *N*-protected amino-acid salts with benzenesulphonoxyacetonitrile (15)[61]. The reagent is better than chloro- or bromo-acetonitrile. Vilsmeier–Haack reagents have been used to prepare aryloxyformimium chlorides (16) which react with *N*-protected amino-acid salts to give activated esters[62]. Eight amino acids were investigated and the whole spectrum of halogeno- and nitro-phenyl esters including succinimdyl esters were prepared. *N*-Hydroxy-*N*,*N*′-diphenylurea (17; R = H) was *O*-acylated with a Z-Phe mixed anhydride and the *O*-acyl compound (17; R = Z-Phe) was shown to acylate glycine and alanine methyl esters[63].

$PhSO_2OCH_2CN$ (15)   PhN(OR)–C(=O)–NHPh (16) $(CH_3)_2\overset{+}{N}=C(OAr)H$ Cl^- (17)   N–OH, =O (18)

Another *O*-acyl hydroxylamine active ester was formed when the chlorosulphite of 1-hydroxy-2(1*H*)-pyridone (18) reacted with thallium salts of *N*-protected amino acids[64]. Only one dipeptide was prepared, but several amides were made. A rate study in which several active esters were allowed to couple with glycine *o*-nitroanilide was performed[65]. The results showed that succinimidyl esters reacted some 400 times faster than the *N*-piperidyloxy and cyanomethyl esters, while hydroxylamine reacted with each of these esters at approximately the same rate. The authors also showed that the addition of *N*-hydroxysuccinimide to a Dcc coupling not only reduced racemisation but also increased the yield of peptide. A recent report by Romanovskii[66] indicated that *O*-acylated oximes are also activated esters. The Z-Gly ester of pyridine-4-carboxaldehyde oxime was shown to couple rapidly with glycine ester.

6.3.2 *N*-Carboxyanhydrides

N-Carboxyanhydrides (Nca) are also readily isolable carboxyl-activated amino-acid derivatives useful in peptide synthesis. They have been known for a long time as intermediates in the synthesis of polypeptides, but were not used successfully in controlled peptide synthesis. In 1967, Hirschmann[67]

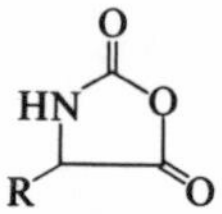

Nca 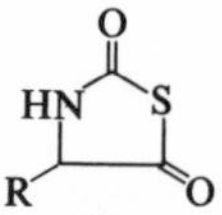 Nta

showed that, under carefully controlled conditions of pH, time and agitation, Ncas could be used in aqueous solution to prepare dipeptides in *ca.* 90% yield. Later, the procedure was adapted to the synthesis of tri- to nonapeptide fragments which were combined by other methods to form the *S*-protein of RNase[68] having a 104 amino-acid sequence. More recently[69], the Merck workers have published procedures for the synthesis of the Ncas from polyfunctional amino acids such as aspartic and glutamic acids, serine, histidine, and ε-Boc lysine. *N*-Thiocarboxyanhydrides (Nta), have also been studied by the Merck group[70], are generally more stable toward alkaline hydrolysis and undergo coupling in aqueous solution at lower pH than Ncas. However, in contrast to Ncas, some Ntas underwent 1–20% racemisation during peptide bond formation. It was concluded that the Ntas most useful in peptide synthesis would be those of glycine, alanine and histidine. The first two gave higher peptide yields than the corresponding Ncas and, in contrast to histidine Nca, histidine Nta could be successfully used in controlled peptide synthesis. Iwakura[71] has very recently reported the synthesis of several di- and tripeptides using Ncas in a two-phase acetonitrile–water mixture at low temperatures. Using this method, it was not necessary to control the pH other than to have sodium carbonate present. Some peptides of lysine and cysteine have also been prepared by Preobrazhenskii[72] using the Nca method. The preparation of Ncas has recently been simplified[73] by the reaction of *N*-trimethylsilyl amino-acid trimethylsilyl esters with thionyl chloride, phosphorus halides or phosgene. This method allows the direct synthesis of Ncas having silylated alcoholic or phenolic hydroxyl groups from the completely silylated amino acids. This is possible because the silylamino function reacts preferentially with the reagent leaving the silyloxy groups intact.

6.3.3 Coupling agents

There are many compounds available which, in the presence of a carboxylic acid and an amine, will bring about dehydration with the formation of an amide linkage. We shall discuss the most recent developments related to several of these reagents.

6.3.3.1 Dicyclohexylcarbodi-imide

Dicyclohexylcarbodi-imide (Dcc) has become the reagent of choice in the years since its use began. It was discovered some time ago that racemisation

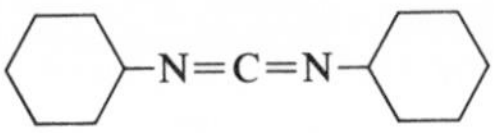

Dcc

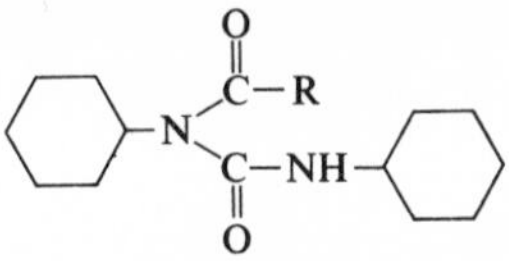

(19)

in varying amounts occurred during Dcc promoted couplings depending on solvent and temperature. The formation of acylurea (19) by-products also created difficulties. Later work showed that the addition of hydroxylamine derivatives to the reaction mixture in molar amount suppressed both racemisation and acylurea formation. *N*-Hydroxysuccinimide (HSu), is an excellent additive, but was found[74] to lead to the formation β-alanine containing by-products. As a result of the extensive studies of Koenig and Geiger[75], the additive of choice now appears to be 3-hydroxy-4-oxo-3,4-dihydro-1,2,3-benzotriazine (HOBt).

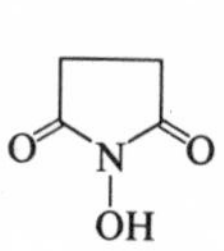

HSu 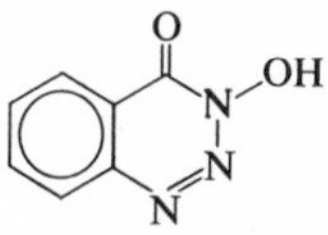 HOBt 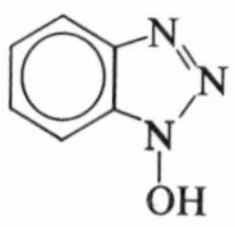 HBt

These workers had found earlier that 1-hydroxybenzotriazole (HBt) was also an excellent additive, but later work showed that it allowed some five to six times as much racemisation during the coupling of TFA-Pro-Val-OH with Pro·OBu^t as did the triazine, HOBt. A by-product formed by ring opening of HOBt was also found which may make this reagent less useful than HBt in cases where racemisation is not a serious problem. Since these reports, Jorgensen[76] has found that HBt suppressed the racemisation of Boc-His(Bzl) during its coupling with dibenzyl-L-glutamate with Dcc. Less than 0.1% of the D-His-L-Glu was found in the presence of HBt as compared to 1.8% in its absence. When this coupling was carried out using the solid-phase method, 0.3% of the racemised peptide was obtained when HBt was present, but 11% was found when it was omitted. These workers found also that HSu suppressed racemisation during Dcc coupling both in solution and by solid phase, but the yield[77] of dipeptide was only *ca.* 70% as compared to quantitative when HBt was used.

6.3.3.2 Miscellaneous

Since the discovery that isoxazolium salts could be used as peptide-bond forming reagents, the search has continued in several laboratories for the best isoxazolium salt for this purpose. Kemp[78] has described in some detail the use of 7-hydroxy-2-ethylbenzisoxazolium fluoroborate (20) in fragment couplings. This compound yields the phenyl ester (21) when treated with an *N*-protected amino-acid salt and this is the intermediate which reacts with

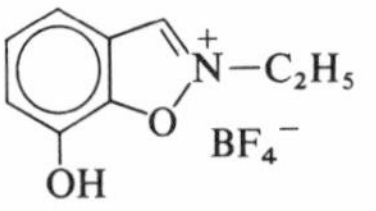

$\overset{+}{N}-C_2H_5$ BF_4^- OH

(20)

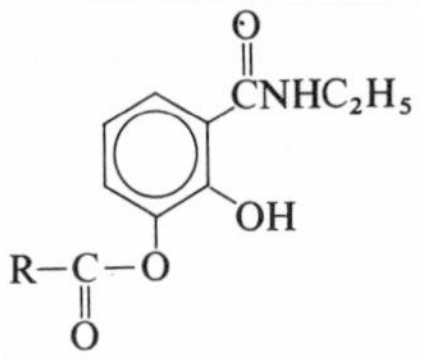

$CNHC_2H_5$ (C=O) OH $R-C(=O)-O$

(21)

$\overset{+}{N}-C_2H_5$ BF_4^-

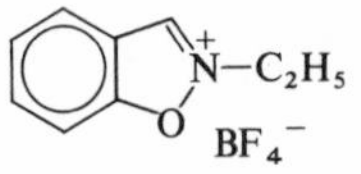

(22)

the amine fragment. Coupling yields were good and at low temperatures the amount of racemisation was very low (<0.1%). Olofson[79] has reported the synthesis of several phenyl- and benzisoxazolium salts and has shown spectroscopically that the benzo salts (22) give essentially no azlactonisation (racemisation) during coupling.

Earlier reports have shown that amides could be prepared using trivalent phosphorus compounds. Yamada[80] applied this reaction to peptide synthesis and showed that dipeptides could be prepared in 60–85% yeilds with *ca.* 3–5% racemisation. The method consists of allowing a phosphine $(C_4H_9)_3P$ or phospinamide like $[(CH_3)_2N]_3P$ to react with carbon tetrachloride or bromide forming a salt (23) which then reacts with an *N*-protected amino

$R_3\overset{+}{P}-CX_3$ X^-

(23)

$R^1_3\overset{+}{P}-OC(=O)R^2$ X^-

(24)

$R_3P \rightarrow O$

(25)

acid giving the acyloxy salt (24) plus the haloform. The acyloxy compound

(24) reacts with an amino ester in the presence of a base like morpholine to give the peptide and a phosphine oxide (25). The method seems very convenient and can possibly be improved to eliminate the racemisation which was found. Wieland[81] and his co-workers have also investigated this coupling method finding also that yields were good, but, in a coupling of Boc-Ala-Phe with Pro-OMe, almost complete racemisation of Phe was found to occur. Certainly, this method could not be used in fragment coupling. Two other reports in which triethyl phosphite[82] and phosphonitrilic chloride[83] have been used to make di- and tetra-peptides, respectively, have appeared.

Several recent reports indicate that the mixed anhydride (MA) procedure still very definitely has a place in peptide synthesis. Meienhofer[84] found that the MA procedure using isobutyl chloroformate was the best way to prepare oligo-γ-glutamic acid having two to seven residues. He attempted the synthesis also by the solid phase, active ester and Dcc methods and found the MA couplings to give the highest yield of crystalline products. The coupling[85] of Z-Phe-Val with Phe·OBu^t was carried out at low temperature using both isobutyryl and isovaleryl chloride in the MA procedure and no racemisation was observed. Another investigation[86] using the sterically hindered amino acid valine showed that the *p*-methoxybenzoyl MA gave better yields (88 %) than the isopropoxycarbonyl MA or the standard Dcc procedure.

The new 'oxidation–reduction condensation' of amino acids reported by Mukaiyama[87] shows promise. Equimolar amounts of acid and amine components are combined with triphenylphosphine and 2,2′-pyridyldisulphide in methylene chloride at room temperature. The coupling proceeds in 90 % yield without racemisation to form triphenylphosphine oxide and 2-pyridinethione as by-products. It was postulated that an acyloxyphosphonium salt (26) was formed, which underwent an anchimerically assisted aminolysis (27) giving the peptide. The intramolecular base catalysis is similar to that postulated for the 2-pyridylthio esters (10) previously discussed and is considered responsible for the lack of racemisation in these couplings.

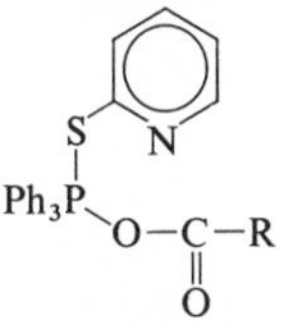

(26)

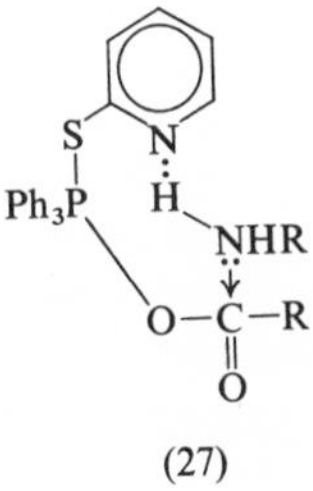

(27)

6.4 RACEMISATION

The present theories on the mechanism of racemisation during aminolysis of variously activated amino-acid derivatives has been critically reviewed by Goodman[88]. Two pathways for racemisation are envisioned: 1) rate determining oxazolone formation followed by rapid racemisation[89] and subsequent ring aminolysis (Figure 6.1); 2) direct base-catalysed abstraction of the α-proton and epimerisation (Figure 6.2). These two mechanisms may be

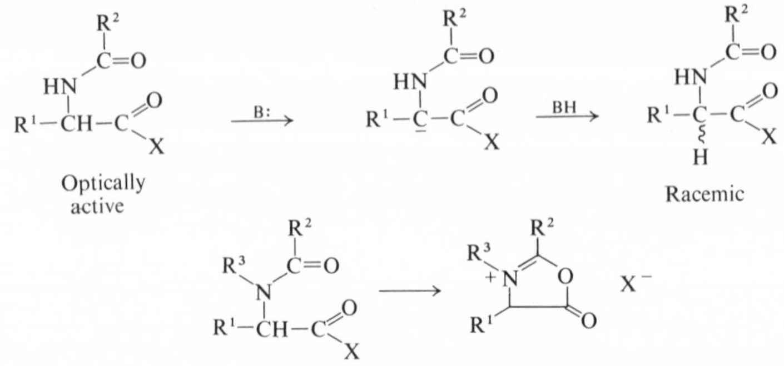

Optically active → B: → BH → Racemic

(28) → (29) X^-

X = leaving group

Figure 6.1

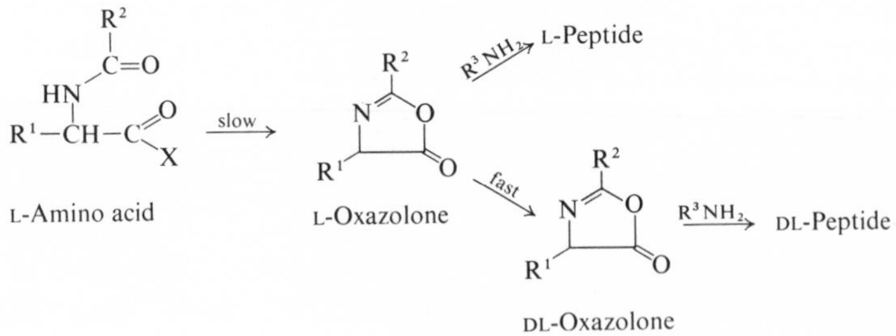

L-Amino acid →(slow) L-Oxazolone →(R^3NH_2) L-Peptide

L-Oxazolone →(fast) DL-Oxazolone →(R^3NH_2) DL-Peptide

Figure 6.2

occurring simultaneously for any given amino-acid derivative under any given set of conditions, but it is generally accepted that oxazolone formation contributes much the larger portion of racemised product. When R^2 is an alkyl or aryl group, the formation of oxazolone is much more rapid than when it is alkoxy. This explains why the urethane protecting groups allow less racemisation than acyl-protecting functions. Proline and *N*-alkylated amino-acids (28) cannot, for structural reasons, undergo oxazolone formation, although there is some speculation as to the possible formation of oxazolonium salts (29). Consequently, these amino acids must undergo racemisation via the α-anion (Figure 6.2).

A method for distinguishing between these two racemisation mechanisms has been described recently by Kemp[90], in which the kinetic isotope effect of α-deuterioamino acid derivatives was determined during various coupling procedures. An isotope effect of 1.0 should indicate that oxazolone formation was the sole racemisation mechanism, while if the α-anion mechanism contributed to the racemisation, k_H/k_D should be substantially larger. When benzoyl-L-leucine was coupled with ethyl glycinate (Young[91] test), the isotope effects were unity for mixed anhydride, azide and isoxazolium salt couplings under various conditions. When applied to the Anderson test[92], in which Z-Gly-L-Phe is coupled with ethyl glycinate, azide and mixed anhydride methods gave isotope effects of 1.5–2.9, indicating that some α-anion formation was occurring. Kemp has also increased the sensitivity of the Young and Anderson tests tremendously by using[93] isotopically tagged reactants. This allows determination of racemic contents in the 1.0–0.001% range. Using these methods, Kemp[94] has evaluated the improved Anderson[95] mixed anhydride procedure, the azide procedure and various active ester couplings. He concluded that the mixed anhydride method rivals the azide process when used under the carefully controlled conditions described. The 3-acyloxy-2-hydroxy-*N*-ethylbenzamides (21), which Kemp has investigated, can also provide peptides of better than 99.99% chiral purity. Further work in this crucial area is continuing.

Among available tests for racemisation besides those of Anderson[92], Young[91] and Weygand[96], is the recently described Izumiya test[97], which allows coupling of Z-Gly-L-Ala with L-Leu·OBzl and, after hydrogenolysis, separates the diastereomers by chromatography. Stewart[98] has modified this test by using the 2,4,6-trimethylbenzyl ester of leucine so that deblocking can be accomplished by hydrogen bromide–acetic acid instead of hydrogenolysis making the procedure more convenient.

The amount of D-amino acid in a peptide can often be determined by hydrolysis followed by separation of various derivatives of the constituent amino acids. Manning[99] has reported a method which determines the amount of racemisation occurring during peptide hydrolysis by using tritiated hydrochloric acid and measuring the amount of incorporated tritium. The method is good only for those amino acids having no replaceable protons in the side chain. The difference between the total amount of D-amino acid in the peptide and the amount formed during hydrolysis is a measure of the quantity formed during synthesis.

6.5 SOLID-PHASE PEPTIDE SYNTHESIS

The standard method of solid-phase peptide synthesis consists of the attachment of the *C*-terminal end of the growing chain to an insoluble resin while step-wise addition of amino acids is accomplished at the *N*-terminal. This sequence is open to a very large number of variations in the methods used at each step and a great deal of work is underway to maximise the yields obtained. Since the intermediate peptides are never purified in a solid-phase synthesis, coupling and deblocking reactions must ideally give quantitative yields without side reactions, if a reasonable amount of a desired, say, eicosapeptide is to be obtained. Modifications of the solid support, coupling methods, protecting groups and general strategy are being investigated. We shall discuss some of the latest developments.

6.5.1 The solid support

Probably the simplest variable to investigate is the pore-size of the solid support. The degree of cross-linking and the exact conditions under which the styrenedivinyl benzene copolymer is prepared largely determines this.

Macroporous[100] resins have been shown to react faster[101] with the *C*-terminal amino-acid giving better yields of aminoacyl resin. The rate of removal of the final product was also increased. The initial velocity of esterification onto the resin was inversely proportional to the swellability of the resin. A new support[102] prepared by attaching a *p*-bis(hydroxymethyl)benzene to Biopak through a silicon–oxygen bond (30) affords a benzyl alcohol function to

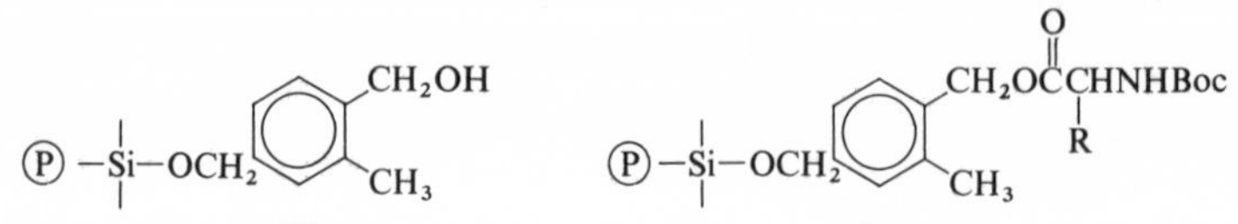

(30) (31)

Ⓟ = Polymer

which a Boc-amino acid can be attached (31). The physical character of Biopak does not allow swelling so that this support can, thus, be used in a column. This cuts down the time necessary for each operation and the large excess of reagents usually used. The coupling efficiency[103] of a new support can be estimated by determining the amounts of Leu-Leu and Ala-Ala in a hydrolysate of $(Leu\text{-}Ala)_6$ prepared on that support. The more of the symmetrical dipeptides found, the more 'failure sequences' present and, consequently, the lower the efficiency. The Biopak support was excellent in this test. A polyethylene glycol of molecular weight 20 000 has been used as a soluble resin support[104]. After each coupling step, the polymer support was dissolved in water and separated from excess reagents and by-products by ultracentrifugation. Coupling yields, even to the hindered valine were excellent. The attachment of a bromoacetyl group to a polystyrene polymer (32)

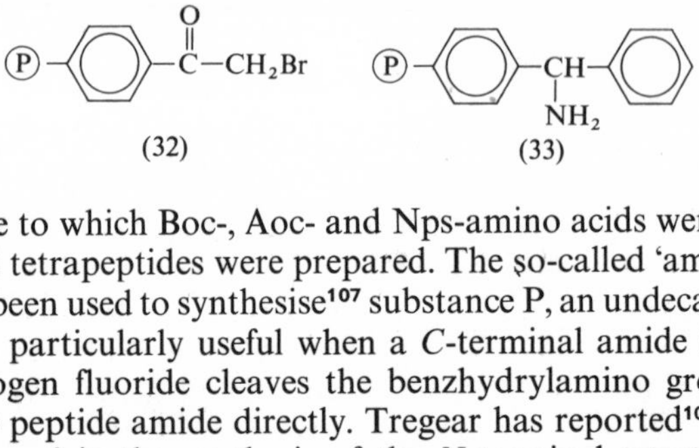

(32) (33)

affords a handle to which Boc-, Aoc- and Nps-amino acids were attached[105]. Several tri- and tetrapeptides were prepared. The so-called 'amine resin' (33) of Pietta[106] has been used to synthesise[107] substance P, an undecapeptide. This kind of resin is particularly useful when a *C*-terminal amide is the desired product. Hydrogen fluoride cleaves the benzhydrylamino group from the resin giving the peptide amide directly. Tregear has reported[108] the use of a 'graft[109] copolymer' in the synthesis of the *N*-terminal tetratricosapeptide sequence of parathyroid hormone. A graft copolymer is a solid support consisting of a central solid core of Kel-F beads with a polystyrene polymer 'grafted' onto the surface. Characteristically, this kind of support has a low capacity, but the polymer strands stand out into the solvent allowing easy access to chemical reagents. Excellent results have been obtained with this kind of solid phase[110]. Kenner[111] has used a solid support having sulphonamido ($—SO_2{\cdot}NH_2$) groups present. An *N*-protected amino acid can replace one of the acidic sulphonamido hydrogen atoms (34) and the peptide chain can then be built up in the usual manner. To cleave the product from the resin, it is treated with diazomethane to methylate the sulphonamido function (35) followed by alkaline saponification, ammonolysis or hydrazinolysis depend-

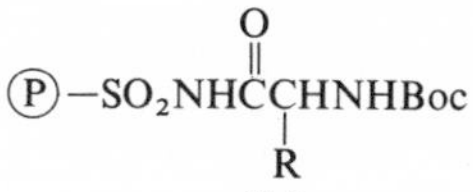

$$\text{Ⓟ}-SO_2NHC(=O)CH(R)NHBoc \quad (34) \qquad \text{Ⓟ}-SO_2N(CH_3)COCH(R)NHBoc \quad (35)$$

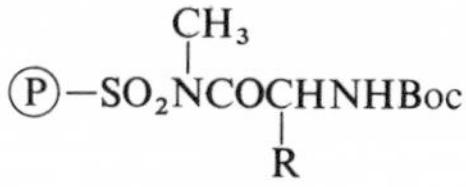

ing on whether the *C*-terminal acid, amide or hydrazide is desired. The concept of having a chemically-stable moiety, the $-SO_2\cdot NH\cdot COR$ function in this case, which can be converted into a labile group, by diazomethane methylation in this case, has been labelled[111] the 'catch principle'. Several peptides up to heptapeptide in size, were synthesised in reasonable yield by this procedure. A polystyrene resin modified by the addition of a 6-hydroxy-n-hexyl group was used by Bayer[112] to synthesise a pentapeptide having valine at the *C*-terminal position. Even with the sterically-hindered amino acid in this position, the peptide was readily ammonolysed from the resin.

6.5.2 Other uses for solid supports

Insoluble solid supports are also being used to carry active esters and coupling agents. A polymer support carrying an *o*-nitrophenol group (36) can be

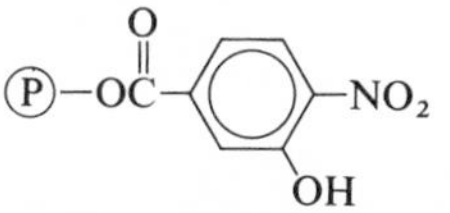

(36) Ⓟ–OC(=O)–C₆H₃(OH)–NO₂ (37) [–CH₂CH₂–(N-hydroxysuccinimide ring, N–OH)]ₙ

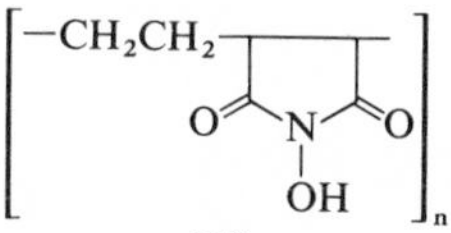

esterified[113] with an *N*-protected amino acid and used to couple that amino acid to a carboxyl-blocked amino acid in solution. A protected tetrapeptide was prepared by this method in yields somewhat lower than that obtained by conventional solution techniques. Active esters of Boc-amino acids were also prepared[114] by esterification of a poly(ethylene-co-*N*-hydroxymaleimide) (37) resin. A heptapeptide fragment of bovine carboxypeptidase A was prepared by this method in 48% yield. The sterically hindered dipeptide Val-Val was prepared in quantitative yield. A variation of this theme was reported by Marshall[115] in which a *p*-hydroxyphenyl sulphide resin (38) was prepared and

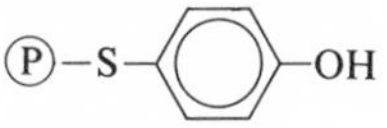

$$\text{Ⓟ}-S-C_6H_4-OH \quad (38) \qquad \text{Ⓟ}-SO_2-C_6H_4-OC(=O)CH(R)NHR^1 \quad (39)$$

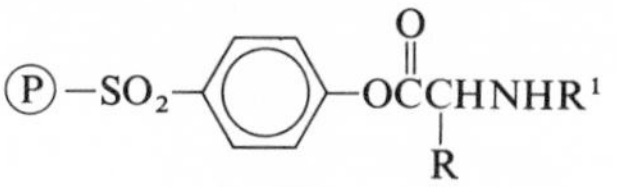

esterified with the penultimate *C*-terminal amino acid of a desired pentapeptide sequence. The tetrapeptide chain was formed in the usual stepwise way and the product was oxidised to the sulphone (39) and treated with the sodium salt of the *C*-terminal amino acid. The pentapeptide product was obtained in *ca.* 40% yield. Even a coupling agent can be incorporated into an insoluble polymer. Brown[116] has copolymerised styrene, divinylbenzene and 6-isopropenylquinoline giving a polymer which was converted into an insolubilised Eedq (40). One gramme of this resin was capable of coupling *ca.* 0.20 mmol of amino acid and was readily reconvertible back to the coupling agent

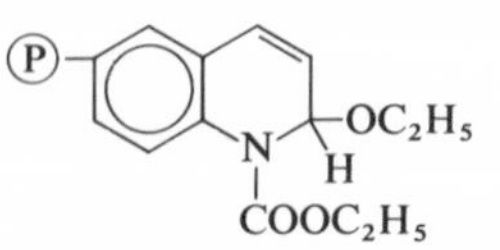

(P)–quinoline–OC_2H_5, N–H, $COOC_2H_5$

(40)

$$\text{(P)}-CH_2-O\overset{O}{\overset{\|}{C}}Cl$$

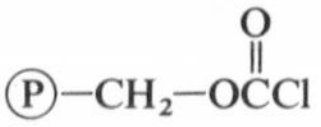

(41)

after each use. In Brown's hands, the Young[91] test showed that 8% racemisation occurred using polymeric Eedq as compared to 5% for monomeric Eedq and 9% for Dcc-HSu mixture.

Solid supports can also be used in a somewhat standard way, but with variations. Meienhofer[117] has attached the ε-amino group of the dipeptide Boc-Lys-Gly-NH_2 to a chloroformoxymethyl resin (41) and proceeded to synthesise Lys^8-vasopressin by standard methods. The product was cleaved from the support with a hydrogen bromide–trifluoroacetic acid mixture. In 1970, Merrifield[118] described a 'reverse' solid-phase method in which an amino acid Boc hydrazine was coupled to a chloroformoxymethyl resin giving the intermediate (42). Removal of the Boc group was followed by conversion to the azide allowed coupling of the next amino acid BOC

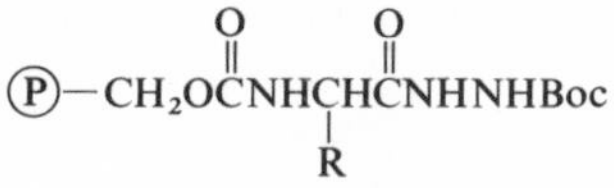

$$\text{(P)}-CH_2O\overset{O}{\overset{\|}{C}}NH\underset{R}{\underset{|}{C}}H\overset{O}{\overset{\|}{C}}NHNHBoc$$

(42)

hydrazide. This cycle was repeated until the C-terminal amino acid was added as its t-butyl ester. Hydrogen bromide-trifluoroacetic acid mixture deblocked and removed the peptide from the resin. The tetrapeptide Leu-Ala-Gly-Val was synthesised in an overall purified yield of 30% by this method. Fragment coupling of (P) -Leu-Ala-N_3 with Gly-Val-OBu^t followed by cleavage and purification gave the same product in 60% yield. This procedure appears to have considerable promise and will undoubtedly be used in the future.

6.5.3 Protecting groups and coupling methods

Some work on protecting groups has been specifically aimed at solid phase synthesis; in particular, side chain protecting groups which are very stable to the conditions used for amine deblocking, yet removable under mild conditions. The blocking of the imidazole ring of histidine has presented a problem which several workers have studied. Merrifield[119] has synthesised N^{α}-BOC-N^{im}-2,4-dinitrophenyl-L-His and used it in the solid phase synthesis of the β-chain of human haemoglobin. The dinitrophenyl group is very stable during BOC removal but can be removed by treatment with mercaptoethanol at the end of the synthesis. Sakakibara[18] found N^{α}-AOC-N^{im}-Tosyl-L-His useful in his synthesis of Ile^5-angiotensin II. The tosyl group was removed in hydrogen fluoride. Several other studies[120] of imidazole protecting groups have also been made. Protection of the mercapto group of cysteine during solid phase work has also been the subject of special study. Zahn[121] found that *p*-methoxybenzyl, diphenylmethyl and benzylthiomethyl groups were most suitable during the synthesis of an insulin β-chain monopeptide fragment. These workers also found[122] that when cystine was coupled with an aminoacyl resin followed by reduction with thiophenol and reoxidation with Fe^{+3}, the peptide chain could be built in the conventional way while the desired cysteine moiety was protected as the disulphide. Reduction and cleavage from the resin gave the desired cysteine peptide.

A preliminary report by Wieland[123] indicated that symmetrical anhydrides of BOC amino acids might be useful in solid phase synthesis. These were prepared by treatment of the BOC amino acid sodium salt with phosgene at low temperature. The linear decapeptide precursor of antamanide was prepared by this method in 96 % yield.

6.5.4 Special problems

While overcoming some of the difficulties encountered in the solution synthesis of peptides, the solid-phase method has other problems peculiar to it. The major problem is the formation of 'failure sequences'. As defined by Bayer[124], when attempting to couple any given amino acid to the *N*-terminal of a peptide chain, a 'truncated sequence' is formed when the amino group of some number of peptide chains fails to react. This truncated sequence is converted into a failure sequence when it couples with an activated amino acid in a later step, thus being converted into a growing chain which is missing one or more amino acids *within* the chain. Since they differ only slightly, separation of these failure sequences from the desired peptide is extremely difficult. With 99 % coupling efficiency, a decapeptide could be obtained in a theoretical 89.5 % yield, while myoglobin, having a 155-amino-acid sequence, would be obtained in only 21.7 % yield. Bayer[124] concluded that by acylating truncated sequences as they are formed to prevent their conversion into failure sequences, homogeneous peptides having 20–30 amino-acid sequences may be synthesised in solid phase. A study[125] of the acetylation of truncated sequences showed that it improved the yield of desired peptide, but that the yield of coupling to form each new peptide bond decreased at each step. The best 'terminating agent' was found[126] to be *N*-acetylimidazole, but a clever use[127] of the anhydride (43) converted truncated sequences into sulphonic

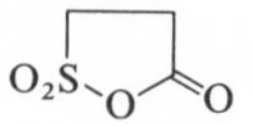

O_2S (cyclic anhydride, ring O and C=O) (43) $\quad HO_3SCH_2CH_2CONH{-}R$ (44)

acids (44) which are readily separated from the desired product on a weakly basic ion-exchange resin. New methods for the determination of free unreacted amino groups have also been reported. Conversion[128] of these into a Schiff base with 2-hydroxy-1-naphthaldehyde allowed their spectroscopic estimation. More elegantly, Bayer[129] has used ^{19}F n.m.r. spectroscopy to estimate free amino groups. A sample of resin is removed after each coupling step and the peptide is cleaved, trifluoroacetylated with trifluoroacetic anhydride and the ^{19}F spectrum determined. Knowledge of the exact chemical shift position of the ^{19}F peak for each trifluoroacetyl amino acid allowed determination of the quantity of each present by integration. The method was applied to the synthesis of the tetrapeptide sequence shown (45). The yields at each coupling stage are shown on the peptide bonds as determined after each coupling. In the older Dorman procedure[130], pyridine hydrochloride

$$\text{Ile} \xrightarrow{85\%} \text{Ala} \xrightarrow{70\%} \text{Val} \xrightarrow{90\%} \text{Gly}$$

(45)

was allowed to react with the uncoupled amino groups and the salt formed was neutralised by a triethylamine solution. Potentiometric titration of the chloride ion present in the triethylamine solution was a direct measure of the free amino groups on the resin. This method is nondestructive of the resin and gives excellent results. The Nps protecting group cannot be used, since pyridine hydrochloride cleaves it at an appreciable rate. A ninhydrin test of high sensitivity[131] has also been described.

The development of methods designed to maximise yields is one approach to the solution of the 'failure sequence' problem. Another solution emerging from the efforts of many workers is a tactical one; i.e. the synthesis of fragments having only 5–15 amino-acid sequences followed by the condensation of these into the larger final product. The coupling methods which allow the absolute minimum amount of racemisation must be used in fragment couplings, since it is well known[88] that activation of a peptide carboxyl group leads to more racemisation than activation of an *N*-protected amino acid. The methods of choice are the azide procedure and Dcc-hydroxylamine additive (HOBt, HBt, HSu) methods previously discussed. The fragment condensation procedure allows relatively easy purification of intermediates so that the final more difficultly purifiable product is easier to obtain pure. This method has been reported by Hofmann[132] in his approach to ribonuclease T_1 synthesis. Three peptides of 10–12 residues each were coupled by the azide procedure in *ca.* 50% yields. Bradykinin analogues[133], insulin fragments[134] and the *N*-terminal tridecapeptide sequence of bovine pancreatic ribonuclease[135] have also been synthesised recently by the azide fragment coupling method. The 12–27 fragment of sekretin was reported[136] synthesised by the condensation of a decapeptide with a tetrapeptide using the Dcc-*N*-hydroxysuccinimide method in reasonable yield. The fragment condensation tactic is presently the method of choice for the synthesis of large peptides and proteins.

6.6 CONCLUSION

The preceding summary of results reported in the 1969–1972 period is a small indication of the tremendous amount of work under way in the field of peptide synthesis. The enthusiasm and dedication of the people involved in this research has now brought this field to the threshold of *protein* synthesis.

There is every promise that we shall soon be preparing pure proteins and variations thereof in the near future. Medical science and man in general will benefit tremendously.

References

1a. Marglin, A. and Merrifield, R. B. (1970). *Ann. Rev. Biochem.*, **39,** 841
1b. Kapoor, A. (1970). *J. Pharm. Sci.*, **59,** 1
1c. Geiger, R. (1971). *Angew. Chem. Int. Ed. Engl.*, **10,** 152
1d. Jones, J. H. (1971). *Amino Acids, Peptides and Proteins,* Vol. 3, 219. (G. T. Young, editor). (London: The Chemical Society)
2. Bergmann, M. and Zervas, L. (1932). *Chem. Ber.*, **65,** 1192
3. Meienhofer, J. (1963). *Proc. of Sixth European Peptide Symposium,* 55. (Oxford: Pergamon)
4. Carpino, L. A. (1957). *J. Amer. Chem. Soc.*, **79,** 98; (1960). **82,** 2725
5. Schwyzer, R., Sieber, P. and Kappeler, H. (1959). *Helv. Chim. Acta,* **42,** 2622
6. Miyoshi, M. and Onishi, T. (1971). *Jap. Pat.* 71 00 010
7. Ito, M. (1970). *Jap. Pat.* 70 36 729
8. Schnabel, E., Stoltefuss, J., Offe, H. A. and Klauke, E. K. (1971). *Justus Liebig's Ann. Chem.*, **743,** 57
9. Wuensch, E. and Wendlberger, G. (1968). *Chem. Ber.*, **101,** 3659
10a. Schnabel, E., Klostermeyer, H. and Berndt, H. (1971). *Justus Liebig's Ann. Chem.*, **749,** 90
10b. Hiskey, R. G., Beacham, L. M., Matl, V. G., Smith, J. N., Williams, E. B., Thomas, A. M. and Walters, E. T. (1971). *J. Org. Chem.*, **36,** 488
11. Loffet, A. and Dernier, C. (1971). *Experientia,* **27,** 1003
12. Alakhov, Yu. B., Kiryushkin, A. A., Lipkin, V. M. and Milne, G. W. A. (1970). *Chem. Commun.*, 406
13. Sieber, P. and Iselin, B. (1968). *Helv. Chim. Acta,* **51,** 622
14. Su-sun, W. and Merrifield, R. B. (1969). *Int. J. Protein Res.*, **1,** 235
15. Klauke, E., Schnabel, E. and Schmidt, G. (1971). *Ger. Pat.* 19 34 783
16. Haas, W. L., Krumkalns, E. V. and Gerzon, K. (1966). *J. Amer. Chem. Soc.*, **88,** 1988
17. Wuensch, E., Wendlberger, G. and Spangenberg, R. (1971). *Chem. Ber.*, **104,** 3854
18. Fujii, T. and Sakakibara, S. (1970). *Bull. Chem. Soc. Jap.*, **43,** 3954
19. Sakakibara, S., Fukuda, T., Kishida, Y. and Honda, I. (1970). *Bull. Chem. Soc. Jap.*, **43,** 3322
20. Noda, K., Terada, S. and Izumiya, N. (1970). *Bull. Chem. Soc. Jap.*, **43,** 1883
21. Southard, G. L., Zaborowsky, B. R. and Pettee, J. M. (1971). *J. Amer. Chem. Soc.*, **93,** 3302
22. Stevenson, D. and Young, G. T. (1969). *J. Chem. Soc.*, 2389
23. Carpino, L. A. and Han, G. Y. (1970). *J. Amer. Chem. Soc.*, **92,** 5748
24. Zervas, L., Borovas, D. and Gazis, E. (1963). *J. Amer. Chem. Soc.*, **85,** 3660
25. Savrda, J. and Veyrat, D. H. (1970). *J. Chem. Soc. C,* 2180
26. Wuensch, E. and Spangenberg, R. (1972). *Chem. Ber.*, **105,** 740
27. Steglich, W. and Batz, H. G. (1971). *Angew. Chem. Int. Edn. Engl.*, **10,** 75
28. D'Angeli, F., DeBello, C. and Filira, F. (1971). *J. Org. Chem.*, **36,** 1818
29. Patchornik, A., Amit, B. and Woodward, R. B. (1970). *J. Amer. Chem. Soc.*, **92,** 6333
30. Stelakatos, G. C. and Argyropoulos, N. (1970). *J. Chem. Soc. C,* 964
31. Maclaren, J. A. (1971). *Aust. J. Chem.*, **24,** 1695
32. Stewart, F. H. C. (1971). *Aust. J. Chem.*, **24,** 2193
33. Turner, D. L. and Baczynski, E. (1970). *Chem. Ind. (London),* 1204
34. Hendrickson, J. B. and Kandall, C. (1970). *Tetrahedron Letters,* 343
35. Camble, R., Garner, R. and Young, G. T. (1969). *J. Chem. Soc. C,* 1911
36a. Burton, J., Fletcher, G. A. and Young, G. T. (1971). *Chem. Commun.*, 1057
36b. Young, G. T., Schafer, D. J., Elliot, D. F. and Wade, R. (1971). *J. Chem. Soc. C,* 46
37. Johnson, B. J. and Jacobs, P. M. (1968). *Chem. Commun.*, 73
38. Muraki, M. and Mizoguchi, T. (1971). *Chem. Pharm. Bull.* (*Tokyo*), **19,** 1708

39. Wieland, T. and Sieber, A. (1969). *Justus Liebig's Ann. Chem.*, **727,** 121
40. Wuensch, E. and Spangenberg, R. (1969). *Ger. Pat.* 19 23 480
41. Hiskey, R. G., Thomas, A. M., Smith, R. L. and Jones, W. C. (1969). *J. Amer. Chem. Soc.*, **91,** 7525; Hiskey, R. G., Davis, G. W., Safdy, M. E., Inui, T., Upham, R. A. and Jones, W. C., Jr. (1970). *J. Org. Chem.*, **35,** 4148
42. Zervas, L., Photaki, I., Taylor-Papadimitriou, J., Sakarellos, C. and Mazarakis, P. (1970). *J. Chem. Soc. C*, 2683
43. Hiskey, R. G. and Sparrow, J. T. (1970). *J. Org. Chem.*, **35,** 215
44. Yajima, H., Watanabe, H. and Okamoto, M. (1971). *Chem. Pharm. Bull.* (*Tokyo*), **19,** 2185
45. Jaeger, G. and Geiger, R. (1970). *Chem. Ber.*, **103,** 1727
46. Fridkin, M. and Shaltiel, S. (1971). *Arch. Biochem. Biophys.*, **147,** 767
47. Weygand, F., Steglich, W. and Bjarnason, J. (1968). *Chem. Ber.*, **101,** 3642
48. Pietta, P. G., Cavallo, P. and Marshall, G. R. (1971). *J. Org. Chem.*, **36,** 3966
49. Koenig, W. and Geiger, R. (1970). *Chem. Ber.*, **103,** 2041
50. Kapoor, A., Kang, S. M. and Trimboli, M. A. (1970). *J. Pharm. Sci.*, **59,** 129
51. Kisfaludy, L., Roberts, J. E., Johnson, R. H., Mayers, G. L. and Kovacs, J. (1970). *J. Org. Chem.*, **35,** 3563
52. Kovacs, J., Mayers, G. L., Johnson, R. H., Cover, R. E. and Ghatak, U. R. (1970). *J. Org. Chem.*, **35,** 1810 and *Chem. Commun.*, 53
53. ( ) indicates this ester's ratio relative to NP as 1. Dnp-2,4=2,4-dinitrophenyl; Tbp-2,4,6 = 2,4,6-tribromophenyl; Su = *N*-succinimidoxy; Tcp-2,4,6 = 2,4,6-trichlorophenyl; Tcp-2,4,5=2,4,5-trichlorophenyl; Dnp-2,6=2,6-dinitrophenyl; NP=*p*-nitrophenyl.
54. Kovacs, J., Cortegiano, H., Cover, R. E. and Mayers, G. L. (1971). *J. Amer. Chem. Soc.*, **93,** 1541
55. Lloyd, K. and Young, G. T. (1971). *J. Chem. Soc. C*, 2890
56. Morley, J. S. and Dutta, A. S. (1971). *J. Chem. Soc. C*, 2896
57. Nakamizo, N. (1971). *Bull. Chem. Soc. Jap.*, **44,** 2006
58. Jones, J. H. and Cowell, R. D. (1971). *J. Chem. Soc. C*, 1082
59a. Johnson, B. J. and Trask, E. G. (1968). *J. Org. Chem.*, **33,** 4521
59b. Johnson, B. J., *J. Org. Chem.*, **34,** 1178
60. Johnson, B. J. and Ruettinger, T. A. (1970). *J. Org. Chem.*, **35,** 255
61. Liplawy, M. and Zabrocki, J. (1971). *Z. Chem.*, **11,** 16
66. Romanovskii, P. Ya., Muiznieks, V. and Cipens, G. (1971). *Zh. Obshch. Khim.*, **41,** 2110
63. Sarantakis, D., Light, W. W., Craig, A. R. and Weinstein, B. (1971). *Synthesis*, 328
64. Taylor, E. C., Kienzle, F. and McKillop, A. (1970). *J. Org. Chem.*, **35,** 1672
65. Merz, D. and Determann, H. (1969). *Justus Liebigs Ann. Chem.*, **728,** 215
62. Itoh, M. (1970). *Chem. Pharm. Bull.* (*Tokyo*), **18,** 784
67. Hirschmann, R., Strachan, R. G., Schwam, H., Schoenewaldt, E. F., Joshua, H., Barkemeyer, B., Veber, D. F., Paleveda, W. J., Jr., Jacob, T. A., Beesley, T. E. and Denkewalter, R. G. (1967). *J. Org. Chem.*, **32,** 3415
68. Hirschmann, R., Nutt, R. F., Veber, D. F., Vitali, R. A., Varga, S. L., Jacob, T. A., Holly, F. W. and Denkewalter, R. G. (1969). *J. Amer. Chem. Soc.*, **91,** 507
69. Hirschmann, R., Schwam, H., Strachan, R. G., Schoenewaldt, E. F., Barkemeyer, H., Miller, S. M., Conn, J. B., Garsky, V., Veber, D. F. and Denkewalter, R. G. (1971). *J. Amer. Chem. Soc.*, **93,** 2746
70. Dewey, R. S., Schoenewaldt, E. F., Joshua, H., Paleveda, W. J., Jr., Schwam, H., Barkemeyer, H., Arison, B. H., Veber, D. F., Strachan, R. G., Milkowski, J., Denkewalter, R. G. and Hirschmann, R. (1971). *J. Org. Chem.*, **36,** 49
71. Iwakura, Y., Uno, K., Oya, M. and Katakai, R. (1970). *Biopolymers*, **9,** 1419
72. Skalaban, T. D., Nazimov, J. M., Pankova, S. S., Zvonkova, E. N., Evstigneeva, R. P. and Preobrazhenskii, N. A. (1971). *Zh. Org. Khim.*, **7,** 47
73. Kricheldorf, H. R. (1971). *Chem. Ber.*, **104,** 87
74. Weygand, F., Steglich, W. and Chytil, N. (1968). *Z. Naturforsch.*, **23b,** 1391
75. Koenig, W. and Geiger, R. (1970). *Chem. Ber.*, **103,** 788, 2024, 2034
76. Windridge, G. C. and Jorgensen, E. C. (1971). *J. Amer. Chem. Soc.*, **93,** 6318
77. Windridge, G. C. and Jorgensen, E. C. (1971). *Intra-Sci. Chem. Rep.*, **5,** 375
78. Kemp, D. S. (1970). *Peptides: Chemistry and Biochemistry*, 33. (S. Lande and B. Weinstein, editors). (New York: Dekker)
79. Olofson, R. A. and Marino, Y. L. (1970). *Tetrahedron*, **26,** 1779

80. Yamada, S. and Takeuchi, Y. (1971). *Tetrahedron Letters,* 3595
81. Wieland, T. and Seeliger, A. (1971). *Chem. Ber.,* **104,** 3992
82. Mitin, Yu. V. and Vlasov, G. P. (1971). *Zh. Obshch. Khim.,* **41,** 427
83. Das, K. C., Lin, Y. Y. and Weinstein, B. (1969). *Experientia,* **25,** 1238
84. Meienhofer, J., Jacobs, P. M., Godwin, H. A. and Rosenberg, I. H. (1970). *J. Org. Chem.,* **35,** 4137
85. Tazner, E., Smulkowski, M. and Lubiewska-Nakomieczna, L. (1970). *Justus Liebigs Ann. Chem.,* **739,** 228
86. Birr. C., Lochinger, W. and Wieland, T. (1969). *Justus Liebigs Ann. Chem.,* **729,** 213
87a. Mukaiyama, T., Matsueda, R. and Maruyama, H. (1970). *Bull. Chem. Soc. Jap.,* **43,** 1271
87b. Mukaiyama, T., Matsueda, R. and Suzuki, M. (1970). *Tetrahedron Letters,* 1901
88. Goodman, M. and Glaser, C. (1970). Reference 78, p. 267
89. Grahl-Nielson, O. (1971). *Chem. Commun.,* 1588
90. Kemp, D. S. and Rebek, J., Jr. (1970). *J. Amer. Chem. Soc.,* **92,** 5792
91. Williams, M. W. and Young, G. T. (1963). *J. Chem. Soc.,* 881
92. Anderson, G. W. and Callahan, F. M. (1958). *J. Amer. Chem. Soc.,* **80,** 2902
93. Kemp, D. S., Wang, S. W., Busby, G., III and Hugel, G. (1970). *J. Amer. Chem. Soc.,* **92,** 1043
94. Kemp, D. S., Bernstein, Z. and Rebek, J., Jr. (1970). *J. Amer. Chem. Soc.,* **92,** 4756
95. Anderson, G. W., Zimmerman, J. E. and Callahan, F. M. (1967). *J. Amer. Chem. Soc.,* **89,** 5012
96. Weygand, F., Prox. A., Schmidhammer, L. and Koenig, W. (1963). *Angew. Chem.,* **75,** 282
97a. Izumiya, N. and Muraoka, M. (1969). *J. Amer. Chem. Soc.,* **91,** 2391
97b. Izumiya, N., Muraoka, M. and Aoyagi, H. (1971). *Bull. Chem. Soc. Jap.,* **44,** 3391
98. Stewart, F. H. C. (1970). *Aust. J. Chem.,* **23,** 1073
99. Manning, J. M. (1970). *J. Amer. Chem. Soc.,* **92,** 7449
100. Tilak, M. and Hollinden, C. S. (1971). *Org. Prep. Proced. Int.,* **3,** 183
101. Losse, A. (1971). *Z. Chem.,* **11,** 386
102. Bayer, E., Jung, G., Halasz, J. and Sebestrian, J. (1970). *Tetrahedron Letters,* 4503
103. Bayer, E. and Koenig, W. (1969). *J. Chrom. Sci.,* **7,** 95
104. Mutter, M. and Hagenmaier, H. (1971). *Angew. Chem. Int. Ed. Engl.,* **10,** 811
105. Mizoguchi, T., Shigezane, T. and Takamura, N. (1970). *Chem. Pharm. Bull.,* **18,** 1465
106. Pietta, P. G. and Marshall, G. R. (1970). *Chem. Commun.,* 650
107. Tregear, G. W., Niall, H. D. and Potts, J. T. (1971). *Nature (London), New Biol.,* **232,** 87
108. Tregear, G. W., Niall, H. D., Keutmann, H. T., Sauer, R., Deftos, L. J., Dawson, B. F., Hogan, M. L. and Aurbach, G. D. (1971). *Proc. U.S. Nat. Acad. Sci.,* **68,** 63
109. Battaerd, H. A. J. and Tregear, G. W. (1967). *Graft Copolymers.* (New York: Wiley–Interscience)
110. Tregear, G. W. (1972). *Graft Polymers as Insoluble Supports in Peptide Synthesis.* Third American Peptide Symposium, Boston, Mass., June 19, 1972
111. Kenner, G. W., McDermott, J. R. and Sheppard, R. C. (1971). *Chem. Commun.,* 636
112. Bayer, E., Breitmaier, E., Jung, G. and Parr, W. (1971). *Hoppe-Seyler's Z. Physiol. Chem.,* **352,** 959
113. Panse, G. T. and Laufer, D. A. (1970). *Tetrahedron Letters,* 4181
114. Fridkin, M., Patchornik, A. and Katchalski, E. (1972). *Biochemistry,* **11,** 466
115. Marshall, D. L. and Liener, I. E. (1970). *J. Org. Chem.,* **35,** 867
116. Brown, J. and Williams, J. E. (1971). *Can. J. Chem.,* **49,** 3765
117. Meienhofer, J. and Trzeciak, A. (1971), *Proc. Nat. Acad. Sci. U.S.,* **68,** 1006
118. Felix, A. M. and Merrifield, R. B. (1970). *J. Amer. Chem. Soc.,* **92,** 1385
119. Chillemi, F. and Merrifield, R. G. (1969). *Biochemistry,* **8,** 4344
120a. Losse, G. and Krychowska, U. (1971). *Tetrahedron Letters,* 4121
120b. Guenter, L. and Krychowska, U. (1970). *J. Prakt. Chem.,* **312,** 1097
121. Hammerstroem, K., Lunkenheimer, W. and Zahn, H. (1970). *Makromal. Chem.,* **133,** 41
122. Lunkenheimer, W. and Zahn, H. (1970). *Justus Liebig's Ann. Chem.,* **740,** 1
123. Wieland, T., Birr, C. and Flor, F. (1971). *Angew. Chem. Int. Ed. Engl.,* **10,** 336
124. Bayer, E., Eckstein, H., Haegele, K., Koenig, W., Bruening, W., Hagenmaier, H. and Parr, W. (1970). *J. Amer. Chem. Soc.,* **92,** 1735
125. Hagenmaier, H. (1970). *Tetrahedron Letters,* 283
126. Markley, L. D. and Dorman, L. C. (1970). *Tetrahedron Letters,* 1787

127. Wissmann, H. and Geiger, R. (1970). *Angew. Chem. Int. Ed. Engl.,* **9,** 908
128. Esko, K. and Karlsson, S. (1970). *Acta Chem. Scand.,* **24,** 1415
129. Bayer, E., Hunziker, P., Mutter, M., Sievers, R. E. and Uhmann, R. (1972). *J. Amer. Chem. Soc.,* **94,** 265
130. Dorman, L. C. (1969). *Tetrahedron Letters,* 2319
131. Kaiser, E., Colescott, R. L., Bossinger, C. D. and Cook, P. I. (1970). *Analyt. Biochem.,* **34,** 595
132. Beacham, J., Dupuis, G., Finn, F. M., Storey, H. T., Yanaihara, C., Yanaihara, N. and Hofmann, K. (1971). *J. Amer. Chem. Soc.,* **93,** 5526
133a. Arold, H. and Feist, H. (1970). *J. Prakt. Chem.,* **312,** 1145
133b. Arold, H. and Reissmann, S., *J. Prakt. Chem.,* 1130
133c. Arold, H. and Stibenz, D., *J. Prakt. Chem.,* 1161
134. Zahn, H. and Schmidt, G. (1970). *Justus Liebig's Ann. Chem.,* **731,** 91
135. Visser, S., Raap, J., Kerling, K. E. T. and Havinga, E. (1970). *Rec. Trav. Chim. Pays-Bas,* **89,** 865
136. Wuensch, E., Wendlberger, G. and Thamm, P. (1971). *Chem. Ber.,* **104,** 2445

7
New Syntheses of Naturally Occurring Peptides and Analogues

R. WADE
CIBA Laboratories, Horsham, Sussex

7.1 INTRODUCTION

This chapter is a report covering work appearing in the literature during 1970 and 1971 on the synthesis of natural peptide sequences and related compounds. In some cases material published before 1970 has been included where its omission would have seriously unbalanced a section (e.g. calcitonin, Section 7.4). Compounds have been given separate sections either because of intense activity in that area during the period under review or because they represent areas in which work is likely to develop in the near future. The choice is admittedly arbitrary and many compounds albeit important, which have received relatively little attention recently, will be mentioned only briefly. The work on glucagon carried out by Wünsch and collaborators falls into this category.

7.2 ABBREVIATIONS

The abbreviations used in this survey follow those given by the IUPAC–IUB Commission on Biochemical Nomenclature (1967), *Biochemistry*, 6, 362 and IUPAC Information Bulletin No. 26, p. 11. Other abbreviations are defined at the time of usage. All amino acid residues can be assumed to be of the natural L configuration unless stated otherwise. In order to conserve space structures of analogues have been presented in tabular form. Any amino acid in a sequence which differs from that in the parent sequence (top line of any table) is written in, a horizontal line indicates no change from the parent sequence. Thus compound (7) in Table 7.1 represents the sequence

Asn-Arg-Val-Tyr(Me)-Val-His-Pro-Phe

A vertical line cutting across a peptide bond in a table or formula indicates a point at which fragment condensation was made during synthesis.

7.3 ANGIOTENSIN II

The amino acid sequence of angiotensin was first reported in 1956 and confirmed by synthesis shortly afterwards. The early work on the elucidation of structure and synthesis has been comprehensively reviewed by Schröder and Lübke[17]. However, interest in this peptide has been maintained and a large number of analogues has been synthesised principally for the purpose of elucidating structure–activity relationships. More recently speculation has arisen as to whether such a relatively small molecule might possess sufficient intramolecular interactions to maintain an ordered structure in solution and if so whether this is maintained on interaction with its biological receptor.

The naturally occurring biologically active peptide is an octapeptide (1), species variations occurring in position 5. The bovine local tissue hormone has valine in this position which in man, horse and pig is occupied by isoleucine. Analogues have been synthesised since 1957[18] and the products of new syntheses appearing during the period under consideration are collated in Table 7.1. Many of these are the result of synthesis using the solid-phase method which was first applied to angiotensin by Marshall and Merrifield some years previously[19]. The integrity of the products of this method of synthesis frequently rests upon characterisation of the final product only. This becomes more difficult to achieve in a satisfactory manner as the chain length of the peptides is increased. As a result of this, chemical and physical differences between compounds differing for example by one amino acid residue become vanishingly small and separation consequently becomes more difficult. Angiotensin II however is of a size at which such problems are not normally encountered and normal analytical criteria such as electrophoresis, chromatography and also residue and elemental analysis would be expected, when taken in combination, to afford adequate characterisation of such a product. Compounds prepared by the conventional solid-phase procedure[19] in Table 7.1 include Nos. (2) to (10). Here t-butoxycarbonyl was used for α-protection and deprotection used HCl in acetic acid at each stage. Dicyclohexylcarbodi-imide was used as coupling agent and the peptide was removed from the resin by the action of hydrogen bromide in trifluoroacetic acid. Benzyl protection was used for the β-carboxyl group of aspartic acid, the phenolic hydroxyl group of tyrosine and the imidazole ring of histidine and was removed by hydrogenolysis. The products were purified on Sephadex G-25 by partition using butanol–acetic acid–water. The same basic synthetic procedure was used in the preparation of (11) but here the presence of *N*-methylalanine at position 7 caused a marked reduction in the rate of the dicyclohexylcarbodi-imide mediated coupling of Boc-N^{im}-benzylhistidine. Purification in this case was by gel filtration on Sephadex G-25 using 0.2 mol l^{-1} acetic acid. Compound (13)[4] was prepared similarly, the *N*-terminal Gly-Gly residue being incorporated as a single unit. In this preparation the synthesis was monitored by sampling at each stage and, where coupling had been found to be incomplete, namely after the incorporation of histidine and isoleucine, residual unreacted amino groups were blocked by acetylation. Purification in this case was by column chromatography on sulphoethylcellulose using an ammonium acetate gradient. The heptapeptide (45) served as a common source from which (15)–(17) were prepared using *p*-nitrophenyl active esters of the relevant benzyloxycarbonyl amino acid. (14) Was obtained from the resin-bound heptapeptide using succinic anhydride. Fission was followed by hydrogenation of protecting groups

Table 7.1 Angiotensin II analogues

	1	2	3	4	5	6	7	8
(1)	Asp	Arg	Val	Tyr	Val Ile	His	Pro	Phe
(2)[1]								
(3)[1]					Ile			Tyr
(4)[1]					Ile			Tyr(Me)
(5)[1]					Val			Tyr(Me)
(6)[1]				Tyr(Me)	Ile			
(7)[1]				Tyr(Me)	Val			
(8)[1]	Asn			Tyr(Me)	Val			
(9)[1]					Ile		Pip	Tyr
(10)[1]					Ile			APB
(11)[2]					Ile			APIB
(12)[3]	Z-Gly	Gly	Gly		Val		MeAla	
(13)[3, 4]	Gly	Gly	Val		Ile			OMe
(14)[5]	Suc	Gly			Ile			
(15)[5]	Sucm	Gly			Ile			
(16)[5]	Asn	Gly			Ile			
(17)[5]	Asp	Gly			Ile			
(18)[6]	Asn				Ile			
(19)[6]	Asn				Ala			
(20)[6]	Asn				Abu			
(21)[6]	Asn				Nle			
(22)[6]	Asn				aIle			
(23)[6]	Asn				Aev			
(24)[6]	Asn				Cpg			
(25)[7]	Asn				Chg			
(26)[7]	Asn				Aib			
(27)[7]	Asn				Sar			
(28)[7]	Asn				Pro			
(29)[7]	Asn				D-Pro			
(30)[8]	Asn				Acpc			
(31)[8]	Asn				His			
(32)[8]	Asn				Arg			
(33)[8]	Asn				Tyr			
					Ser			

(34)[8]		Asn				Pro	
(35)[8]		Asn				Amb	
(36)[9]						Val	
(37)[10]					Phe	Val	Tyr
(38)[11]		Asp				Val	
(39)[12]		Gly	Gly	Gly	Gly	Val	
(40)[12]		Gly	Gly	Gly		Val	
(41)[12]		Gly	Gly			Val	
(42)[12]		Gly	Arg			Val	
(43)[13]	NH_2CO	Gly	Gly			Val	
(44)[13]		Asn	Gly			Val	
(45)[13]			Gly			Val	
(46)[13]		EtCO	Gly			Val	
(47[14]			NH_2CO			Val	
(48)[14]				NH_2CO		Val	
(49)[14]					NH_2CO	Val	
(50)[15]		BuCO				Ile	
(51)[15]		L-Abu				Ile	
(52)[15]		D-Abu				Ile	
(53)[15]		Gly				Ile	
(54)[15]		Ac	Gly			Ile	
(55)[15]			Ahx			Ile	
(56)[15]			Ava			Ile	
(57)[16]			Gly	Gly	Gly	Ile	
(58)[16]				Aoc		Ile	

Pip = pipecolic acid
APB = 3-amino-4-phenylbutyric acid
APIB = 3-amino-3^1-phenylisobutyric acid
MeAla = *N*-methylalanine
Suc = succinyl
Sucm = succinamyl
Abu = α-aminobutyric acid
aIle = allo-isoleucine

Aev = α-amino-β-ethylvaleric acid
Cpg = α-cyclopentylglycine
Chg = α-cyclohexylglycine
Aib = α-aminoisobutyric acid
Acpc = α-aminocyclopentane carboxylic acid
Ahx = ε-aminohexanoic acid
Ava = δ-aminovaleric acid
Amb = *p*-aminomethylbenzoic acid

and anion exchange chromatography. A series of analogues (18)–(29) with variations at position 5 were prepared by the solid-phase procedure[22], the *N*-terminal asparagine again being introduced as the *p*-nitrophenyl ester. In this work both gradient elution with ammonium acetate on sulphoethyl-cellulose and partition chromatography on Sephadex G-25 was used for purification. One interesting observation here was the occurrence of racemisation during the incorporation of butyloxycarbonyl-N^{im}-benzylhistidine as revealed by the enzymic digestion of a hydrolysate of tripeptide-polymer. Diastereoisomeric octapeptides were eliminated during the purification procedure in most cases with the exception of (18), (26) and (27). During the purification of (27) a by-product, from which tyrosine appeared to be absent, was isolated. No other examples of failure sequences were reported, and the authors' alternative suggestion, i.e. that under the conditions used total loss of tyrosine might have occurred by over-reduction seems unlikely. Compounds (30)–(35) were also prepared by the solid-phase procedure and purified using a Sephadex column; full details were not disclosed[8].

[Val^5]-angiotensin II (36) is included in Table 7.1 since it has been prepared in order to test the efficacy of the picolyl ester method of peptide synthesis[9]. This technique, which employs *C*-terminal protection incorporating a basic group to facilitate purification is potentially capable of wider application. Side chain protection by groups capable of removal by reductive conditions is appropriate for this method. Although the deprotected peptide was not purified chromatographically it possessed about $\frac{2}{3}$ pressor activity of International Research Standard A.

Transposition of the hydroxyl group of Tyr^4 to Phe^8, as a result of preparing the analogue (37), gives a compound which is claimed to be a potent inhibitor of angiotensin II. This product was prepared by solid phase using dinitrophenyl to protect the imidazole ring of histidine. This protection was removed by treatment with thiophenol prior to fission of peptide from the resin in anhydrous hydrogen fluoride. The product was purified by partition on Sephadex G-25 but the results of amino acid analysis on acid and enzymic hydrolysates are not impressive.

One further analogue, in which *N*-terminal aspartic acid is linked via its β-carboxyl group (38), has been prepared by solid phase essentially as described above. The α-side chain was protected as the *p*-nitrobenzyl ester and coupling time was extended to a total of 22 h in each cycle. After hydrogenolysis of protecting groups the product was purified by chromatography on Sephadex G-15 using butanol–pyridine–acetic acid–water.

A series of analogues in which up to half of the molecule from the *N*-terminus has been replaced by glycine (39)–(42) has been prepared by a step-wise technique. Some of the intermediates obtained in this work were utilised subsequently by the same group in the preparation of a series of 2-glycine angiotensins II. These, (43)–(45), were assembled by a fragment condensation approach from hexapeptide *p*-nitrobenzyl ester using carbodi-imide couplings when glycine was being added, and *p*-nitrophenyl active ester in the case of asparagine (44). Products were purified by column chromatography on carboxymethyl -Sephadex or -cellulose. Carbamoyl derivatives of *N*-foreshortened angiotensin fragments (47)–(49) have also been reported. These were prepared by carbodi-imide mediated couplings on to the *C*-terminal tetrapeptide of Val^5-angiotensin. Hydroxysuccinimide was present during the reaction to minimise the risk of racemisation[53].

Several more examples of shorter molecules prepared by the solid-phase procedure[22] are represented by (50)–(58). During this work[15] it was noted that the forcing conditions required to remove benzyl protection from histidine were sufficient to cause some reduction of tyrosine and phenylalanine (see also Ref. 23). Chromatography on sulphoethyl Sephadex was used for purification as sulphoethylcellulose which had been used previously[22] failed to remove des-isoleucine by-products. Racemisation of histidine was again noted and could be as high as 10% in these products.

Finally mention must be made of the preparation of an angiotensin II analogue in which tritium has been introduced in a specific manner[20]. [Asn^1-Val^5]-Angiotensin II (Hypertensin-CIBA) was iodinated with iodine chloride to give the [4-di-iodo-tyrosine] analogue. Apparently the histidine ring is unaffected by this treatment. This product was then subjected to catalytic halogen/tritium exchange in the aromatic ring of the tyrosine residue. The labelled analogue has proved useful in a metabolic study[21].

7.4 CALCITONIN

Calcitonin, or thyrocalcitonin, the hormone which appears to be responsible for bone resorption, was first fully characterised chemically in 1968 and preliminary reports of its synthesis appeared at the same time. The porcine hormone, α-thyrocalcitonin, was the first isolated ((59) Table 7.2) and this was synthesised independently by three groups. Reports of the two of these for which details are available will be discussed here.

These two syntheses, which utilise the condensation of fragments assembled initially by step-wise addition, are similar in principle but differ significantly in detail. Positions where fragments were condensed, indicated by the vertical broken lines in (59a) and (59b) are seen to differ generally in the two syntheses. However, both syntheses were designed to incorporate the 1–9 *N*-terminus into the molecule with the disulphide ring formed rather than leave ring closure to the end of the synthesis as is practised in syntheses of other disulphide-containing peptides such as oxytocin or vasopressin. Both syntheses use azide couplings extensively in order to couple peptide fragments but apart from these similarities the syntheses are based on completely opposite approaches.

The strategy used by Rittel and co-workers[24] involved acid-removable t-butyl protection of all seven hydroxyl and the single carboxyl side chain functions. Benzyloxycarbonyl α-protection was used during chain assembly except in the two cases (Ser^{10}, Met^{25}) where protection of an α-amino group had to be removed from sulphur containing molecules with the additional complication of t-butyl ethers requiring to be retained. For this exacting situation, the extremely acid labile 2-(*p*-biphenylyl)isopropyloxycarbonyl (Bpoc) group was adopted. The formation of the disulphide ring used another technique not normally adopted. Sulphur was protected as the *S*-trityl derivative during step-wise synthesis and azide coupling of the 1–4 and 5–9 protected intermediates. Detritylation was then effected using mercury acetate and after decomposition of the mercury mercaptide, oxidative ring

Table 7.2 Calcitonins

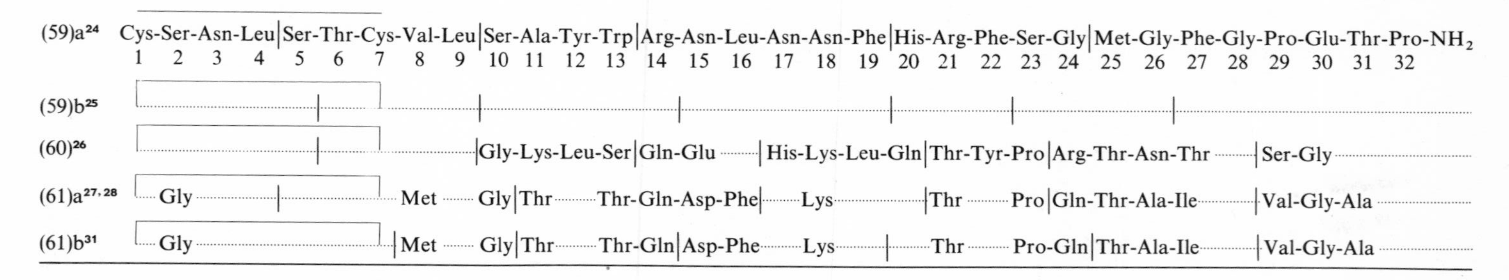

	Sequence
(59)a[24]	Cys-Ser-Asn-Leu|Ser-Thr-Cys-Val-Leu|Ser-Ala-Tyr-Trp|Arg-Asn-Leu-Asn-Asn-Phe|His-Arg-Phe-Ser-Gly|Met-Gly-Phe-Gly-Pro-Glu-Thr-Pro-NH_2
	1 2 3 4 5 6 7 8 9 10 11 12 13 14 15 16 17 18 19 20 21 22 23 24 25 26 27 28 29 30 31 32
(59)b[25]	……|……|……|……|……|……|……
(60)[26]	……|……|Gly-Lys-Leu-Ser|Gln-Glu ……|His-Lys-Leu-Gln|Thr-Tyr-Pro|Arg-Thr-Asn-Thr ……|Ser-Gly ……
(61)a[27,28]	Gly ……|…… Met …… Gly|Thr …… Thr-Gln-Asp-Phe|…… Lys ……|Thr …… Pro|Gln-Thr-Ala-Ile ……|Val-Gly-Ala ……
(61)b[31]	Gly ……|Met …… Gly|Thr …… Thr-Gln|Asp-Phe …… Lys ……|…… Thr …… Pro-Gln|Thr-Ala-Ile ……|Val-Gly-Ala ……

closure was accomplished with di-iodoethane. Otherwise stepwise addition of amino acids to form fragments was achieved by a variety of methods such as mixed anhydride, dicyclohexylcarbodi-imide or nitrophenyl active ester (for asparagine, etc.). Fragments were condensed as azides or using hydroxy-succinimide facilitated dicyclohexylcarbodi-imide[53] to suppress racemisation. Countercurrent distribution was involved at six points to purify fragments and also the protected dotriacontapeptide. Removal of butyl protecting groups was achieved with ice cold concentrated hydrochloric acid for 10 min.

In contrast the synthesis developed by Guttmann and collaborators[25] used no protection of hydroxy side groups of serine and threonine residues. Arginine, which had been used with its guanidine protected by protonation by Rittel and collaborators, was introduced here as nitroarginine. Histidine, which had been unprotected previously, was here incorporated as the bis-(benzyloxycarbonyl) derivative and this was then exchanged for bis-trityl protection. Glutamic acid was again introduced as the t-butyl ester. Cysteine sulphydryl groups were protected by benzylation and after the 1–9 fragment had been assembled by coupling 1–5 azide with 6–9, sodium in liquid ammonia treatment served to remove *S*-benzyl without effecting fragmentation of the molecule.

Fragments were assembled stepwise by active ester couplings or using dicyclohexylcarbodi-imide. No difficulties were apparently encountered arising from undesired acylation of serine or threonine side groups. Fragment 27–32 was constructed using benzyloxycarbonyl as temporary protection for α-amino groups, 23–26 used butyloxycarbonyl and 20–22 which had been constructed from benzyloxycarbonyl derivatives was added to 23–32 as the bistrityl derivative. Fragment 10–19 was constructed from two portions, each built step-wise, generally using active esters and butyloxycarbonyl α-protection. The three large fragments were then combined using azide couplings and the final intermediate required only deprotection of the α-amino group. This was accomplished with trifluoroacetic acid and the free peptide was purified by gel filtration and ion-exchange chromatography.

The hypocalcaemic peptide hormone isolated from the ultimobranchial body of the salmon (60) is of interest because of its high potency relative to the porcine hormone in mammalian species. It has been synthesised by Guttman and collaborators[26] again using the condensation of fragments built up by step-wise methods. The hydroxy groups on serine and threonine side chains were again used unprotected as was histidine on this occasion. Arginine was introduced as nitroarginine but this guanidine protection was removed immediately after incorporation. Other basic side groups and the γ-carboxyl group of glutamic acid were protected by butyloxycarbonyl or t-butyl esters respectively. Benzyloxycarbonyl was used for α-amino protection during build-up of fragments which did not contain sulphur, with the exception of 10-glycine (trityl) and 24-arginine (butyloxycarbonyl). Fragments were condensed by the azide method with the exception of 1–9 which was added as the succinimide ester. After removal of the acid-labile protecting groups with aqueous trifluoroacetic acid the product was purified by gel filtration on Biogel P6 in 0.2 mol l^{-1} acetic acid. No indication was given in this preliminary communication of the occurrence of any by-products.

In contrast the synthesis of the human hypocalcaemic hormone, calcitonin M (61)a has now been published in full and several points of interest were disclosed[27]. From the structure it can be seen that (61) differs from (59) at 18 residues and no fragments common to both could be utilised for synthesis. In this peptide a glycine residue is situated at position 10 (of 59) and advantage was taken of this as a coupling point for the *N*-terminal fragment. As in the synthesis of α-thyrocalcitonin[24], Rittel and collaborators used maximum protection of side chains and were thus able to use carbodi-imide couplings with no chance of side reactions. The *N*-terminal decapeptide was formed from 1–4 and 5–10 fragments by azide coupling[28]. *S*-Trityl groups were then removed and the disulphide ring formed concomitantly by the action of iodine in methanol–dimethylformamide, a technique developed in the same laboratories[29]. After purification by countercurrent distribution, the decapeptide was carefully checked for optical integrity using L-amino acid oxidase treatment of a hydrolysate: thin layer chromatography was used to analyse for the presence of allothreonine but none was detected. The remainder of the molecule was constructed using a similar strategy to that adopted for α-thyrocalcitonin. α-Protection was by benzyloxycarbonyl groups and side chain protection by acid-labile groups derived from t-butanol. Fragments were linked by mixed anhydride (Pro^{23}), hydroxysuccinimide–carbodi-imide (Gly^{10}, Gly^{28}) and azide (His^{20}, Phe^{16}). During this work the occurrence of racemisation while coupling His^{20} azide to 21–28 free peptide was recorded. This was ascribed to the use of triethylamine as base to neutralise the carboxyl terminal of the amino component. Model experiments[30] showed that racemisation could occur if excess triethylamine was present but was much reduced if an equivalent amount of the hindered base ethyl di-isopropylamine was used. In the method finally adopted, no triethylamine was used and ethyl di-isopropylamine was added portion-wise during the first few hours of coupling. In this way, the extent of racemisation was minimised (to 1.5%) and the diastereoisomer was carried through the remaining synthetic stages and separated by countercurrent distribution of the fully protected dotriacontapeptide. The diastereoisomer (about 2%), formed by racemisation of Phe^{16} during a similar azide coupling, was eliminated by countercurrent distribution of fully protected 11–32 docosapeptide. The final stage of the synthesis in which all ten acid-labile protecting groups were removed by a brief treatment with ice-cold concentrated hydrochloric acid was accompanied by two side reactions. One of these, which occurred in trace amounts, was alkylation of methionine sulphur by t-butyl cations to give a sulphonium derivative. This effect was recognised by the enhanced basicity of the by-product on electrophoresis and deduced by analogy with the behaviour of 1–10 decapeptide on treatment with t-butanol–trifluoroacetic acid. A second by-product or mixture of by-products running together was ascribed to an N → O acyl migration having occurred at one of the six β-hydroxyamino acid residues present in the peptide. This product accumulated on increasing the duration of acid treatment and the reverse reaction to restore the normal product could be induced by standing the material in aqueous solution at pH 7.5. All of the by-products formed during acidolysis were removed by countercurrent distribution giving material which was indistinguishable from the natural product. L-Amino acid oxidase digestion of an acid hydrolysate gave residual values for the D isomers differing insignificantly from those obtained from the natural product under the same conditions.

Preliminary reports of a further synthesis of human calcitonin (61)b have appeared in symposium proceedings[31]. Again fragments were condensed by azide or carbodi-imide–hydroxysuccinimide but no protection was applied to side chain hydroxyl groups. Benzyloxycarbonyl was used for α-protection of growing chains in sulphur-free regions of the molecule. S-Protection was by benzyl which was removed by sodium in liquid ammonia and the disulphide ring formed before coupling with methionine in fragment 8–10 which would otherwise have been at risk. Countercurrent distribution and column chromatography on silica gel were used for the purification of intermediates and the absence of D-amino acid residues in fragment 11–32 was checked by L-amino acid oxidase digestion. No less than seven azide couplings were involved in the construction of this fragment. The product possessed the right order of biological activity when compared to Medical Research Council Standard A but comparison of these data with those on synthetic calcitonin M prepared previously is not possible due to a difference in route of administration between the two assays.

7.5 CORTICOTROPIN

Synthetic work achieved before mid-1965 on sequences related to the natural adrenocorticotropins (ACTH) has been reviewed in a comprehensive manner by Schröder and Lübke[49]. At a relatively early point the observation was made that the corticotropic activity of the hormone was associated with the *N*-terminal half of the molecule. Since most of the synthetic work directed towards natural ACTH and its analogues has been motivated by the need to elucidate the contributions made by various portions of the sequence towards bioactivity, the *N*- as distinct from the *C*- terminus has received almost exclusive attention. Species differences reside in the *C*-terminal region and are less than had at first been thought[50]. During the period under review a number of reports of new analogues have appeared. With one exception (63) new compounds described have not extended beyond the *N*-terminal tetracosapeptide portion (62) of the molecule. Table 7.3 lists these in order of decreasing size. Compound (63) was prepared as a biologically active fragment of human ACTH. However, as a result of recent work, the sequence of human ACTH has been revised and the 25–27 region is now considered to be Asn-Gly-Ala[50, 182]. Consequently compound (63), which contains *C*-terminal Asp-Ala-Gly, must now be regarded as an analogue and not a natural sequence.

Although in almost every case the compounds listed in Table 7.3 were prepared with the aim of clarifying structure–activity concepts recognition has also been taken of the relatively rapid breakdown of ACTH in biological systems. As a consequence attempts have been made in many instances to incorporate into the sequence amino acids known to form peptide bonds having a low hydrolytic rate in the presence of the various enzymic systems known to be present in mammalian systems. For example, the replacement of

Table 7.3 Corticotrophin

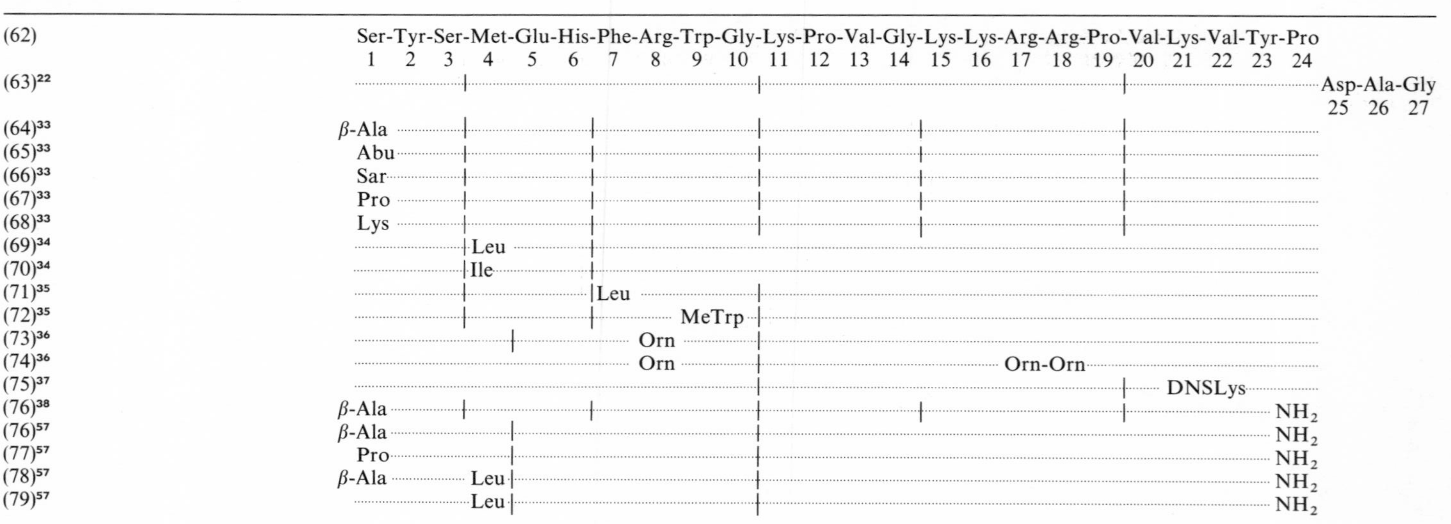

	1	2	3	4	5	6	7	8	9	10	11	12	13	14	15	16	17	18	19	20	21	22	23	24	25	26	27
(62)	Ser	Tyr	Ser	Met	Glu	His	Phe	Arg	Trp	Gly	Lys	Pro	Val	Gly	Lys	Lys	Arg	Arg	Pro	Val	Lys	Val	Tyr	Pro			
(63)[22]																									Asp	Ala	Gly
(64)[33]	β-Ala																										
(65)[33]	Abu																										
(66)[33]	Sar																										
(67)[33]	Pro																										
(68)[33]	Lys																										
(69)[34]				Leu																							
(70)[34]				Ile																							
(71)[35]							Leu																				
(72)[35]									MeTrp																		
(73)[36]								Orn																			
(74)[36]								Orn									Orn	Orn									
(75)[37]																					DNSLys						
(76)[38]	β-Ala																							NH_2			
(76)[57]	β-Ala																							NH_2			
(77)[57]	Pro																							NH_2			
(78)[57]	β-Ala			Leu																				NH_2			
(79)[57]				Leu																				NH_2			

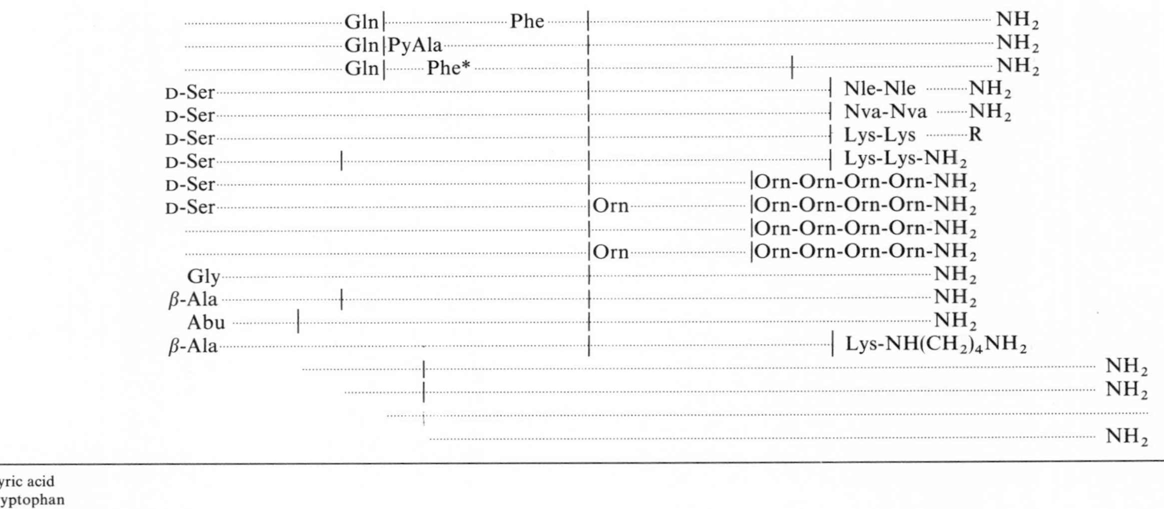

Abu = α-aminobutyric acid
MeTrp = N^{α}-methyltryptophan
DNSLys = N^{ε}-dansyl-lysine
PyAla = β-(pyrazol-3-yl)alanine
Phe* = [^{14}C]-phenylalanine
Nle = norleucine
Nva = norvaline
R = NH_2, $NH(CH_2)_4CH_3$, $NH(CH_2)_9CH_3$, $NH(CH_2)_{15}CH_3$, $N[(CH_2)_7\,CH_3]_2$, $NH(CH_2)_{23}CH_3$, $NHCH[(CH_2)_{16}CH_3]_2$, OH, $O(CH_2)_9CH_3$, $O(CH_2)_{13}CH_3$, $O(CH_2)_{17}CH_3$, $O(CH_2)_{23}CH_3$, $OCH[(CH_2)_{16}\,CH_3]_2$

N-terminal L-serine by β-alanine or better by D-serine would be expected to slow down the degradation of the molecule by aminopeptidase action. If such enzymic action is a limiting factor in determining the biological activity of the molecule, then this activity would be expected to be enhanced or the duration of its action extended by an amino acid replacement of that type. Such a rationale prompted the preparation of the analogue of (62) containing-D-serine at the *N*-terminus. This compound was shown previously to give rise to a doubling of the period of elevated plasma steroid level following subcutaneous injection into rats compared to the all L-peptide[51]. Compounds (83)–(88) were designed with the same objective in mind. A similar line of thought appertained to the use of ornithine as a replacement for lysine. Ornithyl derivatives are split at a much lower rate by trypsin than the corresponding arginyl or lysyl derivatives. On this basis Tesser and co-workers have prepared a series of ornithine-containing corticotropin analogues. By comparing substitutions at position 11 alone, it was observed that replacement of lysine by ornithine did lead to a significant prolongation of enhanced plasma corticosteroid level following injection[44].

The replacement of amino acids in positions 7–9 has usually been prompted by the desire to test the stringency of the requirements of this the critical region of the molecule for steroidogenesis. Replacement of amino acids has generally been found to lead to a significant drop in biological activity. Such considerations do not however apply to changes in position 4 for example. It has long been known[52] that methionine at this site was not essential for biological activity and many analogues have since been prepared with 4-methionine replaced by synthetically less exacting alkyl amino acids at this position. Compounds (69), (70), (78) and (79) are analogues which have been prepared more recently to test the requirements of this site.

Without exception, the compounds listed in Table 7.3 have been synthesised by the classical approach involving the stepwise addition of amino acids from a *C*-terminus to give fragments which may then be condensed. No reports on the application of the step-wise solid-phase technique to this sequence have appeared during the period under review. It is possible that the juxtaposition of several amino acids, which are known to give rise to problems during solid-phase synthesis, in the *N*-terminal nonapeptide coupled with the overall size of ACTH has caused exponents of the technique to look elsewhere (but see Melanotropin below). However, the successful application of classical techniques coupled with purification problems, which are becoming more difficult at this molecular size, has meant that the solid-phase technique does not present such an attractive alternative as in the case of certain smaller peptides. Application of the solid-phase technique to this sequence will undoubtedly be seen in the future, certainly to sections if not to all of the structure.

Inspection of Table 7.3 reveals that the glycine residues at positions 10 and 14 have been favoured sites for condensing fragments which had themselves been built by step-wise methods. There being no racemisation problem, dicyclohexylcarbodi-imide has been used extensively for these couplings, either alone or accelerated with added hydroxysuccinimide[53] or hydroxybenztriazole[54]. Similar remarks apply to proline at position 19. At other points in the sequence, couplings between fragments have usually been effected using azides. It is interesting to note the extension of the use of the manipulatively less demanding Honzl and Rudinger procedure for preparing azides[55]. There have been no reports of any racemisation during the azide couplings used in the preparation of the compounds in Table 7.3. However, only a few reports actually describe experiments designed to detect significant racemisation in products, and it is possible that some may have occurred but have escaped detection.

Protecting groups used have been benzyloxycarbonyl for α-protection during the build up of sequences and groups based on t-butyl for side chains or vice versa, no other groups have found application. Where the final protected intermediate was blocked exclusively by acid-susceptible butyloxycarbonyl and t-butyl residues trifluoroacetic acid has been used for deprotection. Where benzyloxycarbonyl or nitro (on arginine) groups are present as well, the splitting of these has frequently been accomplished by the action of liquid hydrogen fluoride[56]. Histidine and serine have usually been used unprotected without giving rise to additional experimental difficulties. Arginine has generally been incorporated as nitroarginine but in many cases the nitro group has been removed at the next deprotection stage; protonation then served as adequate protection for the remainder of the synthesis.

Compound (75) in which the lysine residue at position 21 carries a dansyl (dimethylaminonaphthalenesulphonyl: DNS) group was prepared by Schwyzer and Schiller[37] in order to obtain information on the conformation of ACTH and also to provide a tagged molecule. Since (75) possesses the biological activity of the parent molecule (62), conclusions based on experiments using it are certainly meaningful. From a synthetic aspect the dansyl residue is an inert irremovable protection for the ε-amino group of lysine and serves only to alter the solubility properties of derivatives somewhat. The *C*-terminal pentapeptide was assembled step-wise, dansyl–lysine being coupled by carbodi-imide and valine added using the *p*-nitrophenyl active ester. The 11–16 hexapeptide azide was then coupled to 17–19 free peptide, care being taken to prevent the solution becoming alkaline and so give rise to the possibility of racemisation. The product was purified by column chromatography on silica gel and then coupled with *C*-terminal pentapeptide using a hydroxysuccinimide accelerated dicyclohexylcarbodi-imide coupling[53]. After deprotection of the *N*-terminus of the 11–24 tetradecapeptide the 1–10 decapeptide was added again using carbodi-imide coupling but using hydroxybenztriazole as the accelerating agent[54]. The protected tetracosapeptide was purified by countercurrent distribution and side-chain protection was then removed using trifluoroacetic acid to give free (75).

Compound (76) is an analogue which has been prepared by two routes. In the preparation described by Fujino and co-workers[38] fragments as indicated were condensed from the *C*-terminus. Coupling at positions 10, 17 and 19 were mediated by pentachlorophenyl active esters themselves prepared from the peptide acids using pentachlorophenyl trichloroacetate. Subsequent couplings at 3 and 6 were by azide. In the earlier preparation by Geiger and co-workers[57], (76) was prepared in conjunction with (77)–(79) and the synthesis designed accordingly. Fragments 2–4 *C*-terminating in methionine or leucine were coupled using the azide method with 5–10 hexapeptide. The nonapeptides thus obtained were then acylated by active ester (nitrophenyl or trichlorophenyl) to give *N*-terminal decapeptide. The reaction of this compound with 11–23 tridecapeptide amide was by a dicyclohexylcarbodi-imide coupling. Acceleration of this coupling by additional pentachlorophenol was more effective than by adding *p*-nitrophenol. It is interesting to note that hydroxy-succinimide was found to have no beneficial effect on this coupling. Purification of (76)–(79) utilised column chromatography on carboxymethylcellulose with an acetic acid gradient.

The analogues (80) and (81) were prepared by Hofmann and collaborators as part of their extensive programme aimed at elucidating the precise structural requirements of the active site of the ACTH molecule. Analogue (82), in which [^{14}C]phenylalanine was introduced, represents the first attempt to prepare a corticotropin radioactively labelled in a known position in the molecule. Further, the molecule does possess full ACTH activity and will be of value for mode of action studies. Unfortunately, the relatively low specific activity associated with ^{14}C, and here it was diluted further down to 0.126 mCi mmol^{-1}, is likely to impose some limitation on the usefulness of this material in view of the extremely low physiological level of the hormone. The synthesis followed a pattern used by the same authors on previous occasions[59]; radioactive phenylalanine was incorporated as the *N*-hydroxysuccinimide active ester. Histidine was then added as azide and fragment 1–5 also added as the azide. Dicyclohexylcarbodi-imide alone was the agent used for coupling with the *C*-terminal decapeptide amide and the protected eicosapeptide was purified by column chromatography on carboxymethylcellulose using a gradient of ammonium acetate in aqueous dimethylformamide (9:1). Deprotection using trifluoroacetic acid was followed by incubation with thioglycollic acid to reduce methionine sulphoxide back to methionine. The optical integrity of the product was checked using an aminopeptidase-M digest.

Of the series of compounds containing D-serine at the *N*-terminus (83)–(88), that containing lysine, replacing arginine at positions 17 and 18 (86) is of considerable interest because of its prolonged steroidogenic activity *in vivo*[43]. The *N*-terminal decapeptide was built using 1–4 tetrapeptide azide, which had been prepared by the Honzl and Rudinger technique[55], coupling with a 5–10 fragment. The *C*-terminal octapeptide was constructed by a 6+2 azide coupling and, after *N*-terminal α-deprotection by hydrogenolysis, was coupled using dicyclohexylcarbodi-imide with *N*-terminal decapeptide. This was purified by countercurrent distribution and was isolated by precipitation using ammonium sulphate. Protecting groups (butyloxycarbonyl and t-butyl) were then removed by the action of 90% trifluoroacetic acid. The product contained only 1.5% sulphoxide (determined by titration).

The analogue (94) in which diaminobutane is attached to a heptadecapeptide can be regarded as a des-carboxy (or des-carbamido) derivative of the corresponding 18-lysine octadecapeptide. Consequently it contains the same pattern of basic groups as (92) and like (92) has high corticotropic activity. This serves to support consideration of the importance of this basic area as a binding site[68]. Synthesis was achieved by coupling the 11–16 hexapeptide with lysine 4-aminobutylamide, the ω-amino groups being blocked by butyloxycarbonyl groups[58]. The coupling agent was dicyclohexylcarbodi-imide–hydroxybenztriazole and the same reagent was used

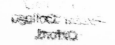

subsequently to add the *N*-terminal decapeptide. As with hydroxysuccinimide[53], hydroxybenztriazole accelerates carbodi-imide couplings and is claimed to avoid racemisation[54]. After removal of protecting groups with trifluoracetic acid and treatment with thioglycollic acid to reduce methionine sulphoxide the final product was purified by chromatography on carboxymethylcellulose. No information on by-products was recorded.

Among other syntheses achieved, Brugger has prepared a series of nonadecapeptide amides and esters (85) in which the *C*-terminal proline carried alkyl amide or alkyl ester groups up to C_{32} in size. These were prepared in the hope that the enhanced lipophilicity conferred upon the molecule thereby would render them absorbable by oral administration. Preparation of these compounds was achieved by coupling various proline esters with a Lys-Lys unit or the requisite amine with *C*-terminal tripeptide by mixed anhydride. Subsequent stages used the same intermediates and methods as had been used previously for (86).

7.6 HUMAN PITUITARY GROWTH HORMONE

In late 1970 a preliminary report appeared in which the synthesis of human pituitary growth hormone (HGH) by the solid-phase procedure was described[60]. Shortly afterwards a series[61–64] of papers from the same laboratory described the synthesis of several protected fragments, prepared by classical methods, which were to be used in an alternative synthesis of the same molecule. A second solid-phase synthesis in this case of two large fragments of HGH, appeared during 1971[65]. Subsequent to these papers being submitted for publication, a report appeared which indicated that the structure used by these two groups in their synthetic work may well be in error to a significant degree. A fragment positioned at 17–31 in the original structural determination[67] would appear to be better located after what had been considered to be Arg^{91}, a Leu-Arg fragment (additional to the original structural content) then follows this transferred fragment and the remainder of the molecule is in accordance with the previous report. In this way, homology with human placental lactogen and sheep prolactin is maintained. The implications of this structural reappraisal on the synthetic work are less profound than on the interpretation placed upon the recorded biological activity of the materials prepared. Sequences which overlap the area in question may no longer be of value for the projected synthesis of the whole molecule.

In the work described, the solid-phase synthesis of a 188 amino acid sequence represents the largest molecule assembled by this technique. An automated apparatus was used, understandably, and conventional solid-phase tactics were adopted in that α-protection was by butyloxycarbonyl and side chain hydroxyl, sulphydryl and carboxyl groups were benzylated, lysine was used with ε-benzyloxycarbonyl protection and arginine as nitroarginine. Histidine was added as the bis(butyloxycarbonyl) derivative and tryptophan was preserved, after introduction, by adding dithiothreitol to the deprotecting reagent trifluoroacetic acid:dichloromethane(1:1). The possibility that the side chain protection of lysine might also be removed to some extent by this treatment[69] was not discussed. Fission from the resin

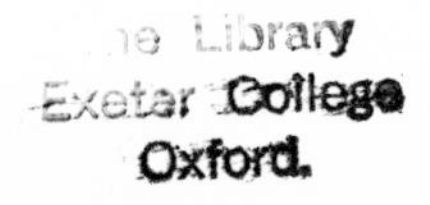

was achieved with hydrogen fluoride–anisole. Although no details are given of the state of the product at this stage, the material was then passed through Sephadex G-25 in 50% acetic acid and then treated with sodium in liquid ammonia to remove *S*-benzyl protection. Inevitably some degradation was brought about by this treatment. Subsequent oxidation was followed by gel filtration on Sephadex G-100 to obtain material behaving analogously to native HGH in this particular system. Analytical data on the product were compared to those determined on native HGH following treatment with hydrogen fluoride and sodium in liquid ammonia, a treatment which the authors assert gave material significantly different from untreated HGH. It would therefore seem unlikely that this approach would be completely satisfactory for the synthesis of HGH unless a more flexible protection for cysteine sulphydryl is used to obviate sodium in liquid ammonia treatment of the large molecule.

In the synthesis of two fragments by solid phase[65], the conventional Merrifield method was used to prepare portions numbered 81–121 and 122–153 of the sequence deduced by Li[67]. The former straddles the portion of the molecule subject to reappraisal but there is currently no dispute concerning the presence of the latter sequence in HGH. Protecting groups and methods of coupling were virtually the same as described above in Li's synthesis. Butyloxycarbonyl was removed from the α-amino group of the growing chain with 1 mol l^{-1} HCl in acetic acid except when glutamine was *N*-terminal and in this event trifluoroacetic acid was used. Peptides were split from the resin using liquid hydrogen fluoride and anisole. Products were purified by countercurrent distribution following removal of dinitrophenyl protection of histidine with mercaptoethanol.

The reports on the synthesis of several HGH fragments (up to nine amino acids in length) by step-wise procedures in solution by Li and co-workers[61–64] are of considerable interest since record is made of difficulty encountered and circumvented at various points. Information of this nature is not always included in communications describing successful syntheses but is of the greatest value to workers in the field. During this work α-protection was by butyloxycarbonyl, side-chain hydroxyl, sulphydryl and carboxyl protection was by benzyl when used and side-chain amino and guanidine by tosyl groups. Histidine was generally unprotected. Couplings were effected by *p*-nitrophenyl or hydroxysuccinimide active ester, dicyclohexylcarbodi-imide, Woodward's reagent, mixed anhydride or azide. Countercurrent distribution was used frequently for purification. Derivatives prepared spanned positions 1–67 and 166–188 (numbering as in Ref. 67) and most would serve for a synthesis of the revised sequence[66]. Difficulties reported include the following:

(a) Boc-Ile + Pro-OBzl $\xrightarrow{\text{DCC}}$ diketopiperazine

while Z-Ile-ONp + Pro-OBu^t ⟶ Z-Ile-Pro-OBu^t;

(b)
```
Boc-Asp-OSu + Thr-Tyr-Glu-Glu
     |            |   |   |   |
    OBzl         Bzl Bzl OBzl OBzl
```

went satisfactorily in dimethylformamide in the presence of *N*-methyl-morpholine but the product when 50% dioxan was solvent and sodium bicarbonate the base showed evidence of $\alpha \rightarrow \beta$ acyl rearrangement;

(c) treatment of

```
Boc-Phe-Glu—Glu-Ala-Tyr-Ile-Pro
        |    |       |
      OBzl OBzl     Bzl
```

with hydrogen bromide in trifluoroacetic acid gave some evidence of glutarnimide formation, while use of trifluoroacetic acid followed by catalytic hydrogenation gave no by-product;

(d) treatment of

```
  Boc-His—Arg-Leu-OMe
     _ |   |
  AcŌ H⁺  Tos
```

with hydrazine hydrate gave 50% yield of a detosylated product. When histidine was not protonated this reaction did not occur.

7.7 INSULIN AND PROINSULIN

The structure of proinsulin was shown by Chance and collaborators[70] to consist of a single peptide chain with disulphide cross-links: it is the subject of a recent review[71]. During insulin biosynthesis the centre portion of the proinsulin chain (the 'connecting peptide') C is cut out leaving the *C*- and *N*-terminal portions. the familiar insulin A and B chains linked by disulphide bridges as shown in (99).

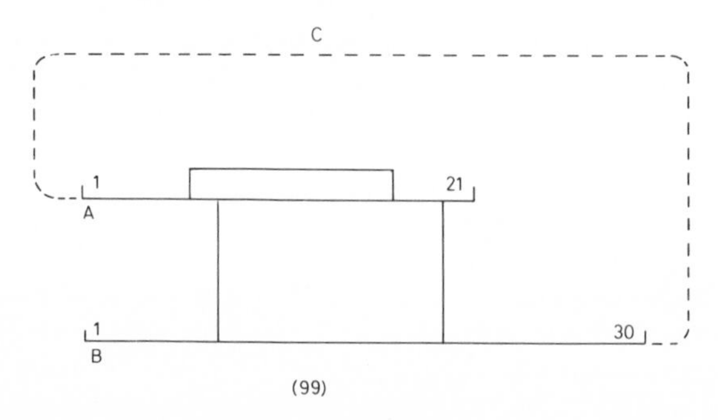

(99)

For convenience, the investigations on the synthesis of the insulin molecule and its analogues, during the period under review, have been subdivided into those involving total synthesis of each chain and those concerned with functional group modification (semisynthesis). As with compounds discussed in previous sections, the driving force instigating studies leading to the preparation of analogues has been the urge to define those portions of the molecule essential for its biological action. Early work on the synthesis of insulin has been reviewed by Schröder and Lübke[72].

7.7.1 Total synthesis

7.7.1.1 A chain

In the single report that has appeared on the synthesis of the insulin A chain several interesting points were discussed[73]. The three sequences given in Table 7.4 were prepared, viz. sheep insulin A chain (100) and two analogues, by the standard Merrifield solid-phase procedure[19]. Side chains were protected by benzyl groups and α-protection was t-butyloxycarbonyl. Cysteine SH groups were also protected by benzyl but synthesis using *S*-benzylcysteine was compared to synthesis using *S*-alkylmercaptocysteine, where alkyl was methyl, ethyl, propyl, isopropyl, t-butyl or benzyl. This protection, in the form of a mixed disulphide was subsequently removed by sulphitolysis without the need for reduction with sodium in liquid ammonia (required to split benzyl thioether links), which has been shown to suffer from side reactions. The A chain *S*-sulphonates were purified by chromatography on DEAE-Sephadex A-25 using a hydrochloric acid gradient in 8 M urea. Combination with B chain sulphonate in the presence of mercaptoethanol in the usual way gave the synthetic insulin. A comparison of the biological activity of the insulin obtained showed that a higher relative activity was obtained following the use of alkylmercapto in place of benzyl for *S*-protection. The issue was however somewhat clouded since failure sequences were present in the products, as evidenced by the biological activity of insulin prepared from synthetic (100) being less than that obtained when natural (100) and natural B chain were recombined.

7.7.1.2 B chain

By far the greatest number of new sequences related to insulin which have been prepared recently are analogous to the B chain. Table 7.5 lists these in order of decreasing size. With the exception of (103), (115), (133), (134), (137) and (138) all of the sequences shown have been prepared by Weitzel and collaborators[74–76] for the purpose of ascertaining the effects of amino acid replacement, or shortening of the B chain, on the biological activity of the hybrid insulin produced from these by combination with natural A chain. The method of synthesis adopted was the basic solid-phase procedure[19] using butyloxycarbonylamino acid *p*-nitrophenyl esters. The side chains of serine, tyrosine, cysteine, histidine and glutamic acid were protected by benzyl groups, of lysine by benzyloxycarbonyl and of arginine by tosyl. Peptide was split from the resin by the action of hydrogen bromide in trifluoroacetic acid and residual side groups were then removed with sodium in liquid ammonia. Following conversion of mercapto groups to the *S*-sulphonate, purification was achieved by ion exchange chromatography on Dowex 50WX-2 using citrate or acetate pH gradients in 8 M urea. After lyophilisation, the products were further purified by gel filtration using Sephadex G-25. Combination with a fourfold excess of natural A chain was by the procedure indicated in Section 7.7.1.1 above.

Full details of the synthesis of human B chain (115) and an improved method

Table 7.4 Insulin A chain

	1	2	3	4	5		7	8	9	10	11	12	13	14	15	16	17	18	19	20	21
(100)[73]	Gly	Ile	Val	Glu	Gln	Cys	Cys	Ala	Gly	Val	Cys	Ser	Leu	Tyr	Gln	Leu	Glu	Asn	Tyr	Cys	Asn
(101)[73]												Ala									
(102)[73]					Glu							Ala						Ala			Ala

Table 7.5 Insulin B chain analogues

	Phe	Val	Asn	Gln	His	Leu	Cys	Gly	Ser	His	Leu	Val	Glu	Ala	Leu	Tyr	Leu	Val	Cys	Gly	Glu	Arg	Gly	Phe	Phe	Tyr	Thr	Pro	Lys	Ala
(103)[77]	1	2	3	4	5	6	7	8	9	10	11	12	13	14	15	16	17	18	19	20	21	22	23	24	25	26	27	28	29	30
(104)[74]										Ala																		Ala		
(105)[74]					Ala																							Ala		
(106)[74]					Ala					Ala																	Ala	Ala		
(107)[74]					Ala				Ala																		Ala	Ala		
(108)[74]					Ala				Ala	Ala																	Ala	Ala		
(109)[75]																													Ala	
(110)[75]																												Ala		
(111)[75]																											Ala			
(112)[75]																											Ala	Ala		
(113)[76]									Ala																		Ala	Ala		
(114)[76]									Ala													Ala					Ala	Ala		
(115)[78]																														Thr
(116)[74]					Ala				Ala	Ala																	Ala	Ala		
(117)[75]									Ala																		Ala			
(118)[75]				Ala					Ala																		Ala	Ala		
(119)[74]									Ala	Ala																	Ala			
(120)[74]					Ala				Ala	Ala																	Ala			
(121)[76]									Ala																		Ala			
(122)[76]									Ala													Orn					Ala			
(123)[76]									Ala													Ala					Ala			
(124)[76]									Ala													His					Ala			
(125)[75]									Ala																		Ala	Ala		
(126)[75]				Ala					Ala																		Ala			
(127)[74]				Ala					Ala	Ala																	Ala			
(128)[74]				Ala	Ala				Ala																		Ala			
(129)[74]				Ala	Ala				Ala	Ala																	Ala			
(130)[75]				Ala					Ala																	Ala				
(131)[75]									Ala																		Ala			
(132)[74]					Ala				Ala	Ala																	Ala			
(133)[79]																														
(134)[80]																														Thr
(135)[76]						Ala			Ala	Ala													Ala							
(136)[76]						Ala			Ala	Ala												Ala								
(137)[82]																														
(138)[83]																														

for the synthesis of sheep (identical with bovine and porcine) insulin have now been reported in a series of papers by Katsoyannis and collaborators[77–81]. This work utilised synthesis by the condensation of fragments which had themselves been prepared by the step-wise addition of amino acid units. Benzyloxycarbonyl was used for α-protection and *p*-nitrophenyl active esters mediated coupling at each stage in the construction of the fragments of (133) and, with only two exceptions, in the synthesis of (134). Fragments comprising sequences 10–14, 15–20 and 21–30 were condensed from the *C*-terminus using the azide method to minimise the risk of racemisation. Peptides (133) and (134) were prepared with cysteine and histidine still protected by benzyl, and arginine and lysine carrying tosyl groups. The last stage in the preparation of these sequences involved saponification of *C*-terminal methyl ester groups. Because of the insolubility of these molecules, it was necessary to use hexamethylphosphoramide alone, or mixed with dimethylformamide as solvent, in the final stage.

In order to construct the full B chains from (133) and (134) the *N*-terminal nonapeptide intermediate (139) (R = H or benzyl) was required. This was synthesised[81] by step-wise condensations involving similar protecting groups and coupling methods to give two fragments which were themselves linked by azide[55]. An alternative preparation used a dicyclohexylcarbodi-imide coupling at this point but the authors assert that significant racemisation did not occur although the possibility was considered. The authors' claim that (139) (R = Bzl), after decarbobenzyloxylation, was completely digested by aminopeptidase-M is not substantiated by the data they have presented. Conversion of (139) to the azide and coupling with partially protected (133) and (134) enabled the respective total B chains (103) and (115) to be obtained.

Protecting groups were removed with the aid of sodium in liquid ammonia. The authors described this stage in some detail in view of its importance regarding the integrity of the Thr-Pro linkage at position 27–28 which is especially susceptible to this reagent. Conditions under which fission at this point is virtually complete are outlined but conditions whereby the reaction may be completely avoided have also been defined. It was found necessary to conduct the reduction in the presence of a 50 molar excess of sodium amide and to avoid any local excess sodium concentration. The mercaptan thus obtained was converted to B-chain *S*-sulphonate by oxidative sulphitolysis using sodium sulphite and tetrathionate and the product was purified by ion exchange chromatography on carboxymethylcellulose using pH 4.0 acetate buffer in urea. Urea was removed by gel filtration on Sephadex G-15 in 5% acetic acid and the product was separated from solution as the picrate salt. The materials were fully characterised and appeared identical with peptide obtained from natural sources. Combination with natural A chains gave insulins of full activity and from sheep B chain the insulin was obtained in crystalline form.

Another approach in which the problem of sulphur protection has been solved without recourse to additional protecting groups is that used by Zahn and Schmidt[82, 83] and involves synthesis of the appropriate cystine peptides. Compounds (137) and (138), suitably protected, were prepared by applying this procedure and were then coupled in concentrated solution by the azide method[55] to give protected B chain (103) as the polymeric sulphide.

In the synthesis of (138) the 1–8 fragment was prepared by step-wise active ester (usually *p*-nitrophenyl) addition, the α-benzyloxycarbonyl protection being removed at each stage with hydrogen bromide in acetic acid. Histidine, which had its imidazole ring unprotected was added as the azide. The *C*-terminal octapeptide was prepared by the condensation of two fragments by azide at the point shown. The Honzl and Rudinger azide technique[55] served to condense the octapeptide fragments to give (138) with the side chains of serine, tyrosine and glutamic acid protected by t-butyl groups. Using similar protecting groups but utilising mixed anhydride and dicyclohexylcarbodiimide couplings for chain building Zahn and Schmidt[82] have synthesised (137) as the disulphide. Problems arising due to the low solubility of cystine-containing products were successfully overcome partly by the use of extensive protection of side chains in (137). After coupling (137) with (138), the high molecular weight polymeric disulphide was deprotected with hydrogen bromide in trifluoroacetic acid and converted by oxidative sulphitolysis to B chain *S*-sulphonate. Compound (140), the *N*-terminal octapeptide disulphide fragment used by Zahn and Schmidt in the preparation of (138), has also been prepared in a similar way by Titov and Ardemasova[84].

The use of trityl (for histidine) and benzhydryl (for cysteine) as protecting groups has been advocated by Bosshard[85] in order to improve the solubility of intermediates during the synthesis of (141).

(139)[81] Z-Phe-Val-Asn-Gln-His-Leu-Cys-Gly-Ser-$NHNH_2$
(His: R; Cys: Bzl)

(140)[83, 84] [Z-Phe-Val-Asn-Gln-His-Leu-Cys-Gly-OMe]$_2$
(Cys–Cys linked)

(141)[85] Boc-Phe-Val-Asn-Gln-His-Leu-Cys-Gly-$NHNH_2$
(His: Trt; Cys: Bzh)

(142)[86] Phe-Phe-Tyr-Thr-Pro-Lys-Ala

(143)[86] Z-Thr-Pro-Lys-Ala
(Lys: Tos)

Compounds (142) and (143) are sequences prepared by Vdovina and collaborators[86] by solid phase.

7.7.1.3 Connecting peptide C

The two syntheses of the connecting peptide which have been reported were based on the original sequence determination (144) by Chance and collaborators[70]. This has since been revised at one point.

Arg-Arg-Glu-Ala-Gln-Asn-Pro-Gln-Ala-Gly-Ala-Val-Glu-Leu-Gly-Gly-
1 2 3 4 5 6 7 8 9 10 11 12 13 14 15 16

Gly-Leu-Gly-Gly-Leu-Gln-Ala-Leu-Ala-Leu-Glu-Gly-Pro-Pro-Gly-
17 18 19 20 21 22 23 24 25 26 27 28 29 30 31

Lys-Arg
32 33

(144)

It now appears that the glutamine residue at position 5 is in fact glutamic acid in the natural material so that the two synthetic products represent *C*-peptide amidated at this point. The first synthesis, by Geiger and collaborators[87], has been described in a preliminary manner but contains several novel features. One is the use of the adamantyloxycarbonyl group for the protection of the guanidine side chains of two of the three arginine residues present. The synthesis of the nonapeptide fragment incorporating this protection has been described in detail more recently[88]. Other protecting groups used were t-butyl for glutamic acid side chains and butyloxycarbonyl for lysine. Fragments 1–9, 10–12, 13–19, 20–23, 24–27 and 28–33 were constructed by step-wise active ester couplings and were then condensed via *C*-terminal hydroxysuccinimide esters. The possibility of racemisation occurring during the carbodi-imide mediated esterification of these peptides does not appear to have been excluded by the data presented.

The second synthesis of (144) by Yanaihara and collaborators employed a more orthodox approach[89]. Fragments, *C*-terminating at glycine residues 10, 19 and 28, were constructed by step-wise active ester (trichlorophenyl or hydroxysuccinimidyl) couplings using benzyloxycarbonyl for α-protection. Side chains of glutamic acid were t-butylated and lysine carried δ-formyl. The optical integrity of each fragment was checked after deprotection by aminopeptidase-M digestion. Fragments, which were obtained as t-butyloxycarbonylhydrazides, were then treated with anhydrous trifluoroacetic acid and linked successively by azide[55] couplings. After hydrogenolysis the tritriacontapeptide, which still carried a formyl on 32-lysine was purified by gel filtration on Sephadex G-25 and by ion exchange chromatography on carboxymethylcellulose.

7.7.2 Semisynthesis

7.7.2.1 A chain

Derivatives of insulin have been prepared by a semisynthetic approach using parts of the natural molecule as starting material. The insulin A chain (100) which carries only a single amino group (at its *N*-terminus) is therefore a suitable intermediate for the preparation of derivatives extended at this point.

Zahn and collaborators[90] have prepared lysyl, histidyl and arginyl derivatives of bovine A chain in the *S*-sulphonate (145)–(147) or disulphide form by allowing the appropriate bisbutyloxycarbonylamino acid active ester to react with whole A chain (145) and (146) on butyloxycarbonyl-arginylglycine *p*-nitrophenyl ester with des-1-glycine-A chain (147). Reaction times were 4–5 days and the products were purified by chromatography on DEAE-Sephadex (in the case of *S*-sulphonates) or on Dowex 50 (disulphide) using gradients prepared in 8 M urea.

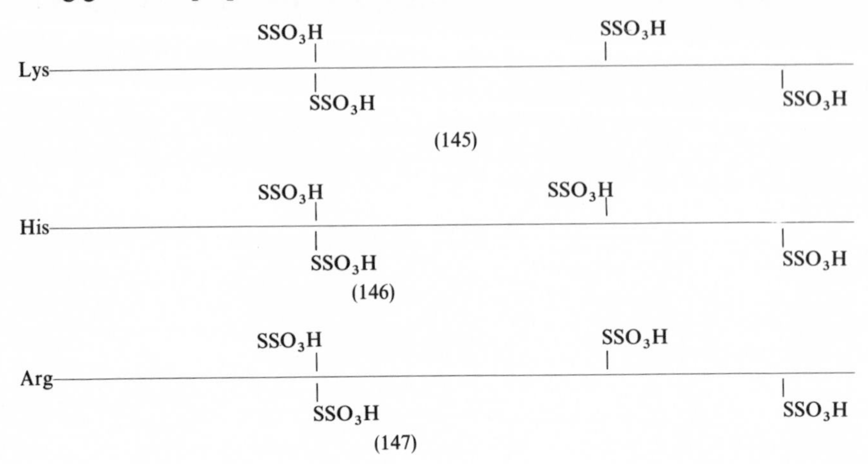

7.7.2.2 *Insulin*

A method of selectively protecting insulin has been described by Geiger and collaborators[91]. These authors have shown that insulin reacts with t-butoxycarbonyl azide in aqueous dimethylformamide containing sodium bicarbonate to give a product which is predominately a bisbutyloxycarbonyl derivative. End group studies showed that the free amino groups of glycine^{A1}

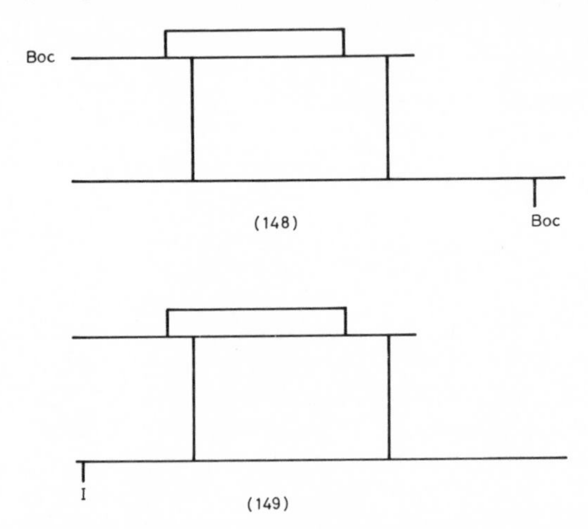

and lysine^{B29} had been acylated (148). Reaction of this derivative with phenylisothiocyanate and treatment of the phenylthiocarbamyl (PTC) derivative with trifluoroacetic acid gave des-Phe^{B1}-insulin. In a second report[92] this material itself was shown to undergo a similar selective A1, B29 acylation.

After purification by preparative gel electrophoresis, bisbutyloxycarbonyl derivative was treated with butyloxycarbonyl-*p*-iodophenylalanine–2,4,5-trichlorophenyl ester. Removal of protecting groups from the product gave an insulin analogue with *p*-iodophenylalanine at the B1 position (149).

A second method whereby insulin may be protected selectively has been developed by Offord[93]. The technique is based upon the selective reaction of the α-amino group of B chain *N*-terminal phenylalanine with phenylisothiocyanate. The mono-PTC derivative was then purified by chromatography on DEAE-cellulose and the two remaining amino groups were

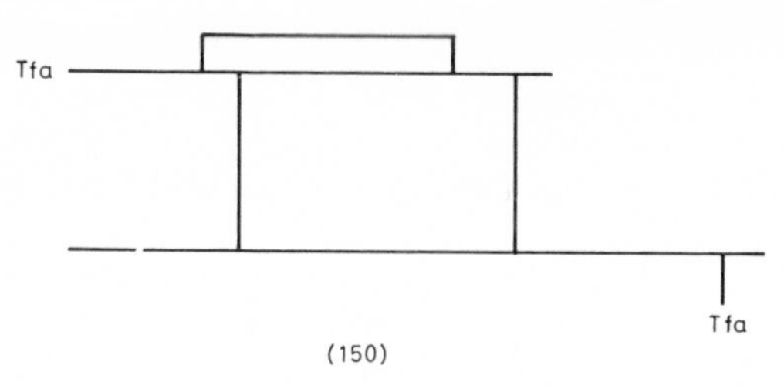

(150)

protected by conversion to the trifluoroacetyl derivatives. Treatment with trifluoroacetic acid then eliminated the PTC-amino acid residue giving bistrifluoroacetyl-des-Phe^{B1}-insulin (150).

7.8 MELANOTROPIN

Recent work on the two melanocyte stimulating hormones (MSH) of the intermediate lobe of the pituitary comprises the synthesis of α-MSH and [5-glutamine]-α-MSH (152) by solid phase and accounts of the synthesis of human-β-MSH and its 10-lysine analogue by classical methods.

7.8.1 α-Melanotropin (α-MSH)

The various syntheses of α-MSH (151) by classical methods, first achieved in 1958, have been reviewed by Schröder and Lübke[94]. The sequence of amino acids in this hormone coincides with the first 13 amino acids in the corticotropins and consequently, synthetic work directed towards MSH is of value in the synthesis of the larger molecule.

Of the recent work, the report[95] on the preparation of compound (152) described the use of Merrifield type resin, butyloxycarbonyl to protect α-amino groups, benzyl groups to protect the side chains of histidine and serine, and tosyl groups for arginine and lysine side chains. During coupling four equivalents of the protected amino acid and of dicyclohexylcarbodi-imide were employed and α-protection was removed using 3.6 mol l^{-1} HCl in dioxan containing β-mercaptoethanol to protect the tryptophan ring. Glutamine was added in the form of its *p*-nitrophenyl ester and after coupling was α-deprotected using trifluoracetic acid: dichloromethane (1:1). After completing the synthetic steps, fission of peptide from the resin by ammonolysis was very slow, presumably due to steric hindrance, and it was found advantageous to prepare the ester by fission with methanol:triethylamine: dimethylformamide and then use ammonolysis of the methyl ester to prepare the amide. After removal of side chain benzyl groups with sodium in liquid ammonia, the product was purified by carboxymethylcellulose chromatography using an ammonium acetate gradient. Subsequently, the same authors, Li and collaborators, reported the synthesis of α-MSH (151)[96] (Table 7.6) again by solid phase but differing in several aspects from the synthesis of (152). Because of the difficulty mentioned above at the stage of ammonolysis of the peptide from the resin a more labile linkage was used. Merrifield resin was esterified with *p*-hydroxyphenylacetic acid and the phenolic group of this acid used to bind butyloxycarbonylvaline to the support. An acetylation step was used at each coupling to mask unreacted amino groups. Side chain protection was as above and the γ-carboxyl of glutamic acid was masked as the t-butyl ester. Serine, tyrosine, methionine and γ-butyl glutamate were added as the biphenylylisopropyloxycarbonyl (Bpoc) derivatives and were α-deprotected using hydrogen chloride in chloroform. Glutamic acid γ-carboxyl was deprotected prior to ammonolysis of the peptide from the resin. It was found necessary to conduct the ammonolysis at −20 °C as *C*-terminal valine was racemised extensively if the reaction was carried out at room temperature.

Subsequent treatment with sodium in liquid ammonia and purification were as in the previous synthesis. Characterisation of the product was exhaustive and confirmed its identity with natural material chemically and biologically.

7.8.2 β-Melanotropin β-MSH

The synthesis of human β-MSH (153) was achieved by Yajima and collaborators[97] following their synthesis[98] of the β-MSH of the monkey. The monkey hormone comprises sequence 5–22 of the human hormone and a convenient common intermediate was the 8–22 pentadecapeptide protected only at 21-lysine as the ε-formyl derivative. Sequences 1–4 and 5–7 were each built step-wise using nitrophenyl active esters and coupled as 1–4 azide using the Honzl and Rudinger·technique[55]. The heptapeptide was converted to its pentachlorophenyl ester using dicyclohexylcarbodi-imide and on reaction with *C*-terminal pentadecapeptide for 3 days a product was obtained which was applied to a column of carboxymethylcellulose in 70% aqueous methanol. Addition of 2% acetic acid to this mixture eluted the required product. t-Butyloxycarbonyl and t-butyl ester protecting groups were removed by treatment with anhydrous trifluoroacetic acid containing anisole and methionine as t-butyl cation scavengers. Purification by ion-exchange on carboxymethylcellulose was followed by fission of the formyl group by hydrazine hydrate and repeat ion exchange purification to give the required human β-MSH.

A semisynthetic approach has been used by Burton and Lande[99] in their synthesis of [Lys^{10}]-β-human MSH (154) from porcine β-MSH. Like the monkey hormone, porcine β-MSH is an octadecapeptide covering positions

Table 7.6 Melanotropin

Compound	Sequence
(151)[96]	Ac-Ser-Tyr-Ser-Met-Glu-His-Phe-Arg-Trp-Gly-Lys-Pro-Val-NH_2
	1 2 3 4 5 6 7 8 9 10 11 12 13
(152)[95]	............Gln............
(152)[97]	Ala-Glu-Lys-Lys|Asp-Glu-Gly|Pro-Tyr-Arg-Met-Glu-His-Phe-Arg-Trp-Gly-Ser-Pro-Pro-Lys-Asp
	1 2 3 4 5 6 7 8 9 10 11 12 13 14 15 16 17 18 19 20 21 22
(154)[99]	............|............Lys............

5–22 of the human sequence but contains a single replacement of lysine for arginine at position 10.

```
   OBuᵗ Boc Boc
    |    |   |
Boc-Ala-Glu-Lys-Lys-OMe
```

(155)

Fully protected *N*-terminal tetrapeptide (155) was prepared by step-wise synthesis using benzyloxycarbonyl for α-protection and dicyclohexylcarbodiimide for coupling. The ester (155) was converted first to the hydrazide and then to the azide and this was allowed to react with unprotected octadecapeptide in cold pyridine for 4 hours. The authors' hope that, due to p*K* differences, such conditions would result in the selective acylation of the α-amino group leaving the ε-amino group of lysine residues unreacted was not realised. Chromatography on DEAE Sephadex using an ammonium acetate gradient led to the recovery in low yield of required material and of a branched chain by-product in proportions of approximately 2:1. Protecting groups were removed with aqueous formic acid to give the free peptide.

7.9 OXYTOCIN AND VASOPRESSIN

The synthesis of oxytocin (156) carried out by du Vigneaud and collaborators in 1954[100] marked the beginning of a period of development in peptide synthesis. At that time new reactions and techniques applicable to peptide synthesis had become available and were becoming exploited. Among the latter, column chromatography, particularly on cellulose ion-exchange media, and countercurrent distribution were beginning to be applied widely. Since that time there have been many syntheses of oxytocin and the structurally closely related vasopressin and hundreds of analogues have been prepared but nearly all have been purified by one or both of these techniques. Schröder and Lübke[137] give comprehensive literature coverage of the early work and more recently the survey of Pettit[138] affords a useful source of references covering the period until that of the present review.

New compounds and old compounds prepared by new methods are indicated in Tables 7.7 and 7.8. Analogues have been arranged in an arbitrary fashion with progressive amino acid replacement followed by functional group modification and finally ring changes (both in composition and size). Vasotacin analogues having ring structures analogous to oxytocin are included in Table 7.7. In Table 7.8 species differences account for the two natural vasopressins. [8-Arginine]-vasopressin (199) is the pressor hormone of most mammalian species, [8-lysine]-vasopressin (200) is the porcine hormone. The amino acid at position 8 is therefore always specified in Table 7.8. The overlap of the oxytocic and pressor actions of the two hormones, which is modulated in analogues, has been a source of interest to investigators and has provided the incentive for the synthesis of new related molecules.

The solid-phase method of peptide synthesis has been used extensively in

Table 7.7 Oxytocin analogues

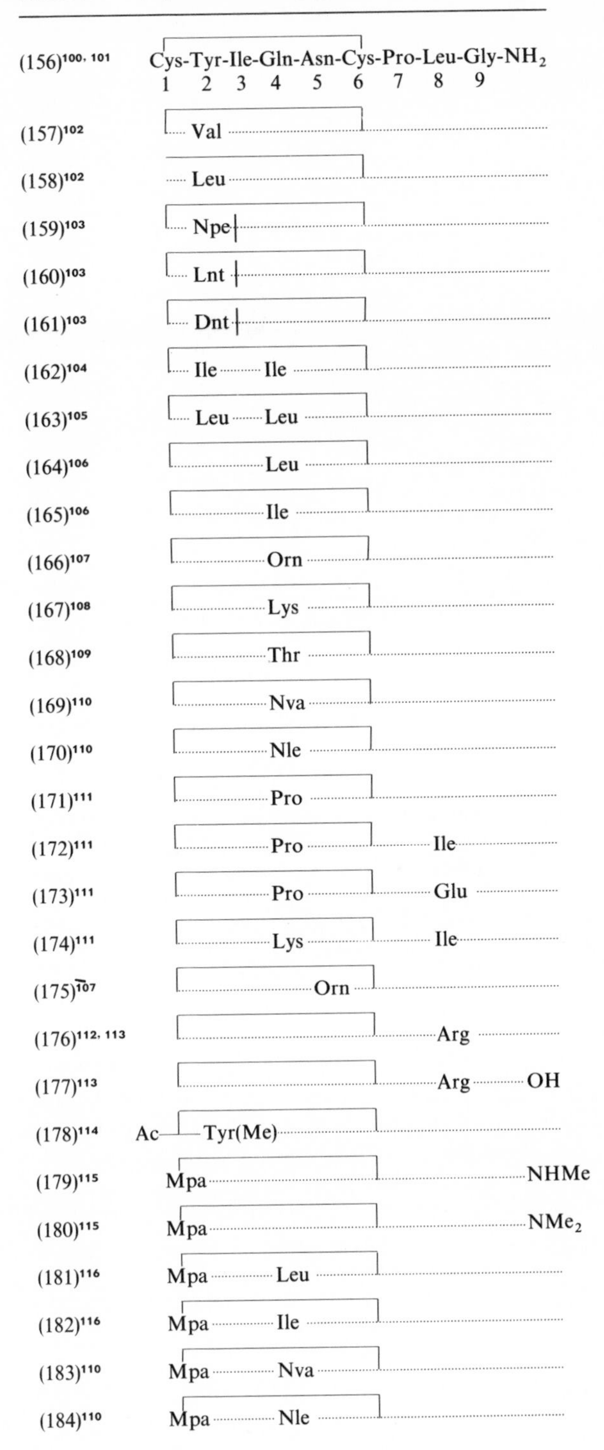

(156)100, 101
Cys-Tyr-Ile-Gln-Asn-Cys-Pro-Leu-Gly-NH2
1 2 3 4 5 6 7 8 9
(157)102 Val
(158)102 Leu
(159)103 Npe
(160)103 Lnt
(161)103 Dnt
(162)104 Ile Ile
(163)105 Leu Leu
(164)106 Leu
(165)106 Ile
(166)107 Orn
(167)108 Lys
(168)109 Thr
(169)110 Nva
(170)110 Nle
(171)111 Pro
(172)111 Pro Ile
(173)111 Pro Glu
(174)111 Lys Ile
(175)107 Orn
(176)112, 113 Arg
(177)113 Arg OH
(178)114 Ac Tyr(Me)
(179)115 Mpa NHMe
(180)115 Mpa NMe2
(181)116 Mpa Leu
(182)116 Mpa Ile
(183)110 Mpa Nva
(184)110 Mpa Nle

Table 7.7 Oxytocin analogues—*continued*

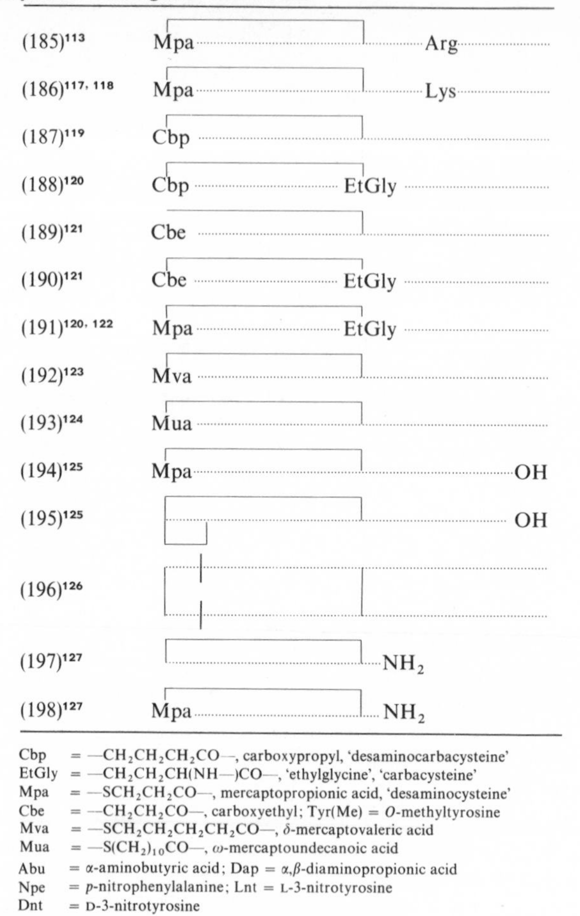

Cbp = $—CH_2CH_2CH_2CO—$, carboxypropyl, 'desaminocarbacysteine'
EtGly = $—CH_2CH_2CH(NH—)CO—$, 'ethylglycine', 'carbacysteine'
Mpa = $—SCH_2CH_2CO—$, mercaptopropionic acid, 'desaminocysteine'
Cbe = $—CH_2CH_2CO—$, carboxyethyl; Tyr(Me) = *O*-methyltyrosine
Mva = $—SCH_2CH_2CH_2CH_2CO—$, δ-mercaptovaleric acid
Mua = $—S(CH_2)_{10}CO—$, ω-mercaptoundecanoic acid
Abu = α-aminobutyric acid; Dap = α,β-diaminopropionic acid
Npe = *p*-nitrophenylalanine; Lnt = L-3-nitrotyrosine
Dnt = D-3-nitrotyrosine

the preparation of the analogues depicted in the Tables. Molecules of this size are sufficiently large for the process to show significant time saving over most classical procedures yet the molecule once cleaved from the resin is not beyond the molecular size range which can be purified effectively by methods currently available (countercurrent distribution, ion-exchange chromatography or gel filtration). Classical methods have however contributed the majority of the analogues given in Tables 7.7 and 7.8. Techniques have evolved from earlier work in the various interested groups and there appears to have been no radical new approach during the period under consideration, although some interesting developments of standard methods have appeared.

The du Vigneaud group has continued to be prolific in the synthesis of new analogues and has made the major contribution to the Tables. The work of this and other groups on analogues prepared by the classical step-wise technique derives to some extent from the synthesis of oxytocin itself using protected amino acid *p*-nitrophenyl esters by Bodanszky and du Vigneaud[139].

In recent work by various authors, benzyl (occasionally *p*-methoxybenzyl) groups have been used for *S*-protection and tyrosine hydroxyl was not normally protected. Benzyloxycarbonyl was used for α-amino group protection and this was removed by the action of hydrogen bromide in acetic acid. *S*-Benzyl protection was removed with sodium in liquid ammonia

Table 7.8 Vasopressin analogues

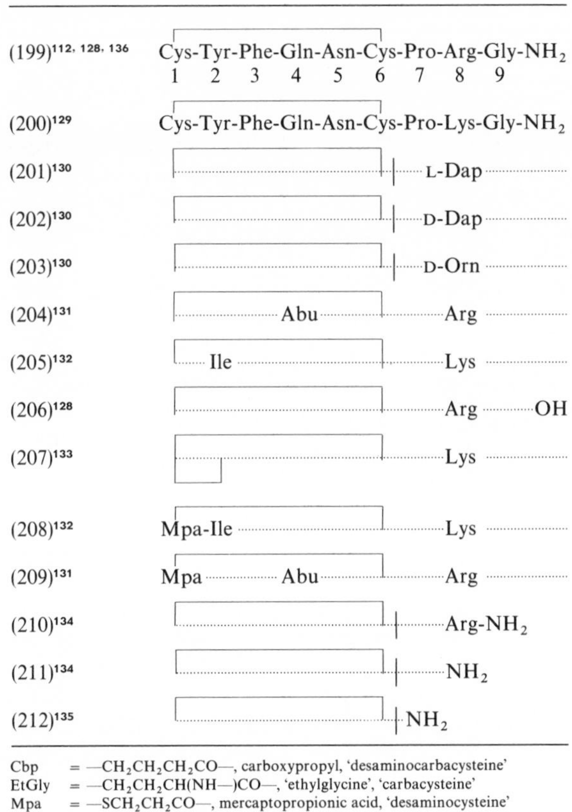

(199)[112, 128, 136]	Cys-Tyr-Phe-Gln-Asn-Cys-Pro-Arg-Gly-NH_2 (positions 1 2 3 4 5 6 7 8 9)
(200)[129]	Cys-Tyr-Phe-Gln-Asn-Cys-Pro-Lys-Gly-NH_2
(201)[130]	… L-Dap …
(202)[130]	… D-Dap …
(203)[130]	… D-Orn …
(204)[131]	… Abu … Arg …
(205)[132]	… Ile … Lys …
(206)[128]	… Arg … OH
(207)[133]	… Lys …
(208)[132]	Mpa-Ile … Lys …
(209)[131]	Mpa … Abu … Arg …
(210)[134]	… Arg-NH_2
(211)[134]	… NH_2
(212)[135]	… NH_2

Cbp = —$CH_2CH_2CH_2CO$—, carboxypropyl, 'desaminocarbacysteine'
EtGly = —$CH_2CH_2CH(NH$—$)CO$—, 'ethylglycine', 'carbacysteine'
Mpa = —SCH_2CH_2CO—, mercaptopropionic acid, 'desaminocysteine'
Cbe = —CH_2CH_2CO—, carboxyethyl
Tyr(Me) = *O*-methyltyrosine
Mva = —$SCH_2CH_2CH_2CH_2CO$—, δ-mercaptovaleric acid
Mua = —$S(CH_2)_{10}CO$—, ω-mercaptoundecanoic acid
Abu = α-aminobutyric acid
Dap = α,β-diaminopropionic acid
Npe = *p*-nitrophenylalanine
Lnt = L-3-nitrotyrosine
Dnt = D-3-nitrotyrosine

and ring closure effected by ferricyanide oxidation. After desalting, purification has been carried out (particularly by the du Vigneaud group) by partition chromatography using a butanol–water–pyridine–acetic acid–benzene system on Sephadex G-25 followed by gel filtration on the same medium in 0.2 mol l^{-1} acetic acid.

Compounds prepared in this way include (157), (158), (162)–(166), (169), (170), (175), (183), (184), (186), (204), (205), (208), (209). Other active esters which have been used include those of hydroxyphthalimide for compounds

(159)–(161) and (167), trichlorophenol (188) and (191) and hydroxysuccinimide (187). Countercurrent purification was used to purify the final free peptides in the case of (159)–(161), (164), (167), (183), (184), (186)–(191) and (201)–(203). As an alternative to benzyloxycarbonyl for α-protection the *o*-nitrophenylsulphenyl protecting group was used in the preparation of compounds (187)–(191), (201)–(203) and was removed by treatment with hydrogen chloride in ether.

Relatively few of the analogues prepared have utilised fragment condensation involving the azide method. Advantage was taken of the availability of the vasopressin 1-6-hexapeptide hydrazide, obtained some years previously[140], to prepare analogues containing variations in the *C*-terminal tripeptide. Compounds (201)–(203), (210) and (211) were prepared from this intermediate. An azide coupling was also used in the preparation of compounds (159)–(161) which have a common *C*-terminal heptapeptide. Ring closure with these compounds was achieved by oxidation with CO_2-free air following removal of *S*-benzyl protecting groups with liquid hydrogen fluoride.

'Carba' analogues (187)–(190) in which the normal disulphide ring is replaced by one containing only one —S— group or none at all deserve comment. The appropriate α-amino dicarboxylic acid was incorporated, in the form of its 5-chloroquinolin-8-yl active ester at position 6 of the peptide chain with the ω-carboxyl group esterified (usually as the methyl ester). Ring closure was then effected by removing this ω-protection and activating the carboxyl to react with the amino group of tyrosine. Nitrophenyl active ester was found to be more effective than mixed carbonic–carboxylic anhydride or than acid azide in achieving this cyclisation.

A distinction between the parallel and antiparallel dimers of oxytocin is now possible as a result of the unequivocal synthesis of the parallel dimer (196). In this preparation bisbutyloxycarbonylcystine was coupled as a mixed anhydride (using isobutyl chloroformate) with the disulphide of oxytocin 2-9-octapeptide amide. The product was purified by partition chromatography on Sephadex G-25 to give a product (albeit in low yield) which corresponded with the α-dimer in that it ran faster than the β-dimer on Sephadex G-25 in 0.2 mol l^{-1} acetic acid. Reduction and reoxidation gave oxytocin.

Application of the solid-phase method of peptide synthesis to the preparation of oxytocin and vasopressin analogues can yield a bonus in comparison with classical procedures, since the -oic acids are as readily available as the amides. Classical syntheses have usually introduced the *C*-terminal amide group as carboxyl protection at the tripeptide stage or earlier and have carried it through the synthesis. A nonapeptide bound to resin at its *C*-terminal may be cleaved by ammonolysis to the amide or by liquid hydrogen fluoride (177) and (206) or hydrogen bromide in trifluoroacetic acid (194) to give the -oic acid. Further, other amines may replace ammonia for cleavage to give amide substituted oxytocins (179) and (180). During solid-phase synthesis tyrosine hydroxyl has frequently but not always been protected, often as the *O*-benzyl derivative; butyloxycarbonyl was used for α-protection and other protecting groups were unaltered (*S*-benzyl and guanidine-tosyl). Coupling has usually been mediated by dicyclohexylcarbodi-imide except in the case of asparagine and glutamine residues when the nitrophenyl esters are preferred to prevent loss of side chain amide function. After cleavage from the resin, deprotection, oxidation and purification usually have followed the same procedures used for the products of classical synthesis.

In two recent cases (199) and (200) Meienhofer and collaborators have been able to prepare protected vasopressins in crystalline form following ammonolysis from the resin support. Subsequent stages of the synthesis involving purification were thereby much easier and vasopressin of a higher pressor activity than hitherto was obtained. [4-Threonine]-oxytocin (168) was prepared in a similar manner, the protected nonapeptide being isolated in crystalline form. Ion exchange (IRC-50) served to effect purification. Manning and collaborators have prepared (171)–(174) similarly. [8-Lysine]-vasopressin (200) has also been prepared by Meienhofer and collaborators as an example of an alternative method of linking peptide to resin. In this, the ε-amino group α-butoxycarbonyllysylglycine ethyl ester was allowed to react with chloroformyloxymethylresin whereby the lysine side chain was bonded to the resin support, virtually as a benzyloxycarbonyl derivative. The α-protecting group was removed using hydrogen chloride in dioxan and the remainder of the sequence built up in the usual way. Fission from the resin with trifluoroacetic acid and conversion of the tyrosine hydroxyl and lysine ε-amino groups to the benzyloxycarbonyl derivatives gave a protected nonapeptide which was induced to crystallise. Deprotection oxidation and purification solely by ion-exchange (IRC-50) afforded pure [Lys^8]-vasopressin[129].

Lübke[112] has described difficulties in applying the solid-phase technique, which had been used effectively for the preparation of [Arg^8]-vasopressin, to the corresponding [Arg^8]-vasotocin in which isoleucine replaces phenylalanine at position 3. Yield was approximately halved and this was attributed to incomplete nitrophenyl active ester coupling of isoleucine. Model experiments using a variety of solvents, reactant ratios and coupling methods confirmed this.

Des-amino analogues (179)–(186), (191)–(194), (208), (209) of oxytocin and vasopressin have been prepared on a number of occasions by classical synthesis and by solid phase, the appropriate benzylmercapto-alkanoic acid being used in the final coupling stage. Other derivatives in which the *N*-terminal amino group is modified include the derivatives formed by interaction with acetone (195) and (207). That the α-amino groups of the two *N*-terminal amino acids are involved in the formation of these derivatives was shown by coupling the product of reacting acetone with *S*-benzyl-L-cysteinyl-L-tyrosine with *C*-terminal heptapeptide. The usual working up procedure gave a product identical with that obtained from [Lys^8]-vasopressin and acetone (207). Desamino-oxytocinoic acid (194) and acetoneoxytocinoic acid (195) have each been coupled with oxytocin with the aid of Woodward's reagent K to give octadecapeptides. Purification was by gel filtration on Sephadex G-25.

7.10 RIBONUCLEASE AND OTHER NUCLEASE ENZYMES

7.10.1 Pancreatic ribonuclease

The observation by Richards[141] that pancreatic ribonuclease A could be split at a single peptide bond by subtilisin into an *N*-terminal eicosapeptide

(*S*-peptide) and a *C*-terminal tetrahectapeptide (*S*-protein) has formed the basis for several investigations. These two fragments may be physically separated and subsequently recombined to give enzymically active ribonuclease S′. A large number of analogues of *S*-peptide have been prepared, mainly by Hofmann and collaborators[142] but also by Scoffone and collaborators[143], in the search for constituents of the active site of the enzyme and in the investigation of conformational aspects of the active site.

7.10.1.1 S-peptide

The majority of the reports on the synthesis of *S*-peptide appeared prior to the period covered by this review but Hofmann and collaborators[142] have recently described the synthesis of a series of eight tetradecapeptides (213) in which 12-histidine was carboxymethylated or replaced by pyrazolylalanine (PyAla) and 10-arginine was replaced by ornithine or δ-formylornithine (ForOrn).

Lys-Glu-Thr-Ala-Ala-Ala-Lys-Phe-Glu-X-Gln-Y-Met-Asp
1 5 10 14

(213)

Y = 1- or 3-HO_2CCH_2-His; 1,3-bis(HO_2CCH_2)-His or PyAla. X = Arg, Orn or ForOrn.

During the synthesis of (213) *C*-terminal aspartic acid was unprotected and carboxymethyl groups were introduced by reacting butyloxycarbonylhistidylmethionylaspartic acid sulphoxide with iodoacetic acid and separating the products (after deblocking) by ion exchange chromatography on AG 1-X2 resin using an acetic acid gradient. Methionine was used as the sulphoxide to prevent *S*-alkylation. This protection was retained throughout and was finally removed by treatment with thioglycollic acid. Synthetic pathways used previously[144] were employed for the remainder of the preparation. Couplings at 7-lysine and 10-glutamine were effected by the azide method. Final purification again used ion exchange on AG 1-X2 resin.

Kerling and collaborators have described a series of tridecapeptide analogues, corresponding to the *N*-terminus of *S*-peptide, which have been prepared by azide condensation of fragments themselves constructed by the Merrifield solid-phase procedure.

(214)[145] Lys-Glu-Thr-Ala-Ala-Ala-Lys-Phe-Glu-Arg-Gln-His-Leu
1 8 9 13

(215)[146]..Harg............................

(216)[147]..Cit............................

The *C*-terminal pentapeptide was built on a resin support using carbodiimide or active ester couplings (for glutamic acid derivatives), reaction times being overnight or several days. Histidine and arginine were protected by benzyl and nitro respectively. After fission from the resin and hydrogenolysis the free pentapeptide was coupled with 1-8-azide derived from a hydrazide prepared again by solid phase and split from the support with hydrazine. The pentapeptide was also prepared by classical step-wise active ester couplings and shown to be identical with the product of solid-phase synthesis by electrophoresis, and paper and thin layer chromatography. The tridecapeptide (214) was purified by gel filtration on Sephadex G-15.

Analogue (215) was prepared in a similar manner to (214) with homoarginine (Harg) replacing arginine. Solid-phase synthesis used mainly active ester (pentachlorophenyl) couplings. A slightly different approach was used in the synthesis of (216). In this case the *C*-terminal citrulline (Cit) containing peptide was left bound to the resin support after selective removal of *o*-nitrophenyl-sulphenyl from the *N*-terminal glutamic acid. Reaction with a 2.5-fold excess of 1-8-azide for 6 days was followed by fission from the resin. The product contained only a small proportion of unreacted pentapeptide and was purified by gel filtration as above.

7.10.1.2 S-protein

In early 1969, Hirschmann and collaborators[148] described the synthesis of material which, when combined with ribonuclease *S*-peptide and incubated with mercaptoethanol, showed ribonuclease S activity. The amount of synthetic *S*-protein was 60 μg and represented the final product of a synthesis of the tetrahectapeptide. As only a preliminary account of this work has appeared, a detailed analysis would be out of place but main points of interest will be mentioned. The tactics for constructing this molecule were designed to make maximum use of *N*-carboxyanhydroamino acids (NCA amino acids) for the step-wise construction of fragments. The NCA coupling method had been developed by the Merck group and possesses significant advantages in time and in the need for protection of side chains during the preparation of short peptide chains. Chains once prepared were then purified prior to condensation using the azide[55] procedure. In order to effect purification of fragments it was necessary to take the usually unprotected intermediate which had been prepared by NCA coupling in an aqueous medium, add an *N*-protected amino acid, usually as the succinimide ester in anhydrous medium and then esterify the product in order that a hydrazide could subsequently be made. The merit of conducting hydrazide formation in dimethylformamide using a large excess of hydrazine for a short time was stressed particularly in the presence of arginine residues. Since the NCAs of asparagine, serine and threonine lead to significant side-products formation during coupling these residues were also added as succinimide esters. Arginine and histidine NCAs react satisfactorily without side chain protection and lysine was used as the ε-benzyloxycarbonyl derivative. In order to avoid problems of treating a large molecule with sodium in liquid ammonia, cysteine, of which 8 are present, was protected as the acetamidomethyl derivative. This protection is stable to treatment with trifluoroacetic acid or liquid hydrogen fluoride but may be removed readily with mercuric salts. In all 19 fragments were prepared and characterised. It is interesting to note that dry column chromatography was of use in the purification of intermediates. The specific enzymic activity of the final product was about 30% but in view of the small amount available a just assessment of the product is impossible.

7.10.1.3 Synthesis of ribonuclease A

Simultaneously with the report on the synthesis of *S*-protein[148] Gutte and Merrifield gave a preliminary account[149] of the synthesis of the whole molecule by the solid-phase procedure. Here again, a full description is not available for critical assessment but the product was obtained in sufficient quantity (85 mg) for a full enzymic characterisation. The product compared favourably with the natural enzyme although the specific activity was low (13%) presumably due to the presence of molecules of closely related structure and of unknown (more active or more probably less active) enzymic character. On peptide mapping, after tryptic digestion of the performic oxidised synthetic product, a good correlation with the natural material was seen. On analysis, amino acid ratios compared more favourably with those on natural material which had been subjected to the deprotection conditions (although differences from untreated natural material were not large). This can be interpreted as a measure of the inadequacy of the purification subsequent to deprotection combined with the degradative character of the deprotection reactions. It would be unfair to defame the purity of the product on this basis, especially since the natural molecule does survive this treatment enzymically.

The synthesis itself used α-butyloxycarbonyl protection removed with trifluoroacetic acid in dichloromethane (1:1). Side chains of all trifunctional amino acids were protected by benzyl or benzyl-derived groups with the exception of arginine (nitro) and methionine which was used as the sulphoxide. Coupling was by dicyclohexylcarbodi-imide in threefold excess for 5 h. Cleavage from the resin and removal of protecting groups was effected with liquid hydrogen fluoride containing anisole and then trifluoroacetic acid for 90 min at 0–15 °C. Conversion to the *S*-sulphonate was followed by chromatography on Dowex 1-X2 followed by gel filtration on Sephadex G-50. Reduction with mercaptoethanol followed by gel filtration and air oxidation caused the four disulphide bridges of the enzyme to form in the requisite pattern, and the product was again chromatographed on IRC-50 to give the material required. Aminopeptidase-M digestion confirmed the optical integrity but about 20% methionine sulphone was observed. Since four methionine residues are present in the molecule concentration of sulphone at one critical site could hazard up to 80% molecules present.

Merrifield and collaborators have also described the synthesis, by solid phase, of the *C*-terminal tetradecapeptide (217) of pancreatic ribonuclease[150]. This material was required for activation studies using *C*-terminal truncated ribonuclease. The method of synthesis was as described above for ribonuclease. The product was purified by free-flow electrophoresis and by gel filtration on Sephadex G-25.

Glu-Gly-Asn-Pro-Tyr-Val-Pro-Val-His-Phe-Asp-Ala-Ser-Val

(217)

7.10.2 Ribonuclease T

Two groups of workers have been engaged upon the synthesis of ribonuclease T_1 [151]. Izumiya and collaborators[152] have described four fragments covering positions 1–30 of the sequence but the major contribution to the synthesis of this tetrahectapeptide has come from the laboratory of Hofmann.

Preliminary communications[153] have disclosed that fragments covering the entire sequence of ribonuclease T_1 have been synthesised and detailed account has been given[154] of the preparation and condensation of three of these to give a protected hexatriacontapeptide covering positions 12–47. The strategy adopted was to divide the sequence into seven fragments (four of which *C*-terminate in glycine residues) and prepare these by a step-wise approach. Each fragment was prepared as the butyloxycarbonylhydrazide; butyl residues were used to protect carbonyl side chains and lysine was blocked as the ε-formyl derivative. Cysteine, which occurs in the first and last fragments, was protected as the ethylcarbamyl derivative[155] which may be removed by mercuric acetate. Benzyloxycarbonyl was used for α-protection and coupling was achieved by active ester (trichlorophenyl or hydroxysuccinimidyl) or by azide couplings. Once prepared in this way the three fragments, covering positions 12–23 (218), 24–34 (219) and 35–47 (220), were deprotected at the *C*-terminus and condensed by the Honzl and Rudinger[55] azide procedure.

```
               OBu
                |
Z-Ser-Ser-Ser-Asp-Val-Ser-Thr-Ala-Gln-Ala-Ala-Gly-R
```

(218)

```
                          OBu OBu     OBu
                           |   |       |
Z-Tyr-Gln-Leu-His-Glu-Asp-Gly-Glu-Thr-Val-Gly-R
```

(219)

```
                                For                   OBu
                                 |                     |
Z-Ser-Asn-Ser-Tyr-Pro-His-Lys-Tyr-Asn-Asn-Tyr-Glu-Gly-R
```

(220)

$R = —NHNHCO_2Bu^t$

This work was characterised by the care taken in examining the various oligopeptides for purity. With increasing size solubility problems became enhanced which is perhaps not unexpected. Of considerable value in the purification of large fragments was the technique of ion-exchange chromatography on AG 1-X2 resin using butanol–methanol–aqueous acetic acid mixtures, the normality of acetic acid being varied to provide resolution.

7.10.3 Staphylococcal nuclease

Digestion of staphylococcal nuclease with trypsin in the presence of calcium ion and 3′, 5′-deoxythymidine dephosphate causes fission into only three fragments. The two larger, P_2 and P_3 may subsequently be recombined under appropriate conditions to give an enzymically active structure known as nuclease-T. P_2, which comprises the 6–48 sequence of the enzyme, may be further split by removal of the *C*-terminal lysine without loss of nuclease-T activity on recombination with P_3. Anfinsen and collaborators[156] have synthesised a peptide covering positions 6–47 of fragment P_2 (221) and several analogues and partial sequences in order to examine their binding and reactivating effect on fragment P_3.

Lys-Leu-His-Lys-Glu-Pro-Ala-Thr-Leu-Ile-Lys-Ala-Ile-Asp-
6 10 15

Gly-Asp-Thr-Val-Lys-Leu-Met-Tyr-Lys-Gly-Gln-Pro-Met-Thr-
20 25 30

Phe-Arg-Leu-Leu-Leu-Val-Asp-Thr-Pro-Glu-Thr-Lys-His-Pro
35 40 45 47

(221)

Synthesis was achieved by the normal Merrifield procedure using butyloxycarbonyl for α-protection, removed with 4 mol l^{-1} hydrogen chloride in dioxan for 30 min. Side chains of aspartic and glutamic acids, threonine and tyrosine were blocked by benzyl groups, histidine by im-benzyloxycarbonyl, arginine by nitro and lysine by trifluoroacetyl. Coupling used 3–4 equivalents of acid and of dicyclohexylcarbodi-imide for 2 h, glutamine being added as the *p*-nitrophenyl ester with a reaction time of 12 h. Cleavage from the resin was by hydrogen fluoride/anisole and trifluoroacetyl groups were subsequently removed with aqueous piperidine containing 8 M urea. The product was subjected to gel filtration, then purified by affinity chromatography on a column of nuclease fragment P_3 covalently bound to a Sepharose support. This step alone enhanced the specific activity of the product tenfold and only 7% of the material applied underwent binding to the column. The elution profile obtained following an alternative purification by ion exchange on phosphorylated cellulose indicated considerable heterogeneity. The authors observed that amino acid analysis on both crude and purified material gave similar results; this criterion seemed to have been singularly unhelpful in this

Table 7.9 Staphyloccal nuclease: fragment P_2 analogues

Compound	*Substitution*	*Sequence of P_2 covered*
(222)[158]	Nle^{26}	6–47
(223)[158]	Nle^{32}	6–47
(224)[158]	Gln^{43}	6–47
(225)[159]	Gly^{46}	6–47
(226)[158]	None	9–47
(227)[159]	Gly^{46}	9–47
(228)[160]	None	10–47
(229)[160]	None	11–47
(230)[160]	None	12–47
(231)[160]	None	16–47
(232)[158]	None	18–47
(233)[158]	None	33–47

example. Other comments were on the instability of ε-benzyloxycarbonyl groups on repeated acid treatment[157], which caused these authors to choose trifluoroacetyl for the purpose of protecting lysine, and low coupling efficiency when histidine was added without side chain protection. Peptide mapping after tryptic digests of synthetic (221) and native P_2 showed patterns similar in most respects but extra spots were seen in the digest of (221).

Analogues of the natural sequence were designed on the basis of sequence differences arising from various strains and from consideration of the x-ray crystallographic structures of staphylococcal nuclease and a nuclease–inhibitor complex. These are shown in Table 7.9. The effect of these analogues on binding to fragment P_3 and on the enzymic activity of the nuclease-T obtained has been discussed[156–161].

7.11 OTHER PEPTIDES

7.11.1 Acyl carrier protein

A preliminary report[162] on the synthesis of sequence 1–74 of the *E. coli* acyl carrier protein has appeared. The standard Merrifield procedure[19] was used with the following modifications. Couplings were carried out twice with 2 h reaction times using carbodi-imide and with 12 h reaction times when nitrophenyl esters were used. The solvent for the second coupling was dichloromethane–dimethylformamide (1:1) containing 1.5 M urea. Deprotection of growing chains was also conducted twice, for 20 min each time, using dichloromethane–trifluoroacetic acid and acetylation was used to block unreacted chains on six occasions. A full assessment of this work must await publication of a more complete description.

7.11.2 Bradykinin

Several analogues of bradykinin (234) have been prepared recently, either to clarify structure–activity dependence or to test new synthetic methods. These are shown in Table 7.10. Compound (244) was prepared by solid phase but the remainder were synthesised by classical step-wise methods

Table 7.10 Bradykin analogues

Compound	Sequence
(234)[23]	Arg-Pro-Pro-Gly-Phe-Ser-Pro-Phe-Arg
	1 2 3 4 5 6 7 8 9
(235)[163]	·······Gly-Gly·······Gly-Gly-Gly-Gly·······
(236)[163]	·······Gly-Gly··············Gly-Gly············
(237)[163]	·······Gly-Gly··············Gly-Gly-Gly·······
(238)[163]	·······Gly-Gly·······Gly-Gly-Gly············
(239)[23]	·························Cha·············Cha·······
(240)[164]	···Agm
(241)[164]	Gva···Agm
(242)[164]	Gva···
(243)[166]	·······AEG··
(244)[166]	·························TAla·············TAla·······

Agm = agmatine, 'descarboxyarginine'
Gva = 4-guanidinovaleric acid
Cha = β-cyclohexylalanine
AEG = N(β-aminoethyl)glycine
TAla = β-thienylalanine

using (in the main) various active esters. Bradykinin itself (234) was prepared using the picolyl ester method of Young and collaborators. During this synthesis evidence of *O*-acylation of unprotected serine was observed when a glycine trichlorophenyl ester coupling was being effected; subsequently serine was used as the *O*-benzyl ether. Syntheses were otherwise as expected although the use of vigorous conditions to remove nitro protection from arginine was found to lead to significant hydrogenation of phenylalanine side chains. By taking this reaction to completion the biscyclohexylalanine analogue (239) was prepared.

7.11.3 Cholecystokinin-pancreozymin (CCK-PZ)

The tritriacontapeptide CCK-PZ undergoes tryptic hydrolysis to give the *C*-terminal dodecapeptide (245). Both (245) and its *C*-terminal octapeptide (246) possess the full gall bladder contracting activity of

$$\begin{array}{c}SO_3H\\ \text{Ile-Ser-Asp-Arg-Asp-Tyr-Met-Gly-Trp-Met-Asp-Phe-NH}_2\end{array}$$

(245)

CCK-PZ. Both (245)[167] and a series of analogues of (246)[168] have been prepared by classical procedures involving *p*-nitrophenyl active ester condensation to build fragments of 2–6 amino acids which were then coupled by the

Table 7.11 Analogues of CCK-PZ active fragment

		1	2	3	4	5	6	7	8
(246)		Asp-	Tyr(SO_3H)-	Met-	Gly-	Trp-	Met-	Asp-	Phe-NH_2
(247)	Boc	…	…	…	…	…	…	…	…
(248)		⋮ …	…	…	…	…	…	…	…
		Asp							
(249)		Abu	…	…	…	…	…	…	…
(250)		Ala	…	…	…	…	…	…	…
(251)		…	…	…	…	…	…	Ala	…
(252)		Tyr(SO_3H)-	Asp	…	…	…	…	…	…
(253)		…	Met-	Tyr(SO_3H)	…	…	…	…	…
(254)		…	…	Leu	…	…	…	…	…
(255)		…	…	…	…	…	Leu	…	…
(256)		…	…	Leu	…	…	Leu	…	…

azide method. Tyrosine was sulphated in the free octapeptide by the action of concentrated sulphuric acid or by the action of a pyridine–sulphur trioxide complex on butyloxycarbonyl derivatives. Analogues are shown in Table 7.11.

7.11.4 Glucagon

Although the synthesis of glucagon by Wünsch and collaborators was mainly carried out before the period covered by this review it cannot pass unmentioned. The final paper[169], on the crystallisation and characterisation of the synthetic nonacosapeptide, represents comparison of synthetic material and natural product taken to such a degree that identity becomes indisputable. Characterisation to this extent is seen all too seldom.

7.11.5 Melittin

Dorman and Markley[170] have prepared a series of fragments of melittin, the basic peptide which is the main constituent of bee venom. The largest molecule prepared was the *N*-terminal heptadecapeptide (257) and sequences 1–8, 7–17 and 14–17 were also prepared.

Gly-Ile-Gly-Ala-Val-Leu-Lys-Val-Leu-Thr-Thr-Gly-Leu-Pro-Ala-Leu-Ile
1 2 3 4 5 6 7 8 9 10 11 12 13 14 15 16 17

(257)

Synthesis was by normal Merrifield procedure but *C*-terminal amino acids were bonded to the resin by the procedure developed previously by Dorman and collaborators[171] using resin in the sulphonium form. Difficulty was observed during the deprotection of butyloxycarbonyl–isoleucine resin. This was attributed to steric effects and to the influence of microenvironment at that particular concentration of peptide groups in the resin matrix. Use of a higher loading of amino acid on the resin gave no difficulty during deprotection.

7.11.6 Parathyroid hormone

Potts and collaborators[172] have described the synthesis by the solid-phase procedure, of a tetratriacontapeptide having the sequence of the *N*-terminus of bovine parathyroid hormone. It is claimed that the resin used, viz. poly(trifluorochloroethylene-g-chloromethylstyrene) possesses unique advantages. Basic Merrifield procedure was used with benzyl side chain protection of aspartic and glutamic acids and serine, arginine was protected by nitro, histidine by im-dinitrophenyl and lysine by ε-trifluoroacetyl. Deprotection was by 4 mol l^{-1} hydrogen chloride in dioxan (or by trifluoroacetic acid in the case of *N*-terminal glutamine) and mercaptoethanol was added to preserve tryptophan after its incorporation. Coupling was by carbodi-imide except for glutamine and asparagine which were coupled as the *p*-nitrophenyl esters. Couplings were tested for completeness at each stage and either permitted further reaction or free chains were acetylated. Fission from the resin was by hydrogen fluoride–anisole and removal of trifluoroacetyl protection was by piperidine in 8 M urea. Details of the purification were not given save the indication that gel filtration and chromatography on carboxymethyl-cellulose were used. Amino acid analysis of an acid hydrolysate indicated that the product could possess a significant degree of heterogeneity. In particular, the deviation of the histidine ratio (1.70) from theoretical (3.00) was attributed by the authors to incomplete scission of the im-dinitrophenyl group by thiophenol. Biological characterisation of the product was extensive.

7.11.7 Phyllocaerulein

$$\begin{array}{c} SO_3H \\ | \\ \text{Pyr-Glu-Tyr-Thr-Gly-Trp-Met-Asp-Phe-NH}_2 \end{array}$$

(258)

The nonapeptide phyllocaerulein (258) has been synthesised by Bernardi and collaborators[173]. A pentapeptide amide corresponding to the *C*-terminus of (258) was available from these authors' synthesis of the closely related decapeptide caerulein[174], so that synthesis of (258) itself required only a suitably protected *N*-terminal tetrapeptide. This was prepared by a step-wise procedure using mixed anhydride couplings. Threonine was used as the *O*-acetyl derivative and glutamic acid as the γ-benzyl ester. The 4+5 fragments were coupled by the azide technique[55] and tyrosine was then sulphated using pyridine–SO_3 complex. Selective saponification of the acetylthreonine ester gave a product which was purified by chromatography on DEAE–Sephadex using ammonium bicarbonate buffers.

$$\begin{array}{c} SO_3H \\ | \\ \text{Pyr-Glu-Thr-Tyr-Gly-Trp-Met-Asp-Phe-NH}_2 \end{array}$$

(259)

An exactly similar method was used to prepare the analogue (259)[175].

7.11.8 Secretin

The synthesis of porcine secretin by the step-wise addition of nitrophenyl active esters exclusively was described some years ago by Bodanszky and collaborators[176] and is prior to the period of this review. Recently Wünsch and collaborators have described their work on the synthesis of fragments, covering the entire secretin sequence, which when condensed will lead to a second synthesis of the molecule. This alternative approach makes use of hydroxysuccinimide active esters for step-wise addition of amino acids and hydroxysuccinimide–dicyclohexylcarbodi-imide[53] to condense fragments in a racemisation-free manner.

$$\begin{array}{c} \text{Adoc Bu OBu} \quad\quad \text{Bu} \\ |\quad\;\; |\quad\;\; |\quad\quad\quad\;\; | \\ \text{Adoc-His-Ser-Asp-Gly-Thr-Phe-OH} \\ 1 \;\; 2 \;\; 3 \;\; 4 \;\; 5 \;\; 6 \end{array}$$

(260)[180]

Adoc = adamantyloxycarbonyl

```
Bu  Bu  OBu     Bu
|   |   |       |
Thr-Ser-Glu-Leu-Ser-OH
 7   8   9  10  11
```

(261)[179]

```
Leu-Leu-Gln-Gly-Leu-Val-NH2
 22  23  24  25  26  27
```

(262)[177]

```
Arg-Leu-Gln-Arg-Leu-Leu-Gln-Gly-Leu-Val-NH2
 18  19  20  21  22
```

(263)[177]

```
            OBu Bu
            |   |
Arg-Leu-Arg-Asp-Ser-Ala-Arg-Leu-Gln-Arg-Leu-Leu-Gln-Gly-Leu-Val-NH2
 12  13  14  15  16  17  18
```

(264)[178]

Fragments covering positions 1–6 (260) and 7–11 (261) were prepared by the step-wise addition of hydroxysuccinimide esters to *N*-methylmorpholinium salts of free peptides in a suitable solvent. Protection of α-amino was by benzyloxycarbonyl, removed by hydrogenolysis at each stage. Intermediate (263) was prepared by several alternative procedures. The route preferred involved coupling the two dipeptide units (from the *C*-terminus) then step-wise addition of single residues up to arginine (position 21). It was then found that better yields could be obtained overall by preparing the 18–20 tripeptide fragment and adding this by a hydroxybenztriazole accelerated dicyclohexylcarbodi-imide coupling[54] to give protected (263). Sequence 14–17 was then added as a unit followed by leucine and arginine step-wise to give (264).

The synthesis of a secretin analogue, in which all four arginine residues have been replaced by ornithine, has been described by Smithers in a preliminary report[181].

```
      Bu  OBu
      |   |
Z-His-Ser-Asp-Gly-OH
```

(265)

```
  Bu      Bu  Bu  OBu     Bu
  |       |   |   |       |
H-Thr-Phe-Thr-Ser-Glu-Leu-Ser-NHNHBoc
```

(266)

```
  Boc     Boc     Bu
  |       |       |
Z-Orn-Leu-Orn-Asp-Ser-Ala-NHNH2
```

(267)

```
Boc          Boc
 |            |
Orn-Leu-Gln-Orn-Leu-Leu-Gln-Gly-Leu-Val-NH2
```

(268)

Intermediates (265) to (268) were prepared by step-wise active ester condensations; in most cases hydroxysuccinimide esters were preferred over 2,4,5-trichlorophenyl esters (265) and (266) were coupled by mixed anhydride and this 1–11 fragment was condensed with the product of coupling (267) and (268) by the azide procedure. The final product, which was purified by chromatography on Sephadex G-25, was carefully checked to ensure that no $\alpha \rightarrow \beta$ peptide rearrangement had occurred at the susceptible Asp-Gly sequence (3–4).

7.12 CONCLUSION

During the period under consideration peptides have been synthesised by methods which were developed previously and have been adapted to new situations. No fundamentally new concepts or techniques have been introduced. Solution methods of peptide synthesis have provided the majority of the syntheses in this review. The more established methods of step-wise synthesis such as active ester or carbodi-imide have been complemented by that involving the controlled reaction of *N*-carboxyamino acid anhydrides which was introduced by the Merck group several years ago. This method has shown its capabilities in the synthesis of ribonuclease *S*-protein and will no doubt find its place in other applications in the future. The condensation of fragments, which is widely used in the preparation of larger peptides, has latterly been brought about using carbodi-imide couplings accelerated by hydroxysuccinimide[53] or hydroxybenztriazole[54] or more commonly by the azide method using Honzl and Rudinger's improved procedure[55].

As might have been predicted the solid-phase method of peptide synthesis has been applied to even longer sequences than hitherto. The homogeneity of the products of such syntheses has been the subject of much discussion between protagonists and antagonists of the technique. There would seem to have been no precedent for the polarisation of opinion among interested investigators that was generated by the introduction of the solid-phase procedure. In certain cases, claims respecting homogeneity made by authors in their initial publications have later been admitted to be overstated. The method has also come into disrepute as a result of misuse by investigators unskilled in the art. The onus will always be upon the synthesising investigator to substantiate a claim that a particular sequence has been prepared and in a given state of purity. At the present time, objectivity in characterisation falls with increasing molecular size. The solid-phase procedure is of considerable value in the synthesis of peptides of the size of bradykinin or oxytocin and its most effective exploitation in the synthesis of very large molecules could well be the synthesis of fragments of this size. These fragments can be purified effectively by methods currently available and may then be condensed in solution by a suitable procedure.

References

1. Chaturvedi, N. C., Park, W. K., Smeeby, R. R. and Bumpus, F. M. (1970). *J. Med. Chem.*, **13,** 177
2. Andreatta, R. H. and Scheraga, H. A. (1971). *J. Med. Chem.*, **14,** 489
3. Khan, S. A. and Sivanandaiah, K. M. (1971). *Ind. J. Chem.*, **9,** 184
4. Jorgensen, E. C., Windridge, G. C., Patton, W. and Lee, T. C. (1970). *Peptides: Chemistry and Biochemistry,* 113. (New York: Dekker)
5. Jorgensen, E. C., Windridge, G. C. and Lee, T. C. (1971). *J. Med. Chem.*, **14,** 631
6. Jorgensen, E. C., Rapaka, S. R., Windridge, G. C. and Lee, T. C. (1971). *J. Med. Chem.*, **14,** 899
7. Jorgensen, E. C., Rapaka, S. R., Windridge, G. C. and Lee, T. C. (1971). *J. Med. Chem.*, **14,** 904
8. Paruszewski, R. (1971). *Rocz. Chem.*, **45,** 299
9. Garner, R. and Young, G. T. (1971). *J. Chem. Soc. C,* 50
10. Marshall, G. R., Vine, W. and Needleman, P. (1970). *Proc. Nat. Acad. Sci. U.S.*, **67,** 1624
11. Romanovskaya, I. K. and Chipens, G. I. (1971). *Zh. Obshch. Khim.*, **41,** 1856
12. Pavar, A. P. and Chipens, G. I. (1971). *Zh. Obshch. Khim.*, **41,** 467
13. Pavar, A. P., Auna, Z. P. and Chipens, G. I. (1971). *Zh. Obshch. Khim.*, **41,** 1859
14. Pavar, A. P., Avotinya, G. Ya., Indulen, Yu. I., Auna, Z. P. and Chipens, G. I. (1971). *Zh. Obshch. Khim.*, **41,** 2312
15. Jorgensen, E. C., Windridge, G. C. and Lee, T. C. (1970). *J. Med. Chem.*, **13,** 352
16. Jorgensen, E. C., Windridge, G. C. and Lee, T. C. (1970). *J. Med. Chem.*, **13,** 744
17. Schröder, E. and Lübke, K. (1966). *The Peptides,* **2,** 4. (New York: Academic Press)
18. Rittel, W., Iselin, B., Kappeler, H., Riniker, B. and Schwyzer, R. (1957). *Helv. Chim. Acta,* **40,** 614
19. Marshall, G. R. and Merrifield, R. B. (1965). *Biochemistry,* **4,** 2394
20. Morgat, J. L., Hung, L. T. and Fromageot, P. (1970). *Biochim. Biophys. Acta,* **207,** 374
21. Osborne, M. J., Pooters, N., Angles d'Auriac, G., Epstein, A. N., Worcel, M. and Meyer, P. (1971). *Pfluger's Arch.*, **326,** 101
22. Jorgensen, E. C., Windridge, G. C., Patton, W. and Lee, T. C. (1969). *J. Med. Chem.*, **12,** 733
23. Schafer, D. J., Young, G. T., Elliott, D. F. and Wade, R. (1971). *J. Chem. Soc. C,* 46
24. Rittel, W., Brugger, M., Kamber, B., Riniker, B. and Sieber, P. (1968). *Helv. Chim. Acta,* **51,** 924; Kamber, B. and Rittel, W. (1969). *Helv. Chim. Acta,* **52,** 1074; Riniker, B., Brugger, M., Kamber, B., Sieber, P. and Rittel, W. (1969). *Helv. Chim. Acta,* **52,** 1058
25. Guttmann, S., Pless, J., Sandrin, E., Jaquenoud, P. A., Bossert, H. and Willems, H. (1968). *Helv. Chim. Acta,* **51,** 1155
26. Guttmann, S., Pless, J., Huguenin, R. L., Sandrin, E., Bossert, H. and Zehnder, K. (1969). *Helv. Chim. Acta,* **52,** 1789
27. Sieber, P., Riniker, B., Brugger, M., Kamber, B. and Rittel, W. (1970). *Helv. Chim. Acta,* **53,** 2135
28. Kamber, B., Brückner, H., Riniker, B., Sieber, P. and Rittel, W. (1970). *Helv. Chim. Acta,* **53,** 556
29. Kamber, B. and Rittel, W. (1968). *Helv. Chim. Acta,* **51,** 2061
30. Sieber, P., Brugger, M. and Rittel, W. (1971). *Peptides 1969,* 60. (Amsterdam: North-Holland)
31. Greven, H. M. and Tax, L. J. W. M. (1971). *Peptides 1969,* 38. (Amsterdam: North-Holland)
32. Otsuka, H., Watanabe, K. and Inouye, K. (1970). *Bull. Chem. Soc. Jap.*, **43,** 2278
33. Fujino, M., Hatanaka, C. and Nishimura, O. (1970). *Chem. Pharm. Bull.*, **18,** 1288
34. Fujino, M., Nishimura, O. and Hatanaka, C. (1970). *Chem. Pharm. Bull.*, **18,** 1291
35. Fujino, M., Hatanaka, C., Nishimura, O. and Shinagawa, S. (1971). *Chem. Pharm. Bull.*, **19,** 1075
36. Tesser, G. I. and Rittel, W. (1969). *Rec. Trav. Chim. Pays-Bas,* **88,** 553
37. Schwyzer, R. and Schiller, P. W. (1971). *Helv. Chim. Acta,* **54,** 897
38. Fujino, M., Hatanaka, C. and Nishimura, O. (1970). *Chem. Pharm. Bull.*, **18,** 771
39. Hofmann, K., Andreatta, R., Bohn, H. and Moroder, L. (1970). *J. Med. Chem.*, **13,** 339
40. Moroder, L. and Hofmann, K. (1970). *J. Med. Chem.*, **13,** 839
41. Brugger, M., Barthe, P. and Desaulles, P. A. (1970). *Experientia,* **26,** 1050

42. Brugger, M. (1971). *Helv. Chim. Acta,* **54,** 1261
43. Riniker, B. and Rittel, W. (1970). *Helv. Chim. Acta,* **53,** 513
44. Tesser, G. I. and Buis, J. T. (1971). *Rec. Trav. Chim. Pays-Bas,* **90,** 444
45. Otsuka, H., Shin, M., Kinomura, Y. and Inouye, K. (1970). *Bull. Chem. Soc. Jap.,* **43,** 196
46. Inouye, K., Tanaka, A. and Otsuka, H. (1970). *Bull. Chem. Soc. Jap.,* **43,** 1163
47. Inouye, K., Watanabe, K., Namba, K. and Otsuka, H. (1970). *Bull. Chem. Soc. Jap.,* **43,** 3873
48. Fujino, M., Hatanaka, C. and Nishimura, O. (1971). *Chem. Pharm. Bull.,* **19,** 1066
49. Schröder, E. and Lübke, K. Reference 17, p. 199
50. Riniker, B., Rittel, W., Sieber, P. and Zuber, H. (1972). *Nature (London),* **235,** 114
51. Desaulles, P. A. and Rittel, W. (1968). *Mem. Soc. Endocrinol.,* **17,** 125
52. Hofmann, K., Wells, R. D., Yajima, H. and Rosenthaler, J. (1963). *J. Amer. Chem. Soc.,* **85,** 1546
53. Weygand, F., Hoffmann, D. and Wünsch, E. (1966). *Z. Naturforsch.,* **21b,** 426
54. König, W. and Geiger, R. (1970). *Chem. Ber.,* **103,** 788
55. Honzl, J. and Rudinger, J. (1961). *Collect. Czech. Chem. Commun.,* **26,** 2333
56. Sakakibara, S. and Shimonishi, Y. (1965). *Bull. Chem. Soc. Jap.,* **38,** 1412
57. Geiger, R., Schröder, H-G. and Siedel, W. (1969). *Ann.,* **726,** 177
58. Geiger, R. (1971). *Ann.,* **750,** 165
59. Hofmann, K., Rosenthaler, J., Wells, R. D. and Yajima, H. (1964). *J. Amer. Chem. Soc.,* **86,** 4991
60. Li, C. H. and Yamashiro, D. (1970). *J. Amer. Chem. Soc.,* **92,** 7608
61. Li, C. H. and Chung, D. (1971). *Int. J. Protein Res.,* **3,** 73
62. Danho, W. and Li, C. H. (1971). *Int. J. Protein Res.,* **3,** 81
63. Kovacs, K., Kovacs-Petres, Y. and Li, C. H. (1971). *Int. J. Protein Res.,* **3,** 93
64. Danho, W. and Li, C. H. (1971). *Int. J. Protein Res.,* **3,** 99
65. Chillemi, F. and Pecile, A. (1971). *Experientia,* **27,** 385
66. Niall, H. D. (1971). *Nature New Biology,* **230,** 90
67. Li, C. H., Dixon, J. S. and Liu, W. K. (1969). *Arch. Biochem. Biophys.,* **133,** 70
68. Hofmann, K. (1960). *Brookhaven Symp. Biol.,* **13,** 184
69. Schnabel, E., Klostermeyer, H. and Berndt, H. (1971). *Ann.,* **749,** 90
70. Chance, R. E., Ellis, R. M. E. and Bromer, W. M. (1968). *Science,* **161,** 165
71. Grant, P. T. and Coombs, T. L. (1970). *Essays in Biochemistry,* **6,** 69
72. Schröder, E. and Lübke, K. (1966). Reference 17, p. 379
73. Weber, U., Herzog, K. H., Grossmann, H., Hartter, P. and Weitzel, G. (1971). *Z. Physiol. Chem.,* **352,** 419
74. Weitzel, G., Weber, U., Eisele, K., Zollner, H. and Martin, J. (1970). *Z. Physiol. Chem.,* **351,** 263
75. Weitzel, G., Eisele, K., Zollner, H. and Weber, U. (1969). *Z. Physiol. Chem.,* **350,** 1480
76. Weitzel, G., Weber, U., Martin, J. and Eisele, K. (1971). *Z. Physiol. Chem.,* **357,** 1005
77. Katsoyannis, P. G., Zalut, C., Tometsko, A., Tilak, M., Johnson, S. and Trakatellis, A. C. (1971). *J. Amer. Chem. Soc.,* **93,** 5871
78. Katsoyannis, P. G., Ginos, J., Zalut, C., Tilak, M., Johnson, S. and Trakatellis, A. C. (1971). *J. Amer. Chem. Soc.,* **93,** 5877
79. Katsoyannis, P. G., Tilak, M. and Fukuda, K. (1971). *J. Amer. Chem. Soc.,* **93,** 5857
80. Katsoyannis, P. G., Ginos, J. and Tilak, M. (1971). *J. Amer. Chem. Soc.,* **93,** 5866
81. Katsoyannis, P. G., Tilak, M., Ginos, J. and Suzuki, K. (1971). *J. Amer. Chem. Soc.,* **93,** 5862
82. Zahn, H. and Schmidt, G. (1970). *Ann.,* **731,** 101
83. Zahn, H. and Schmidt, G. (1970). *Ann.,* **731,** 91
84. Titov, M. I. and Ardemasova, Z. A. (1971). *Zh. Obshch. Khim.,* **41,** 1403
85. Bosshard, H. R. (1971). *Helv. Chim. Acta,* **54,** 951
86. Vdovina, R. G., Gracheva, A. K., Poznyak, M. G. and Shvachkin, Yu. P. (1971). *Zh. Obshch. Khim.,* **41,** 239
87. Geiger, R., Jäger, G., König, W. and Volk, A. (1969). *Z. Naturforsch.,* **24b,** 999
88. Jäger, G. and Geiger, R. (1970). *Chem. Ber.,* **103,** 1727
89. Yanaihara, N., Hashimoto, T., Yanaihara, C. and Sakura, N. (1970). *Chem. Pharm. Bull.,* **18,** 417
90. Weinert, M., Brandenburg, D. and Zahn, H. (1969). *Z. Physiol. Chem.,* **350,** 1556
91. Geiger, R., Schöne, H.-H. and Pfaff, W. (1971). *Z. Physiol. Chem.,* **352,** 1487

92. Krail, G., Brandenburg, D., Zahn, H. and Geiger, R. (1971). *Z. Physiol. Chem.*, **352**, 1595
93. Borras, F. and Offord, R. E. (1970). *Nature (London)*, **227**, 716
94. Schröder, E. and Lübke, K. Reference 17, p. 161
95. Blake, J., Crooks, R. W. and Li, C. H. (1970). *Biochemistry*, **9**, 2071
96. Blake, J. and Li, C. H. (1971). *Int. J. Protein Res.*, **3**, 185
97. Yajima, H., Kawasaki, K., Minami, H., Kawatani, H., Mizokami, N., Kiso, Y. and Tamura, F. (1970). *Chem. Pharm. Bull.*, **18**, 1394
98. Yajima, H., Okada, Y., Kinomura, Y. and Minami, H. (1968). *J. Amer. Chem. Soc.*, **90**, 527; Yajima, H., Okada, Y., Kinomura, Y., Mizokami, N. and Kawatani, H. (1969). *Chem. Pharm. Bull.*, **17**, 1237
99. Burton, J. and Lande, S. (1970). *J. Amer. Chem. Soc.*, **92**, 3746
100. du Vigneaud, V., Ressler, C., Swan, J. M., Roberts, C. W. and Katsoyannis, P. G. (1954). *J. Amer. Chem. Soc.*, **76**, 3115
101. Papsuevich, O. S. and Chipens, G. I. (1970). *Zh. Obshch. Khim.*, **40**, 2768
102. Hruby, V. J. and du Vigneaud, V. (1969). *J. Med. Chem.*, **12**, 731
103. Auna, Z. P., Kaurov, O. A., Martinov, V. F. and Morozov, V. B. (1971). *Zh. Obshch. Khim.*, **41**, 674
104. Hruby, V. J., du Vigneaud, V. and Chan, W. Y. (1970). *J. Med. Chem.*, **13**, 185
105. Hruby, V. J. and Chan, W. Y. (1971). *J. Med. Chem.*, **14**, 1050
106. Hruby, V. J., Flouret, G. and du Vigneaud, V. (1969). *J. Biol. Chem.*, **244**, 3890
107. Havran, R. T., Schwartz, I. L. and Walter, R. (1969). *J. Amer. Chem. Soc.*, **91**, 1836
108. Kaurov, O. A., Martinov, V. F. and Mikhailov, Yu. D. (1971). *Zh. Obshch. Khim.*, **41**, 1413
109. Manning, M., Coy, E. and Sawyer, W. H. (1970). *Biochemistry*, **9**, 3925
110. Flouret, G. and du Vigneard, V. (1969). *J. Med. Chem.*, **12**, 1035
111. Manning, M., Baxter, J. W. M., Wuu, T. C., Smart-Abbey, V., Morton, K., Coy, E. J. and Sawyer, W. H. (1971). *J. Med. Chem.*, **14**, 1143
112. Lübke, K. (1971). *Peptides 1969*, 154. (Amsterdam: North-Holland)
113. Havran, R. T., Meyers, C., Schwartz, I. L. and Walter, R. (1971). *Peptides 1969*, 161. (Amsterdam: North-Holland)
114. Jost, K. and Sorm, F. (1971). *Collect. Czech. Chem. Commun.*, **36**, 297
115. Takashima, H., Fraefel, W. and du Vigneaud, V. (1969). *J. Amer. Chem. Soc.*, **91**, 6182
116. Takashima, H., Hruby, V. J. and du Vigneaud, V. (1970). *J. Amer. Chem. Soc.*, **92**, 677
117. Rimpler, M. and Schöberl, A. (1969). *Naturwissenschaften*, **56**, 638
118. Rimpler, M. (1971). *Ann.*, **745**, 8
119. Jost, K. (1971). *Collect. Czech. Chem. Commun.*, **36**, 218
120. Jost, K. and Sorm, F. (1971). *Collect. Czech. Chem. Commun.*, **36**, 234
121. Jost, K. and Sorm, F. (1971). *Collect. Czech. Chem. Commun.*, **36**, 2795
122. Kabayashi, A., Hase, S., Kiyai, R. and Sakakibara, S. (1970). *Bull. Chem. Soc. Jap.*, **42**, 3491
123. Fraefel, W. and du Vigneaud, V. (1970). *J. Amer. Chem. Soc.*, **92**, 1030
124. Fraefel, W. and du Vigneaud, V. (1970). *J. Amer. Chem. Soc.*, **92**, 4426
125. Takashima, H. and du Vigneaud, V. (1970). *J. Amer. Chem. Soc.*, **92**, 2501
126. Aanning, H. L. and Yamashiro, D. (1970). *J. Amer. Chem. Soc.*, **92**, 5214
127. Hruby, V. J., Ferger, M. F. and du Vigneaud, V. (1971). *J. Amer. Chem. Soc.*, **93**, 5539
128. Meienhofer, J., Trzeciak, A., Dousa, T., Hechter, O., Havran, R. T., Schwartz, I. L. and Walter, R. (1971). *Peptides 1969*, 157. (Amsterdam: North-Holland)
129. Meienhofer, J. and Trzeciak, A. (1971). *Proc. Nat. Acad. Sci. U.S.*, **68**, 1006
130. Zaoral, M., Kolc, J. and Sorm, F. (1970). *Collect. Czech. Chem. Commun.*, **35**, 1716
131. Gillesen, D. and du Vigneaud, V. (1970). *J. Med. Chem.*, **13**, 346
132. Havran, R. T. and du Vigneaud, V. (1969). *J. Amer. Chem. Soc.*, **91**, 3626
133. Havran, R. T. and du Vigneaud, V. (1969). *J. Amer. Chem. Soc.*, **91**, 2696
134. Papsuevich, O. S. and Chipens, G. I. (1970). *Zh. Obshch. Khim.*, **40**, 709
135. Papsuevich, O. S. and Chipens, G. I. (1969). *Latv. PSR Zinat. Akad. Vestis. Kim. Ser.*, (6), 751
136. Meienhofer, J., Trzeciak, A., Havran, R. T. and Walter, R. (1970). *J. Amer. Chem. Soc.*, **92**, 7199
137. Schröder, E. and Lübke, K. (1966). Reference 17, p. 281
138. Pettit, G. R. (1970). *Synthetic Peptides Vol. I*, 339. (New York: Van Nostrand Reinhold)
139. Bodanszky, M. and du Vigneaud, V. (1959). *J. Amer. Chem. Soc.*, **81**, 5688

140. Zaoral, M. (1965). *Collect. Czech. Chem. Commun.*, **30,** 1853
141. Richards, F. M. (1958). *Proc. Nat. Acad. Sci. U.S.*, **44,** 162
142. Hofmann, K., Visser, J. P. and Finn, F. M. (1970). *J. Amer. Chem. Soc.*, **92,** 2900 and references cited therein
143. Rocchi, R., Marchiori, F., Moroder, L., Borin, G. and Scoffone, E. (1968). *J. Amer. Chem. Soc.*, **91,** 3927 and references cited therein
144. Hofmann, K., Visser, J. P. and Finn, F. M. (1969). *J. Amer. Chem. Soc.*, **91,** 4883
145. Visser, S., Raap, J., Kerling, K. E. T. and Havinga, E. (1970). *Rec. Trav. Chim. Pays-Bas*, **89,** 865
146. Visser, S., Kerling, K. E. T. and Havinga, E. (1970). *Rec. Trav. Chim. Pays-Bas*, **89,** 876
147. Visser, S. and Kerling, K. E. T. (1970). *Rec. Trav. Chim. Pays-Bas*, **89,** 880
148. Denkewalter, R. G., Veber, D. F., Holly, F. W. and Hirschmann, R. (1969). *J. Amer. Chem. Soc.*, **91,** 502; Strachan, R. G., Paleveda, W. J., Nutt, R. F., Vitali, R. A., Veber, D. F., Dickinson, M. J., Garsky, V., Deak, J. E., Walton, E., Jenkins, S. R., Holly, F. W. and Hirschmann, R. (1969). ibid. p. 503; Jenkins, S. R., Nutt, R. F., Dewey, R. S., Veber, D. F., Holly, F. W., Paleveda, W. J., Lanza, T., Strachan, R. G., Schoenewaldt, E. F., Barkemeyer, H., Dickinson, M. J., Sondey, J., Hirschmann, R. and Walton, E. (1969). ibid. p. 505; Veber, D. F., Varga, S. L., Milkowski, J. D., Joshua, H., Conn, J. B., Hirschmann, R. and Denkewalter, R. G. (1969). ibid. p. 506; Hirschmann, R., Nutt, R. F., Veber, D. F., Vitali, R. A., Varga, S. L., Jacob, T. A., Holly, F. W. and Denkewalter, R. G. (1969). ibid. p. 507
149. Gutte, B. and Merrifield, R. B. (1969). *J. Amer. Chem. Soc.*, **91,** 501
150. Lin, M. C., Gutte, B., Moore, S. and Merrifield, R. B. (1970). *J. Biol. Chem.*, **245,** 5169
151. Takahashi, K. (1965). *J. Biol. Chem.*, **240,** 4117
152. Ohno, M., Kato, T., Mitsuyasu, N., Waki, M., Makisumi, S. and Izumiya, N. (1967). *Bull. Chem. Soc. Jap.*, **40,** 204; Waki, M., Mitsuyasu, N., Kato, T., Makisumi, S. and Izumiya, N. (1968). ibid., **41,** 669; Kato, T., Mitsuyasu, N., Waki, M., Makisumi, S. and Izumiya, N. (1968). ibid., **41,** 2480; Mitsuyasu, N., Waki, M., Kato, T., Makisumi, S. and Izumiya, N. (1970). ibid., **43,** 1556
153. Yanaihara, N., Yanaihara, C., Dupius, G., Beacham, J., Camble, R. and Hofmann, K. (1969). *J. Amer. Chem. Soc.*, **91,** 2184; Hofmann, K. (1971). *Peptides 1969*, 130. (Amsterdam: North-Holland)
154. Beacham, J., Dupuis, G., Finn, F. M., Storey, H. T., Yanaihara, C., Yanaihara, N. and Hofmann, K. (1971). *J. Amer. Chem. Soc.*, **93,** 5526
155. Guttmann, S. (1966). *Helv. Chim. Acta*, **49,** 83
156. Ontjes, D. A. and Anfinsen, C. B. (1969). *Proc. Nat. Acad Sci. U.S.*, **64,** 428
157. Ontjes, D. A. and Anfinsen, C. B. (1970). *Peptides: Chemistry and Biochemistry*, 79. (New York: Dekker)
158. Ontjes, D. A. and Anfinsen, C. B. (1969). *J. Biol. Chem.*, **244,** 6316
159. Chaiken, I. M. and Anfinsen, C. B. (1970). *J. Biol. Chem.*, **245,** 2337
160. Chaiken, I. M. and Anfinsen, C. B. (1970). *J. Biol. Chem.*, **245,** 4718
161. Anfinsen, C. B., Ontjes, D. A. and Chaiken, I. M. (1971). *Peptides 1969*, 121. (Amsterdam: North-Holland)
162. Hancock, W. S., Prescott, D. J., Nulty, W. L., Weintraub, J., Vagelos, P. R. and Marshall, G. R. (1971). *J. Amer. Chem. Soc.*, **93,** 1799
163. Abramson, F. B., Elliott, D. F., Lindsay, D. G. and Wade, R. (1970). *J. Chem. Soc. C*, 1042
164. Johnson, W. H., Law, H. D. and Studer, R. O. (1971). *J. Chem. Soc. C*, 748
165. Atherton, E., Law, H. D., Moore, S., Elliott, D. F. and Wade, R. (1971). *J. Chem. Soc. C*, 3393
166. Dunn, F. W. and Stewart, J. M. (1971). *J. Med. Chem.*, **14,** 779
167. Ondetti, M. A., Pluscec, J., Sabo, E. F., Sheehan, J. T. and Williams, N. (1970). *J. Amer. Chem. Soc.*, **92,** 195
168. Pluscec, J., Sheehan, J. T., Sabo, E. F., Williams, N., Kocy, O. and Ondetti, M. A. (1970). *J. Med. Chem.*, **13,** 349
169. Wünsch, E., Jaeger, E. and Scharf, R. (1968). *Chem. Ber.*, **101,** 3664
170. Dorman, L. C. and Markley, L. D. (1971). *J. Med. Chem.*, **14,** 5
171. Dorman, L. C. and Love, J. (1969). *J. Org. Chem.*, **34,** 158
172. Potts, J. T., Tregear, G. W., Keutmann, H. T., Niall, H. D., Sauer, R., Deftos, L. J., Dawson, B. F., Hogan, M. L. and Aurbach, G. D. (1971). *Proc. Nat. Acad. Sci. U.S.*, **68,** 63

173. Bernardi, L., Bosisio, G., de Castiglione, R. and Goffredo, O. (1969). *Experientia,* **25,** 7
174. Bernardi, L., Bosisio, G., de Castiglione, R. and Goffredo, O. (1967). *Experientia,* **23,** 700
175. De Castiglione, R. and Goffredo, O. (1970). *Il Farmaco Ed. Sc.,* **25,** 40
176. Bodanszky, M. and Williams, N. J. (1967). *J. Amer. Chem. Soc.,* **89,** 685; Bodanszky, M., Ondetti, M. A., Levine, S. D. and Williams, N. J. (1967). *J. Amer. Chem. Soc.,* **89,** 6753
177. Wünsch, E., Wendlberger, G. and Högel, A. (1971). *Chem. Ber.,* **104,** 2430
178. Wünsch, E., Wendlberger, G. and Thamm, P. (1971). *Chem. Ber.,* **104,** 2445
179. Wünsch, E. and Thamm, P. (1971). *Chem. Ber.,* **104,** 2454
180. Wünsch, E., Wendlberger, G. and Spangenberg, R. (1971). *Chem. Ber.,* **104,** 3854
181. Smithers, M. (1971). *Peptides 1969,* 44. (Amsterdam: North-Holland)
182. Graf, L., Bajusz, S., Patthy, A., Barat, E. and Cseh, G. (1971). *Acta Biochim. Biophys. Acad. Sci. Hung.,* **6,** 415

8
Depsipeptides

H. A. JAMES
University College of Swansea

8.1 INTRODUCTION

Peptides, which contain, in addition to the normal amide bonds, other types of linkages, have received much attention in recent years. Depsipeptides, in which the heteromeric linkage is an ester bond, form an extensive group of such compounds. They may be broadly classified into three groups[1], namely, cyclodepsipeptides possessing a regularly alternating array of α-amino and α-hydroxy-acid residues, e.g. enniatin A, (1); cyclodepsipeptides with irregular sequences of amino and hydroxy acids, e.g. sporidesmolide I, (2); depsipeptides with one or more α- or β- or γ-hydroxy-α-amino acid residues, the hydroxyl group of these residues being involved in the ester bond formation, e.g. actinomycin D, (3).

It is of interest to note that a group of closely related depsipeptides is normally produced by a single micro-organism.

The main emphasis in this review will be on those aspects of depsipeptide research which are primarily of a chemical nature. However, from a survey of the literature published on depsipeptides, it becomes very apparent that interest in this subject is not confined to the field of chemistry. It is therefore

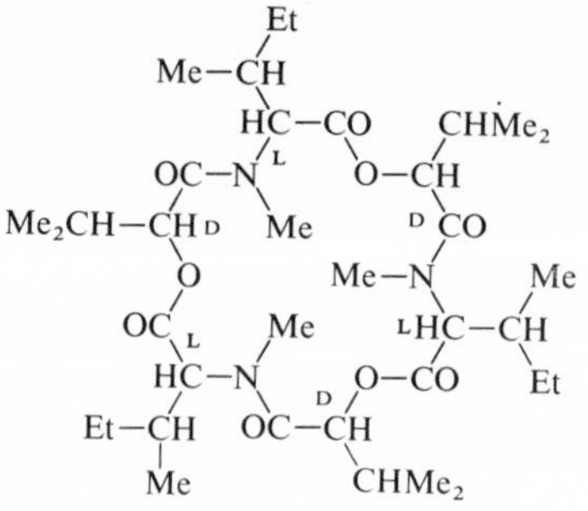

(1)

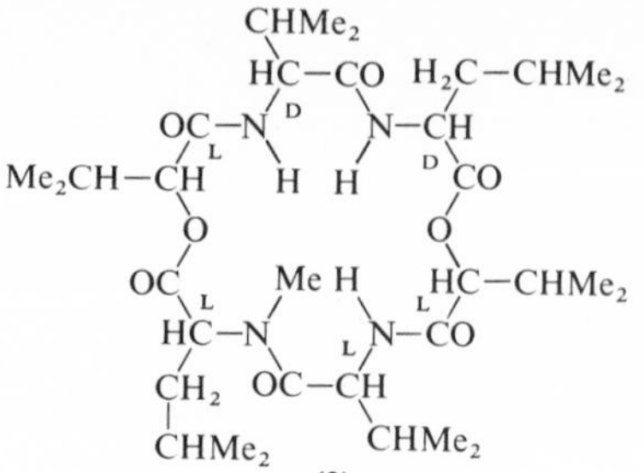

(2)

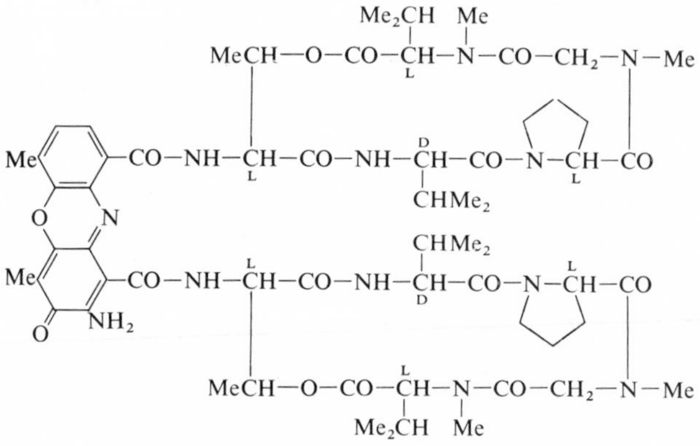

(3)

obvious that the chemical aspects of depsipeptides cannot be discussed in isolation, but any excursions into other fields will, of necessity, be brief. The utilisation of depsipeptides for the elucidation of biologically significant processes, and their applications as models for the more complex systems present in proteins, as well as their potential chemotherapeutic uses, are the main factors which have ensured the continued chemical interest. Reviews of a general nature which have appeared during the period 1970–1971 include an extremely thorough review[2] of peptide lactones, which mainly covers the literature published in the twenty years up to 1970, while the sections on depsipeptides in the reports[3] concerning peptides of abnormal structure continue to provide an exhaustive coverage of recent publications. A review[4] dealing with the chemistry and isolation of peptide antibiotics contains references to a number of depsipeptides, but the majority of these references are pre-1965. Other reviews, which deal either wholly or partly with the various aspects of depsipeptide chemistry, will be referred to in the appropriate sections of this review.

8.2 PRODUCTION AND ISOLATION

The majority of naturally occurring depsipeptides are produced by surface or submerged culture of the appropriate micro-organism. Normally a particular micro-organism produces a closely related group of depsipeptides, but the production of a particular depsipeptide may not be exclusive to a single micro-organism. Actinomycin D, together with other actinomycins, is produced by *S. antibioticus*, *S. chrysomallus* JA 1449 [5], several *S. parvullus* and *S. parvus* strains[6], *A. oliveobrunneus*[7] and various other Streptomyces species. Other depsipeptides which are produced by more than one species of micro-organisms include valinomycin, echinomycin, the enniatins and the ostreogricins.

Generally, once it has been established that a family of depsipeptides is produced by a micro-organism, the conditions for the maximal production of either the group of depsipeptides, or of an individual depsipeptide within that group, are investigated. This involves a study of fermentation parameters such as pH, temperature, medium composition and strain of micro-organism. The most effective conditions for the production of actinomycin D from *A. oliveobrunneus*[8] and of valinomycin (4) from *Streptomyces* species PRL 1642[9],

$$\begin{array}{c} \to \text{D-Val} \to \text{L-Lac} \to \text{L-Val} \to \text{D-Hyiv} \to \text{D-Val} \to \text{L-Lac} \\ \text{D-Hyiv} \leftarrow \text{L-Val} \leftarrow \text{L-Lac} \leftarrow \text{D-Val} \leftarrow \text{D-Hyiv} \leftarrow \text{L-Val} \leftarrow \end{array}$$

(4)

have been described, and it has also been reported[10] that addition of elemental sulphur to the soybean meal medium used for the fermentation of *S. sioyaensis* resulted in an increased production of siomycin. In the case of the mycobactins, iron-chelating growth factors produced by various *Mycobacterium* species, it was found[11] that certain metal ions were essential to their maximum production. The addition of amino acids to the culture medium may result in either an increased production of a particular depsipeptide, or in the production of new depsipeptides. Thus actinomycin $AyX_{1\alpha}$, which contains *N*-methylvaline, *N*-methylisoleucine, D-isoleucine and D-leucine, was produced by the addition of a mixture of L-isoleucine and DL-isoleucine to actinomycin producing strains of *Actinomycetes*[12], and new actinomycins containing *N*-methyl-L-*allo*isoleucine have been produced by the addition of L-isoleucine, D-isoleucine or L-*allo*isoleucine to the culture medium of *S. antibioticus* or *S. chrysomallus*[13].

The isolation of a family of congeneric depsipeptides is usually accomplished with relative ease by extraction of the culture medium with a suitable organic solvent. However, the isolation of a particular congener within that family often requires the application of a number of separation techniques, and is frequently a tedious process. The following examples illustrate methods which are currently used.

Echinomycin (5) which was obtained from a cell free extract of *Streptomyces* culture X-63 in the presence of precursors, ATP, mercaptoethanol and Mg^{2+}, was isolated by extraction of the medium at pH 4 with ethyl acetate followed by counter-current distribution of this extract[14].

A mixture of destruxins A, B, C and D and desmethyldestruxin B, (6), (7), (8), (9) and (10) respectively, produced by *Metarrhizium anisopliae*, was isolated by extraction at pH 3 of the filtrate obtained after removal of mycelia, with ethyl acetate. The previously unknown destruxins C and D and desmethyldestruxin B were isolated from this mixture using a combination of chromatographic techniques[15].

The enduracidins, a depsipeptide family produced by *Streptomyces*

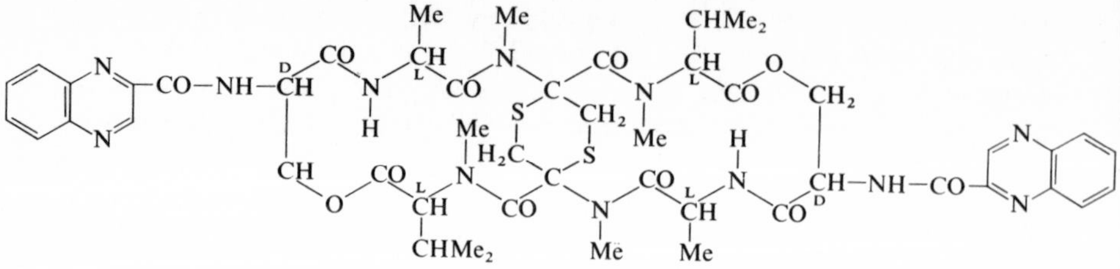

(5)

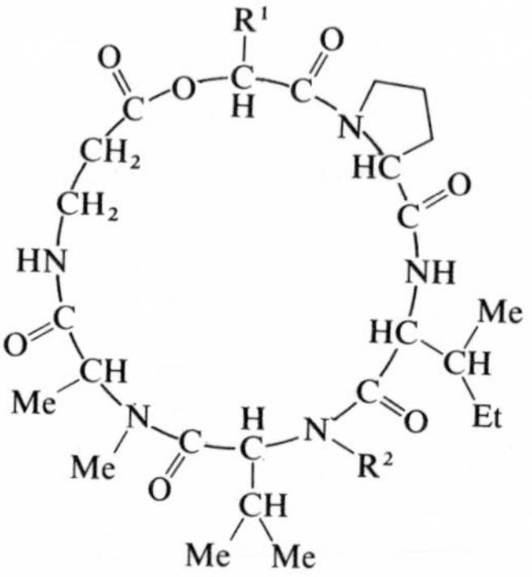

	R^1	R^2
(6)	$-CH_2\cdot CH{:}CH_2$	$-CH_3$
(7)	$-CH_2\cdot CH(CH_3)_2$	$-CH_3$
(8)	$-CH_2\cdot CH(CH_2OH)CH_3$	$-CH_3$
(9)	$-CH_2\cdot CH(CO_2H)CH_3$	$-CH_3$
(10)	$-CH_2\cdot CH(CH_3)_2$	$-H$

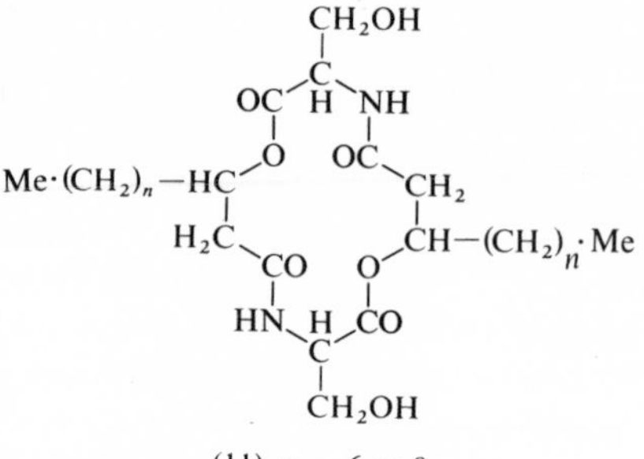

(11) $n = 6$ or 8

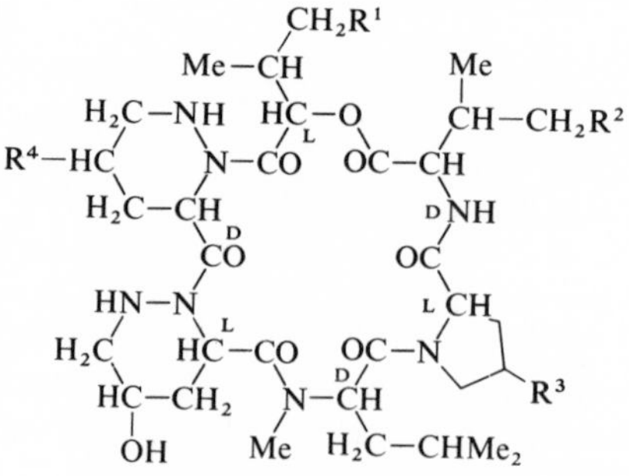

(12)

R^1 = H or Me
R^2 = H or Me
R^3 = H or Me
R^4 = H or Cl

fungicidicus B-5477, have been isolated and separated into two main components by chromatography on Amberlite XAD-2[16], while janiemycin, which is similar in composition to enduracidin, and is produced by *Streptomyces macrosporeus* ATCC 21388, has been separated into three components by electrophoresis at pH 3.3[17]. A closer examination of the acetone extracts of the freeze dried cells of five different strains of *Serratia marcescens*, employing chromatographic techniques, has revealed the existence of a family of depsipeptides (11) which are similar to serratamolide[18].

The monamycin family of depsipeptides (12) produced by *Streptomyces jamaicensis*[19], was isolated as a mixture of 15 congeners by extraction of the culture medium at pH 6.5–7.5 with either light petroleum or diethyl ether. Fractional crystallisation of the crude mixture, followed by extensive countercurrent distribution (up to 5000 transfers) resulted in the isolation of monamycins A, C, D_1, E, F, H_1 and I, from this mixture[20]. However, D_1 was still contaminated with 6% D_2 and H_1 with 3% H_2.

This last example serves to demonstrate that the methods presently available for the isolation of a single depsipeptide from its congeneric depsipeptides often prove to be inadequate. In some cases, the difficulties inherent in such isolations have only become apparent due to the improvements in physico-chemical techniques, since the use of these techniques has shown that certain depsipeptides which were believed to be homogeneous are, in fact, mixtures of congeners.

8.3 STRUCTURAL ELUCIDATION

The successful utilisation of physico-chemical techniques for the examination of the structure of depsipeptides has been a major feature of recent research. The procedure normally employed for the structural investigation of a depsipeptide involves an initial acid hydrolysis in order to identify the amino and hydroxy-acid constituents, followed by partial hydrolysis and mass spectral studies to establish the sequence of these constituents. Nuclear magnetic resonance spectroscopy may then be used, in conjunction with infrared spectroscopy, optical rotatory dispersion and potential energy calculations, to obtain information on the conformation of the depsipeptide. Reviews which illustrate the use of these techniques include one on valinomycin and enniatin B[21] and another on cyclic peptides and depsipeptides[22].

8.3.1 Hydrolysis

Vigorous acid hydrolysis cleaves all the amide and ester bonds in a depsipeptide, whereas mild alkaline hydrolysis usually selectively hydrolyses the ester linkage(s). The disappearance of a band at *c.* 1745 cm^{-1} in the infrared spectrum of a suspected depsipeptide on treatment with dilute alkali is normally taken as conclusive proof of its lactone structure.

The identification of the acid hydrolysis products of depsipeptides may present a number of problems. The presence of any new amino acids, the identification of which may involve mass spectral and n.m.r. studies, and the necessity of isolating the hydroxy-acid(s) present, are the most common difficulties encountered. Known amino acids are readily identified using an amino acid analyser or chromatographic techniques and their configuration may be established by specific rotation measurements or by treatment with D- and L-amino acid oxidase. Comprehensive tables of the amino and hydroxy acids obtained on acid hydrolysis of various depsipeptides may be found in two review articles[2, 23].

Most of the procedures employed for sequence determination by hydrolytic methods were used to establish the amino acid sequence of enduracidins A and B. For this reason, the enduracidins have been chosen as representative examples of the use of hydrolysis techniques. Other depsipeptides mentioned in this section are of interest either due to the isolation of new amino acids from their hydrolysis products or because of some other unusual features.

The enduracidins[16] presented a considerable challenge in sequence determination due both to the number of amino acid residues in the molecule, seventeen, and the presence of three new amino acids together with a previously unknown fatty acid moiety. These new amino acids were identified[24] as α-amino-3,5-dichloro-4-hydroxyphenylacetic acid, α(*S*)-amino-β-4(*R*)-(2-iminoimidazolinyl)propionic acid and α(*R*)-amino-β-4(*R*)-(2-iminoimidazolinyl)propionic acid, and were present in both enduracidin A and enduracidin B. The fatty acid moiety of enduracidin A was shown, using n.m.r. spectroscopy and mass spectrometry, to be 10-methylundeca-2(*cis*)-4(*trans*)-dienoic acid, and in a similar manner, the fatty acid component of enduracidin B was shown to be 10-methyldodeca-2(*cis*)-4(*trans*)-dienoic acid.

Treatment of enduracidin A with dilute alkali at room temperature gave enduracidic acid A. The 1750 cm^{-1} absorption in the infrared spectrum of enduracidin A could not be detected in the spectrum of enduracidic acid A, which, taken together with the fact that no second fragment containing a hydroxyl group could be found in this alkaline hydrolysate, was consistent with a lactone structure for enduracidin A.

Partial acid hydrolysis of enduracidic acid A under various conditions gave several peptide fragments. Edman degradation, dansylation and dinitrophenylation were used for the determination of the *N*-terminal sequences of these fragments, while tritium labelling, reduction with lithium borohydride and hydrazinolysis were employed for the estimation of the *C*-terminal amino acids. By reaction of enduracidin A with aqueous barium hydroxide, it was shown that *N*-(10-methylundeca-2(*cis*)-4(*trans*)-dienoyl)aspartic acid was the *N*-terminal moiety. Hydrolysis of enduracidic acid A with very dilute acid gave, as one of the products, a hexadecapeptide which contained all the amino acid residues, except aspartic acid, and Edman degradation of this hexadecapeptide enabled the sequence of five amino acids from the *N*-terminal end to be ascertained. It was proved that the α-carboxyl group of the aspartic acid residue was involved in peptide bond formation with the neighbouring threonine, by conversion of the free carboxyl of the aspartic acid residue in enduracidic acid A to the corresponding alcohol, which was subsequently detected, as homoserine, in the acid hydrolysates of this modified enduracidic acid A.

The position of the ester bond in enduracidin A was established by lithium borohydride reduction, followed by partial acid hydrolysis, which showed that α-amino-4-hydroxyphenylacetic acid was the *C*-terminus of the lactone and by reaction of both enduracidin A and enduracidic acid A with phenyl isocyanate, threonine was proved to be the *O*-terminus.

As a result of these experiments (13) has been tentatively proposed for the structure of enduracidin A, and identical results obtained in the various experiments with enduracidin B led to the conclusion that the peptide and lactone structures of enduracidin A and B are identical, and that enduracidin B has the structure (14).

$NH{\cdot}CO{\cdot}CH{:}CH{\cdot}CH{:}CH{\cdot}(CH_2)_4{\cdot}CHRMe$

$CH{\cdot}CH_2{\cdot}CO_2H$

CO

$CH_3 \quad NH$

$CH{-}CH$

(13) R = Me
(14) R = Et

The only other depsipeptides which can be compared with the enduracidins with respect to the number of amino acid residues present, are the stendomycin family, which contain the novel basic amino acid stendomycidine[25]. The structure of the dominant congener was determined by partial acid

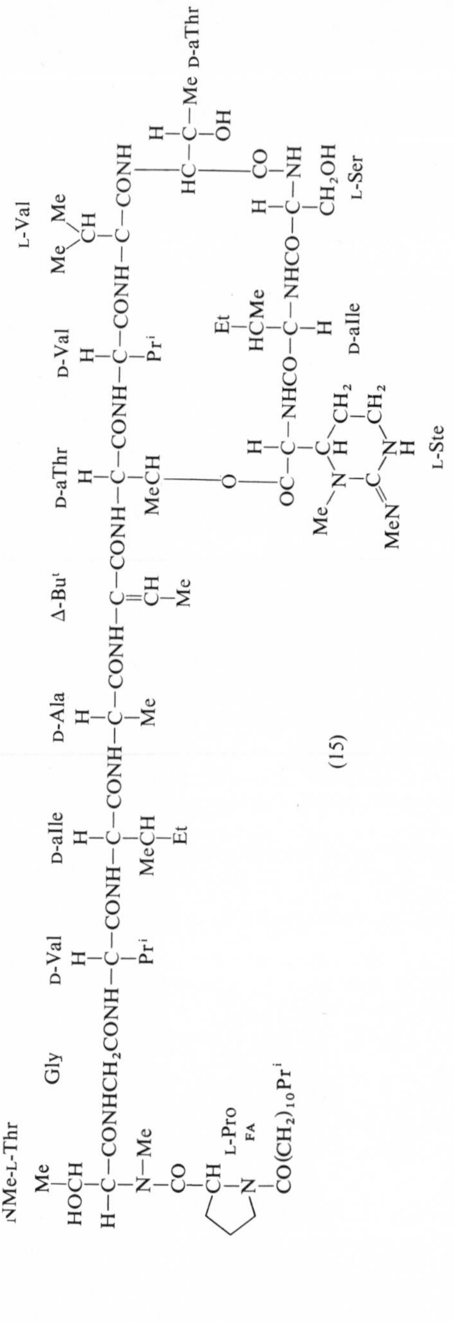

NMe-L-Thr, Gly, D-Val, D-aIle, D-Ala, Δ-Bu^t, D-aThr, D-Val, L-Val, D-aThr, L-Pro, FA, L-Ste, D-aIle, L-Ser

(15)

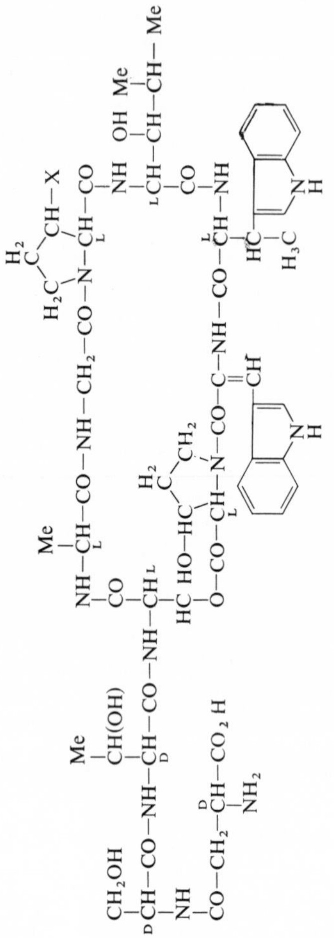

(16) X = OH
(17) X = H

hydrolysis of both stendomycin and dihydrostendomycin[26]. Various combinations of solvent extraction, t.l.c., ion-exchange chromatography, dinitrophenylation, column chromatography and electrophoresis were used to isolate the peptide fragments in the hydrolysates. The other members of the stendomycin family differ from the dominant member (15) in that isomyristic acid is replaced by its lower homologues and alloisoleucine by valine or leucine. This structure was confirmed by mass spectrometry[27]. The determination of the amino acid sequences of the antibiotics A-128-OP and A-128-P, obtained from neotelomycin by chromatography on Sephadex G-25, was accomplished using alkaline hydrolysis, since partial acid hydrolysis proved to be unsuccessful[28]. In each case, a mixture of water soluble di-, tri-, tetra- and penta-peptides and a water soluble hexapeptide was obtained. These fragments were identified using procedures which had been previously used for the determination of the structure of telomycin[29]. A comparison of the structures of telomycin and A-128-OP (16), shows that the only significant difference between these two depsipeptides is in the size of the lactone ring.

In telomycin, the ester bond is formed between the carbonyl of the same 3-hydroxyproline residue and the hydroxyl of the *allo*threonine residue. It has also been suggested (but not positively established) that all the residues in telomycin possess the L-configuration.

A further point of interest concerning the hydrolysis of A-128-OP and A-128-P (17), is that acid hydrolysis of the undecapeptides obtained from them on mild alkaline hydrolysis, did not reveal the presence of the L-threonine residue. This was due to the conversion of threonine to β-methyldehydroalanine during the alkaline hydrolysis and the subsequent destruction of the latter compound in the acid hydrolysis[30].

Acid hydrolysis of the S-520 complex of antibiotics[31] gave four previously unknown amino acids, together with five known amino acids[32]. Three of these new amino acids were identified, using n.m.r. spectroscopy, as α-amino isoheptanoic acid, α-amino isooctanoic acid and α-amino nonanoic acid[33]. The other new amino acid was shown to be L-*threo*-β-hydroxyglutamic acid. It was inferred from o.r.d. studies that the α-carbon possessed the L-configuration and the configuration at the β-carbon was deduced from the n.m.r. coupling constants of the protons of the anhydride (18)[34].

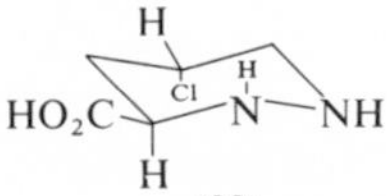

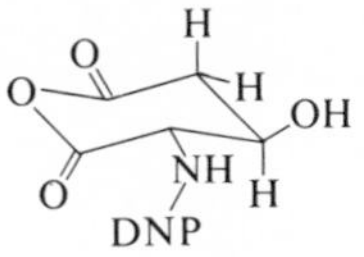

(18)

(19)

(20)

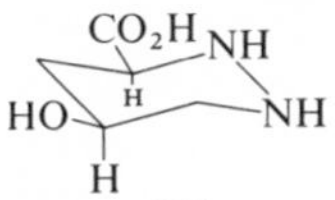

(21)

No structure has yet been proposed for individual constituents of the S-520 complex, but it has been suggested that it is a complex of several depsipeptides, certain parts of which are composed of one of a pair of amino acids which are replaceable by one another, such as valine and isoleucine, and ornithine and lysine.

Three new imino acids, piperazic acid (hexahydropyridazine-3-carboxylic acid), (19) (3*R*,5*S*)-5-chloropiperazic acid (20) and (3*S*,5*S*)-5-hydroxypiperazic acid (21) were obtained from acid hydrolysates of the monamycin family of depsipeptides. Their structures were established using degradative and n.m.r. techniques[35].

Another point of interest was the formation, in good yield, of the dioxopiperazines (22) and (23), the structures of which were determined by a combination of i.r. and n.m.r. spectroscopy[35]. Dioxopiperazines had previously been occasionally observed in the acid hydrolysates of proteins, but the yields were low. Their production, in this case, was attributed to the presence of the protonated form (24) of the combined piperazic acid residue, which could be converted into the dioxopiperazine derivative (25) by an intramolecular process.

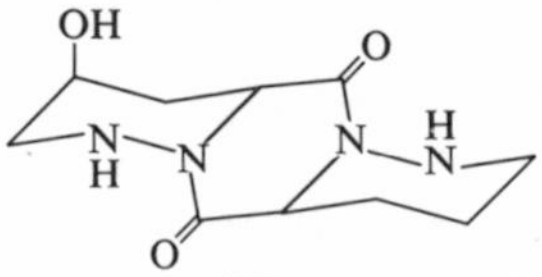

(22)

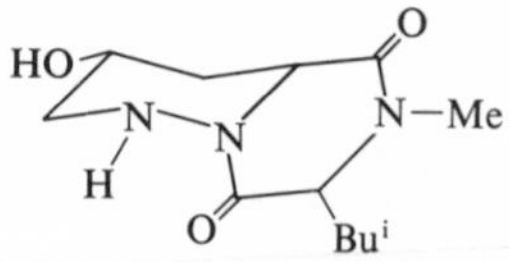

(23)

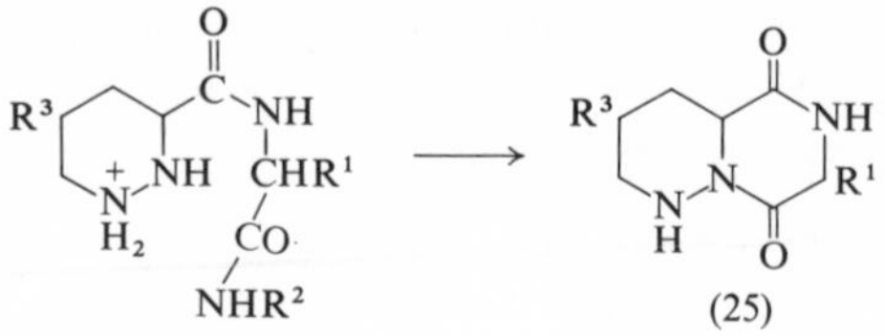

(24)

(25)

8.3.2 Mass spectrometry

As may be seen from the previous section, mass spectral studies are not always necessary for the determination of the primary structure of a depsipeptide. However, the application of mass spectrometry to molecular-weight determination and for the identification of new amino and hydroxy acids, as well as for the confirmation of proposed sequences, has meant that it has been of considerable value in structural elucidation. For sequence determination, mass spectrometry is normally used in conjunction with hydrolysis studies, since the amino acid fragmentation pattern is usually insufficient to determine the sequence; the quantitative amino acid content of the depsipeptide provides information of the volatility of any derivative which may be prepared, and is indicative of the type of fragmentation which will have to be taken into account in the interpretation of the spectrum.

A number of primary fragmentation routes have been suggested for depsipeptides. These include the 'COX' route, in which loss of the elements of an ester or amide bond occurs, the 'morpholine route', which involves loss of a 2,5-dioxomorpholine from adjacent amino and hydroxy acid residues, or a 2,5-dioxopiperazine from two adjacent amino acid residues, and the 'acylaminoketen route' which involves the formation of an acylaminoketen ion as a result of the cleavage of two ester bonds separated by an amide bond. The fragmentation patterns of some depsipeptides are included in two reviews[37, 38]. More recently reported fragmentation patterns are given below.

The initial fragmentation in a series of depsipeptides of the staphylomycin S (26) and etamycin (27) group has been reported to be a CO_2 type fragmentation which gave a linear peptide in which further fragmentation was of the amino acid type, with the splitting off of the amino acid residues in such a way that the positive charge always remained on the fragment carrying the 3-hydroxypicolinic acid residue[39].

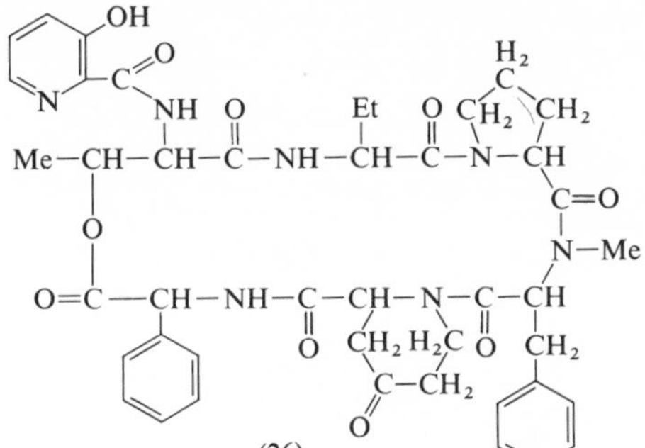

(26)

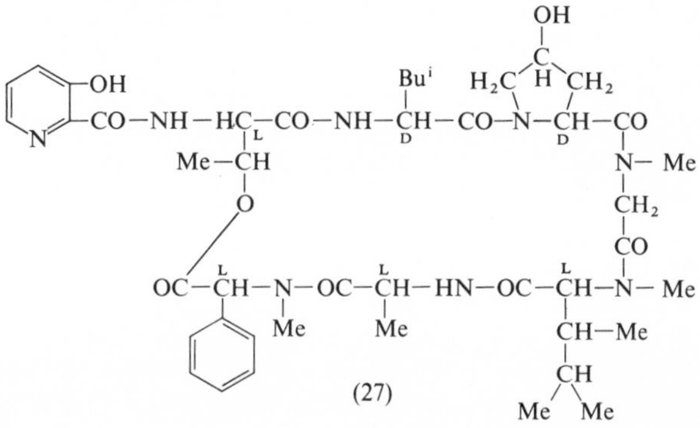

(27)

The side chains of the aromatic amino acids readily split off and were easily identified, but aliphatic side chain fission was only marked in the case of etamycin. This was attributed to the presence of leucine and *N*,3-dimethylleucine in the molecule. A later report, however, suggests that the initial fragmentation in depsipeptides involving only one lactone function, involves elimination of an azomethin molecule[40], and that this process is preferentially triggered by cyclic amino acids such as proline and pipecolic acid derivatives, and that no simple generalisations can be drawn concerning the course of further fragmentations. The azomethin type fragmentation for staphylomycin S is as shown in Figure 8.1.

Staphylomycin S $M^+ = 823$

$m^*640.4$

(a) 726 — m/e

$m^*642.6$

(b) 683

$m^*439.7$ $-CH_2C_6H_5$ $-CO_2$

(c) 548

Figure 8.1 Mass-spectral fragmentation pattern of staphylomycin S (From Compernolle, Vanderhaeghe and Janssen[40], by courtesy of Heyden Associates Ltd.)

In the case of the monamycins, the major fragmentation resulted in the formation of the acylaminoketen unit (28) and a linear ion (29). There was also evidence of a morpholinic type fragmentation leading to the ion (30) but this was a minor process[36].

[5-HyPip-MeLeu-Pro-Val]$^{+\cdot}$

(28) (29) (30)

The structure of monamycin D_1 was confirmed from the fragmentation pattern of the linear ester methyl monamycinate D_1, the major fragmentations of which were associated with sequential cleavage of the peptide bonds from both the *N*- and the *C*-terminus.

The major ion peaks observed in the mass spectra of destruxins A and B, (6) and (7), were proved to be characteristic of the *N*-methylvalyl-*N*-methylalanyl moiety in the ring structure[41]. The amino acid sequence was confirmed by analysis of the mass spectra of the open-chain derivatives. Other depsipeptides to which mass spectrometry has been applied include the griselimycins[42, 43], destruxins C and D, and desmethyldestruxin B[15] and virginiamycin S components[44].

In all of the above cases, the mass-spectral studies were carried out mainly on unmodified depsipeptides and perhaps the only general comment that can be made is that the fragmentation patterns depend on the depsipeptide and on the conditions employed. However, the use of derivatives of depsipeptides, such as permethylated derivatives, results in much simpler fragmentation patterns and in enhanced volatility, two factors which enable a better prediction to be made regarding the amino acid sequence giving rise to a particular fragmentation pattern. Thus, a revised structure (31) has been proposed for esperin, produced by *Bacillus mesentericus*, as a result of the mass-spectral fragmentation of some *N*-permethylated derivatives[45]. Doubt has been expressed regarding the previously accepted structure as a result of synthetic studies[46].

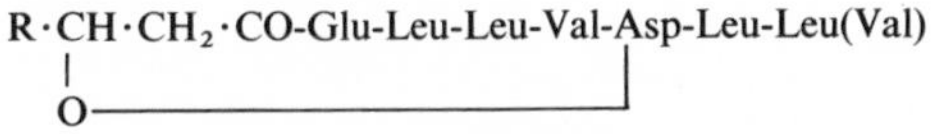

```
R·CH·CH2·CO-Glu-Leu-Leu-Val-Asp-Leu-Leu(Val)
  |                           |
  O----------------------------
```

$R = —C_{12}H_{25}, —C_{11}H_{23}$ or $—C_{10}H_{21}$

(31)

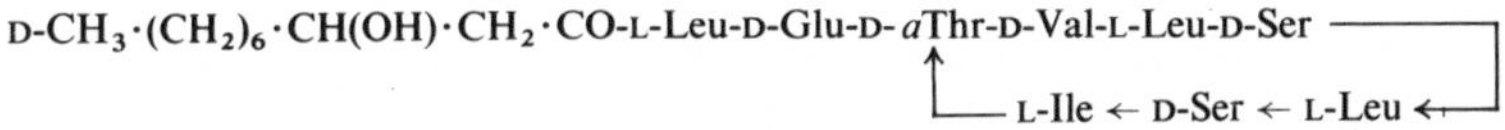

```
D-CH3·(CH2)6·CH(OH)·CH2·CO-L-Leu-D-Glu-D-aThr-D-Val-L-Leu-D-Ser ----
                                          ↑                          |
                                          L-Ile ← D-Ser ← L-Leu ←----
```

(32)

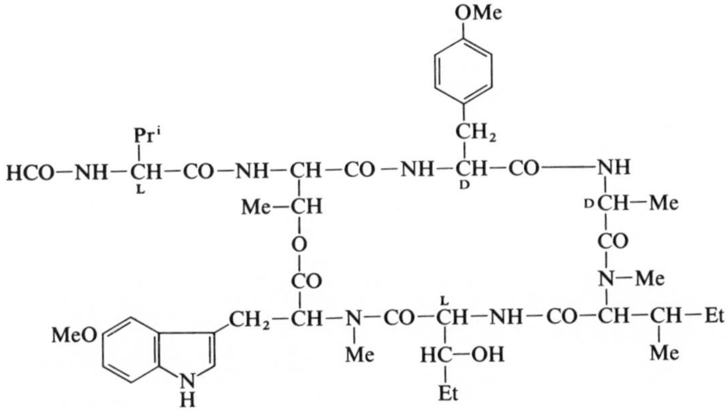

```
                                                OMe
                                                 |
                                               (C6H4)
                                                 |
         Pri                                    CH2
         |                                       |
HCO—NH—CH—CO—NH—CH—CO—NH—CH—CO————NH
        L        |          D            |
              Me—CH                   DCH—Me
                 |                       |
                 O                       CO
                 |                       |
                 CO                      N—Me
                 |            L          |
MeO(indolyl)CH2—CH—N—CO—CH—NH—CO—CH—CH—Et
                   |      |          |
                   Me   HC—OH        Me
                          |
                          Et
```

(33)

Similarly, a revised structure (32), for viscosin, produced by *Pseudomonas viscosa*, has been put forward, on the basis of the mass spectra of permethylated viscosin and viscosic acid[47]. Again, synthetic studies[48, 49] had raised doubts concerning the originally proposed structure.

Other cases in which permethylation has been used include cycloheptamycin[50] (33) and stendomycin[27] (15). In this latter case, permethylation introduced difficulties, since it was impossible to distinguish between proline, *allo*threonine, *N*-methylthreonine and dehydro-α-aminobutyric acid. However, by using perdeuteriomethylation, these difficulties were overcome.

Two recently-developed mass-spectrometric techniques which appear to have applications in the depsipeptide field, particularly for molecular-weight determination, are chemical-ionisation mass spectrometry (which has already been applied to peptides[51], using methane as the reactant gas) and vapour-deposited AgBr plates for measurements of masses up to m/e = 1700[52]. Valinomycin was used as a test compound for this latter technique and gave 42 peaks above m/e = 1000, the breakdown pattern being accounted for by straightforward loss of small alkyl fragments and CO_2. It is not generally accepted that the exclusive use of mass spectrometry provides sufficient evidence to unequivocally establish the primary structure of a depsipeptide. However, its application for the determination of molecular weights and to provide confirmatory evidence on amino acid sequences, as well as its utilisation for detecting the presence of minor congeneric constituents of depsipeptide families[39, 53], has meant that it is firmly established as an indispensable technique for structural studies of depsipeptides.

8.3.3 Other methods for determination of primary structure

The two previously-discussed techniques have found general application in primary structure determination. Methods which are only occasionally employed or which may only find application in special cases, are discussed in this section.

Beauvericin was isolated from cultures of *Beauveria bassania*, and its depsipeptide nature deduced from two carbonyl bands in its infrared spectrum (ester, 1740 cm^{-1}; amide, 1670 cm^{-1})[54]. Elemental analysis, and a molecular weight determination by mass spectrometry, suggested a molecular formula of $C_{45}H_{57}N_3O_9$. The simplicity of its n.m.r. spectrum was indicative of its cyclic symmetrical nature, but the n.m.r. data did not distinguish between two possible structures. An unambiguous assignment of its structure (39) resulted from the identification of the fragment (35) which was produced on lithium aluminium deuteride reduction of beauvericin.

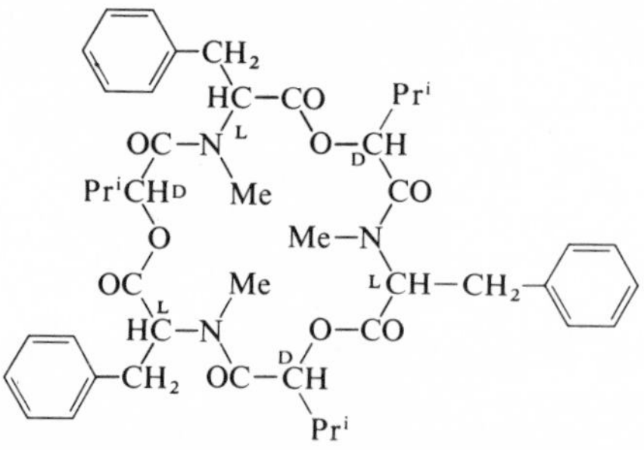

(34)

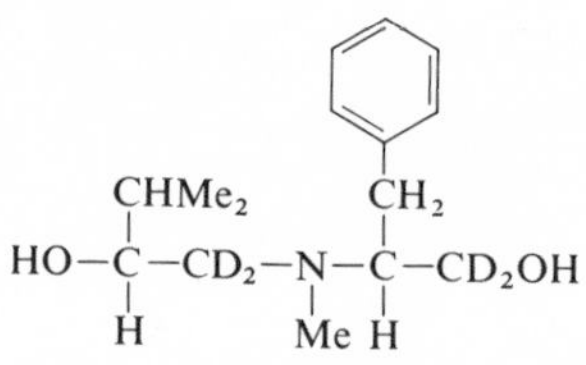

(35)

Single step pyrolysis-gas chromatography studies[55] on four actinomycins support the supposition, originally made in the determination of the amino acid sequence of actinomycin D, that diketopiperazines can only be produced from neighbouring pairs of amino acids in these peptides. Pyrolysis–g.l.c. was also used for the separation of antimycin A components (36)[56], which, because of their thermal instability, could not be identified using straightforward gas liquid chromatography. The pyrolysate of each of the isolated

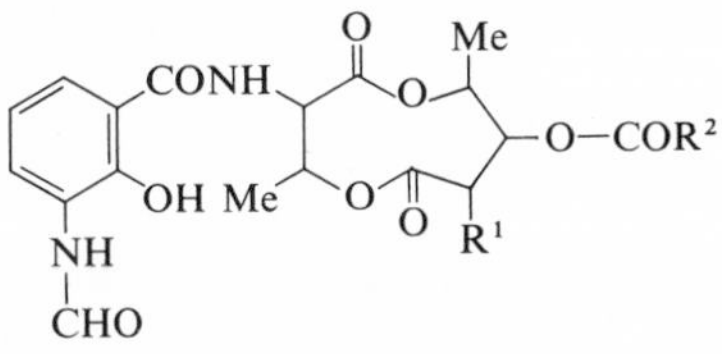

R^1 = Bu^n or n-C_6H_{13}
R^2 = Pr^i, Pr^n, Bu^s or Bu^i

(36)

pure antimycin components gave a characteristic g.l.c. pattern consisting of three major peaks, two of which were common to two different antimycins. The mass spectral fragmentation patterns of individual antimycins suggested that fragmentation occurred due to both electron impact and thermolysis. Molecular ions were only observed if the temperature of the direct inlet system of the mass spectrometer was kept below 155 °C.

Similar conclusions concerning the alkyl and acyl substitutions in antimycin components resulted from an independent study involving g.l.c. on trimethylsilyl ether derivatives of antimycins obtained from blastmycin[57].

Thiostrepton proved to be an extremely difficult problem in structural elucidation, and the proposal of the structure (37) was the culmination of many years of study.

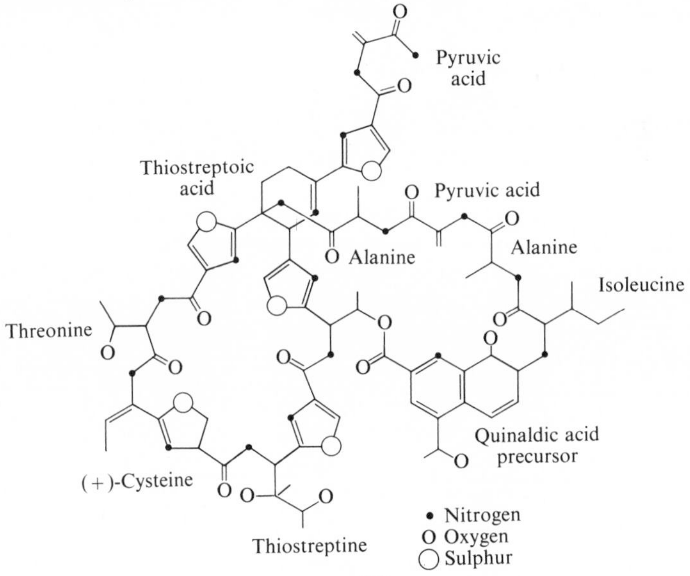

(37)

From Anderson *et al.*[59], reproduced by courtesy of Macmillan Ltd.

Acid hydrolysis gave the products shown in Figure (8.2) and earlier work[58] had suggested that the carboxyl group of the quinaldic acid precursor in thiostrepton was involved in the ester linkage. Structure (37) was proposed after a detailed x-ray study[59], and has been accepted[60] with the reservation

Alanine, Threonine, Isoleucine
Cysteine, Pyruvic acid, Ammonia

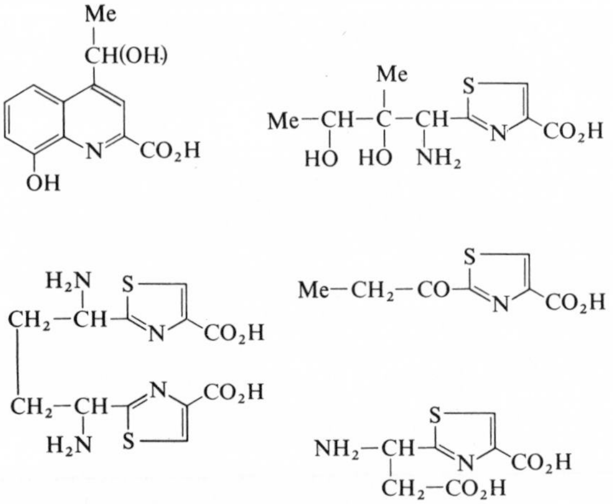

Figure 8.2

that another dehydroalanyl residue must be present in a part of the molecule not yet defined by the x-ray studies.

8.3.4 Conformational aspects

The investigation of the secondary and tertiary structure of depsipeptides, particularly those which are of biological interest, has been, without doubt, the area in which the most spectacular advances in the field of depsipeptide chemistry have occurred. Various techniques, including i.r. spectroscopy, o.r.d. and c.d. data, dipole-moment measurements and potential-energy calculations have been employed for these studies, but the one technique which has proved to be the most fruitful is n.m.r. spectroscopy. Since most of the conformational information has been obtained through the utilisation of more than one of these techniques, it is preferable to discuss conformational

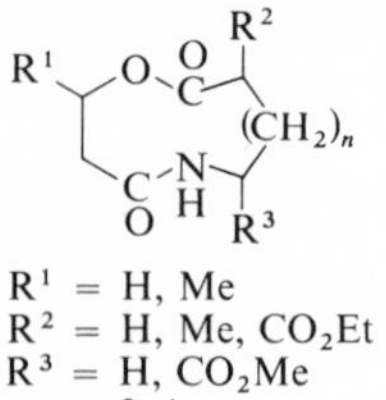

R^1 = H, Me
R^2 = H, Me, CO_2Et
R^3 = H, CO_2Me
n = 2–4

(38)

aspects in terms of individual depsipeptides rather than in terms of single techniques.

The conformational analysis of the ten-, eleven-, and twelve-membered cyclic depsipeptides (38) using vibrational spectra and dipole-moment measurements[61], has shown that the size of the depsipeptide ring and the ring substituents have a marked effect on the ratio of *cis* and *trans* amide bonds, and that this *cis*:*trans* ratio is temperature independent. For the ten membered cyclodepsipeptides, conformations allowing intramolecular interactions between the amide and ester groupings are adopted, but interactions of this type are considerably weaker in the 11- and 12-membered rings. Theoretical calculations on other 12-membered cyclodepsipeptides have resulted in the postulation of *cis* amide and *trans* ester groupings for $(Gly\text{-}OCH_2CO)_2$[62] and $(MeAla\text{-}Lac)_2$* with various configurations of the α-C atoms[63]. Optimum conformations were calculated by varying the dihedral angles and subsequently performing potential-energy minimisation calculations based on the premise that the preferred form of the ring is determined by the non-valence interactions of the closest atoms of the backbone and side chains. An x-ray study[64] of another 12-membered cyclic tetradepsipeptide cyclo-(D-Hyiv-L-MeIle-D-Hyiv-L-MeLeu)*, again confirms that for this particular ring size, the amide bonds are *cis* while the ester bonds are *trans*.

An n.m.r. and i.r. study of the 14-membered cyclic tetradepsipeptide, serratamolide, and related compounds[65], has led to the assignment of conformation (39) in which the amide and ester bonds are *trans*, being suggested. This conformation accounts for the observed coupling constants and is unaffected by temperature changes in the range −85 to 150 °C.

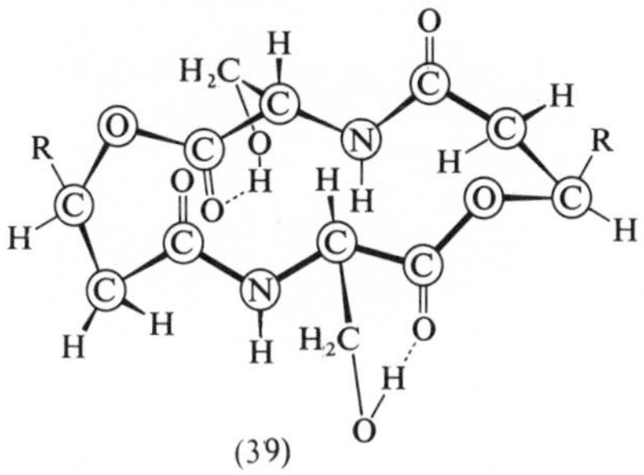

(39)

From Hassall and Thomas[22], reproduced by courtesy of The Chemical Society.

The conformation of the 16-membered pentapeptide lactone rings in actinomycin D, a potent inhibitor of DNA-dependent RNA synthesis and 'the most popular biochemical tool'[66], has been the object of much attention. As yet, none of the proposed structures have met with universal approval, but this is mainly due to differences in the interpretation of the available data. Assignments of the individual resonance in the 60,100 and 220 MHz n.m.r. spectra of actinomycin D in various solvents, involving the use of deuterium exchange, solvent perturbation and spin decoupling techniques, have shown

*Lac = lactic acid. Hyiv = α-hydroxyisovaleric acid.

that the majority of the chemically equivalent protons are not magnetically equivalent[67–71]. The involvement of the D-valine NH protons in hydrogen bonding with the sarcosine carboxyl group has been inferred from their slow exchange rates and low temperature coefficients[68], while the consistently larger deshielding of the higher field threonine NH has been used as evidence for the suggestion that the low-field threonine NH proton is involved in hydrogen bonding with the C_9 phenoxazine ketone[69]. Data obtained from a semi-empirical approach employing n.m.r., i.r., x-ray data and theoretical calculations, is consistent with the NH protons of both D-valine residues being involved in hydrogen bonding[72].

The main differences of opinion have arisen over the orientation of the sarcosine–proline peptide bond. Objections[73] to the suggestions[74, 75] that this is *trans* are based on the steric interactions introduced between the sarcosine *N*-methyl group and the proline ring. However, the fact that the equivalence of the chemical shifts of both sarcosine *N*-methyl groups is

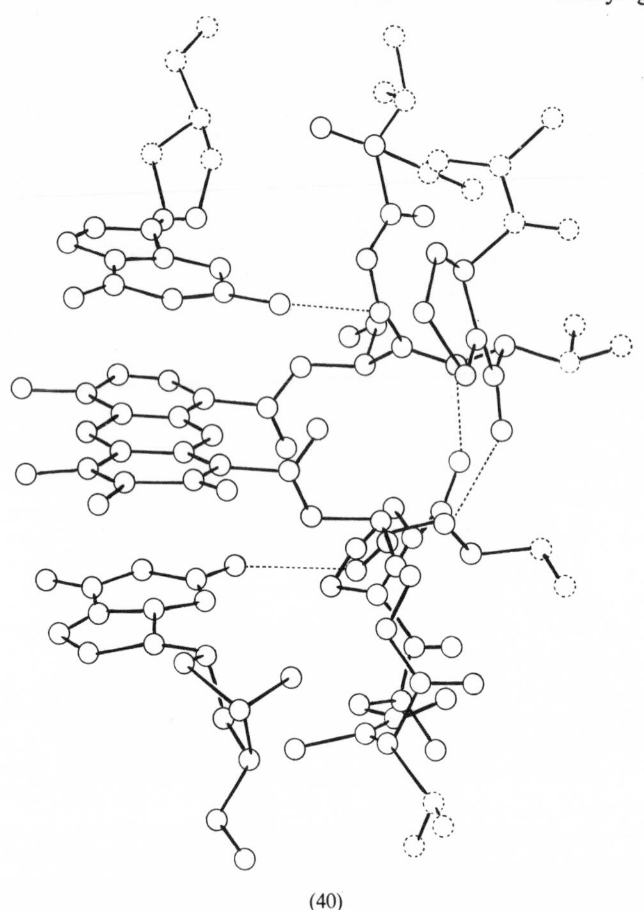

(40)

Illustration of actinomycin-deoxyguanidine complex viewed from a sideways direction
(From Sobell *et al.*[77], reproduced by courtesy of Macmillans)

maintained in a wide variety of solvents over a wide temperature range has led to the suggestion that this may be due to a screening effect exerted by the proline rings being in very close proximity to these methyl groups[67]. It is generally accepted that all other linkages in the lactone rings are *trans*. A further complicating factor regarding the conformation of the lactone rings has been introduced by the report that the sign of the Cotton effects in the c.d. spectra of actinomycin D is solvent dependent[76].

An X-ray study of the actinomycin D–deoxyguanosine complex[77] has shown that the conformation of bound actinomycin D is significantly different to its unbound solution conformation. In the complex (40) both the D-valine—L-proline and L-proline—sarcosine peptide bonds are *cis*, the other bonds are *trans*. The only hydrogen bonding involving the lactone rings occurs between the D-valine residues in different lactone rings.

An n.m.r. study of the actinomycin D-deoxyguanylic acid complex[69], has produced a qualitative picture of this complex, but insufficient data was available to allow a detailed conformational description to be made.

Cyclic hexadepsipeptides are the most common naturally-occurring depsipeptides and it is therefore not surprising that they have received more attention than other depsipeptides. They are also of interest since they are the smallest depsipeptides which act as ionophores, i.e. they may interact with certain alkali metal cations to form lipid-soluble complexes which greatly affect the cation permeability of various natural and artificial membranes. Except in cases where *N*-methylation is a precluding factor, transannular hydrogen bonding and β-turns are characteristic features of their conformations. Thus the conformation of monamycin D_1, which contains only one amide proton, incorporates a β-turn in which the proline residue occupies a similar position to that adopted in certain proline-containing cyclic peptides[22]. It should, however, be pointed out that a recent report[78] has concluded that the transannular hydrogen bonds in cyclic hexapeptides are not important conformational determinants, and that the residues with side chains adopt the energetically most favourable positions.

A 220 MHz n.m.r. study of vernamycin Bα (41) and patricin A (42) in

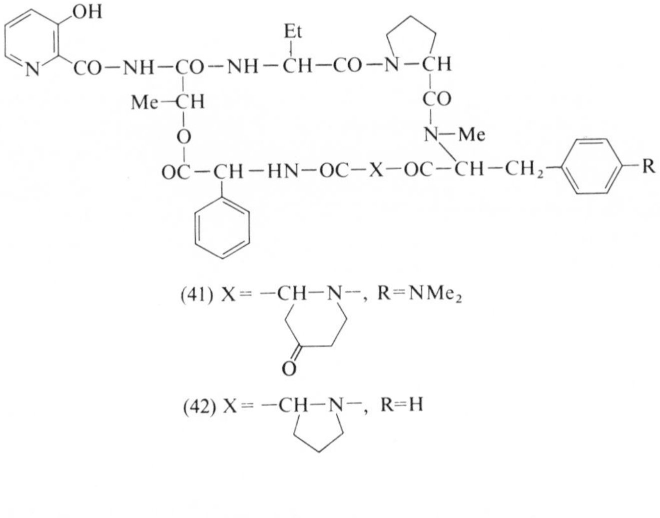

OH; Et; CO—NH—CO—NH—CH—CO—N—CH; Me—CH; CO; O; N—Me; OC—CH—HN—OC—X—OC—CH—CH$_2$—(C$_6$H$_4$)—R

(41) X= —CH—N—, R=NMe$_2$

(42) X= —CH—N—, R=H

DMSO, has shown that although they are closely related members of the streptogramin family of antibiotics, the conformations that they adopt differ markedly[79].

In the case of vernamycin Bα, all of the amide proton resonances were observed and from proton–deuterium exchange data, temperature studies and decoupling experiments, it has been suggested that the L-phenyl-glycinyl amide proton is hydrogen bonded to the peptide carbonyl of the D-amino-butyryl residue, and that the amide proton of the D-aminobutyryl residue is hydrogen bonded to the oxygen of the ester linkage. In the case of patricin A, only two of the three proton resonances were observed. The α-CH–NH coupling constants were comparable with those of the three amide protons of vernamycin Bα, but the proton-deuterium exchange rates were slower and in a reverse order. It has been suggested that one possible explanation for these differences may be due to the fact that non-planar amide bonds are possible in particin A.

The enniatins and beauvericin have been studied in great detail because of their ionophoric behaviour. In these compounds, all of the amide nitrogens are methylated, so that they might be expected to possess considerable conformational flexibility due to the absence of hydrogen bonding. Theoretical conformational analysis[80, 81] of MeCO—*N*(Me)—(L)CH($CHMe_2$)—CO_2Me and MeCO—O—(D)CH($CHMe_2$)—$CONMe_2$, which model the amino and hydroxy acid fragments of enniatin B, strongly suggests that, in solution, enniatin B can assume two principal conformations, designated N and P, which do not differ significantly in their energies. O.R.D. studies[82–84] confirm the existence of two different conformations in non-polar and polar media. A study of the N → P transition in the system dioxan–trifluoroethanol[85] shows that the tendency to adopt conformation P in low polarity solvents increases in the series enniatin B, enniatin A, beauvericin, enniatin C. Information on the 'non-polar' conformation N was obtained from low-temperature n.m.r. studies[84] in CS_2, which indicated that this conformation does not possess a central cavity or symmetry elements. The 'polar' conformation P, on the other hand, has six carbonyl groups oriented towards an interior cavity, so that efficient ion–dipole interaction is provided when the cavity is occupied by an alkali metal ion. It has been demonstrated by X-ray crystallography[86] that this type of conformation is characteristic of the crystal enniatin B–K^+ complex, and it has been proved by n.m.r. studies on (tri-*N*-desmethyl)enniatin B that in the complexes with alkali metal ions, the preferentially adopted conformation is the P conformation.

Both ^1H and ^{13}C n.m.r. studies[87] show that the size of the central cavity is adjusted, by variation of the carbonyl orientation, so that cations of varying sizes may be accommodated. This consequence of the lack of hydrogen bonding is reflected in the lower ionic selectivity of the enniatins compared to valinomycin, which as will be seen below, adopts a relatively rigid conformation in its complexes, due to hydrogen bonding, and is a highly selective ionophore for K^+.

The conformation of the cyclododecadepsipeptide valinomycin and its K^+ complex have come under close scrutiny[79, 80, 82, 83, 87–97]. As in the case of the enniatins, the o.r.d. curves of valinomycin vary with the solvent[80, 82, 83] which is indicative of a conformational equilibrium, and n.m.r. and i.r. studies

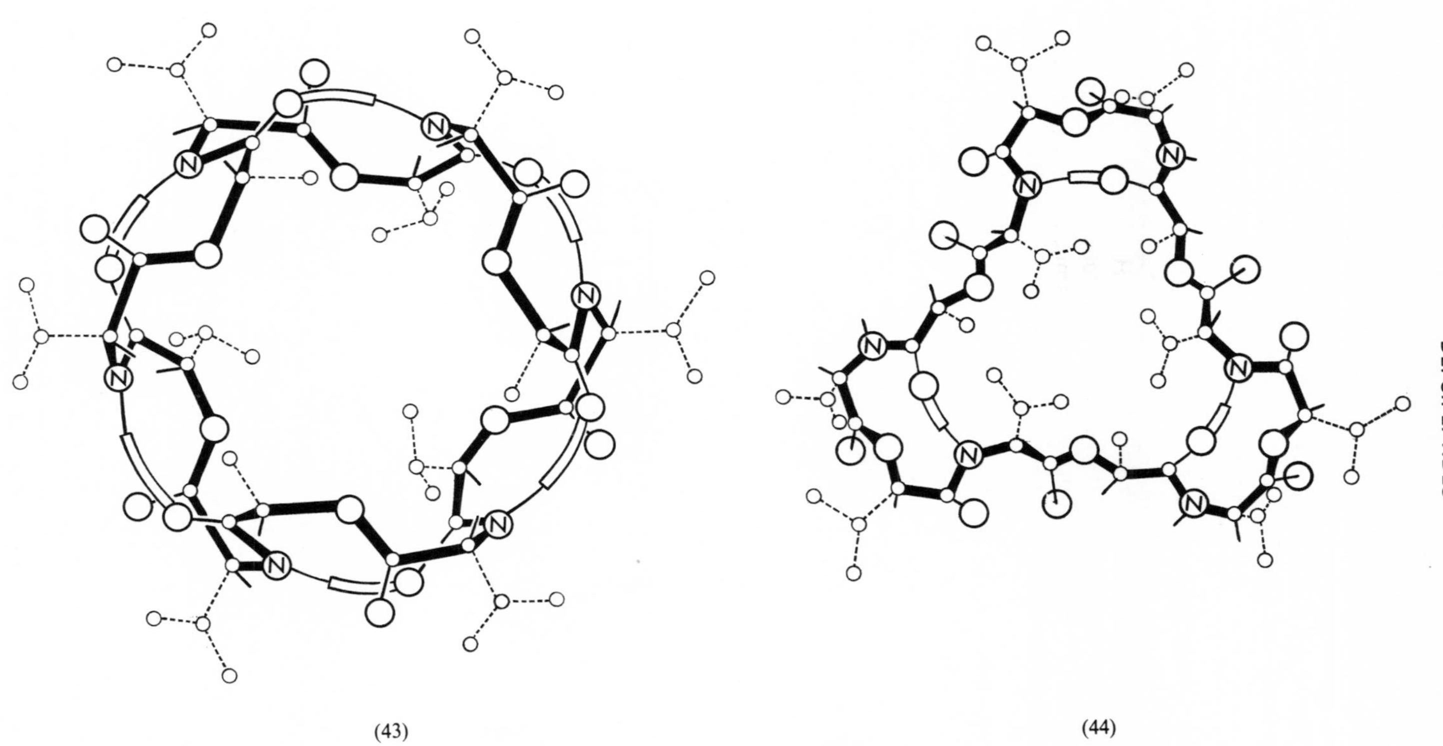

From Ovchinnikov[125], by courtesy of Butterworths.

support the proposition that in low polarity solvents, two conformers are present. Theoretical calculations of the energy of various conformations stabilised by hydrogen bonding are in agreement with the suggestion that in non-polar solvents, valinomycin adopts a bracelet-like conformation (43) 8 Å in diameter and 4 Å in height, in which there are six β-turns stabilised by hydrogen bonds. As the polarity of the solvent is increased, a conformation (44) in which there are three β-turns is adopted.

From a 220 MHz n.m.r. study of valinomycin in DMSO[79], one conformation which is very similar to (44) has been forwarded, and from proton–deuterium exchange studies, it has been suggested that this conformation may be rapidly interconverting with another of similar energy. However, low-temperature studies in CD_3OD indicate that a unique conformation is adopted at low temperature. Polar solvents destroy all the intramolecular hydrogen bonding, so that a number of conformations of similar energies are populated.

An examination of the solution conformation of the valinomycin–K^+

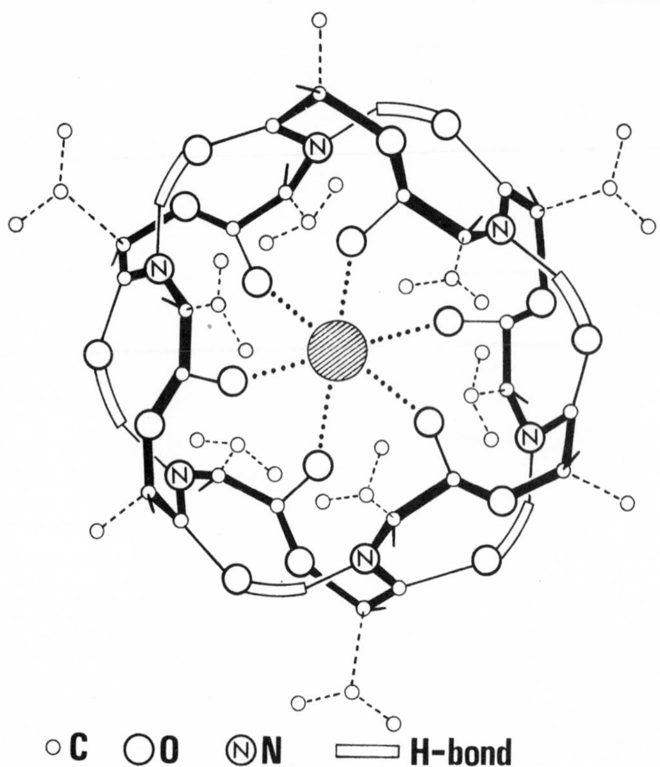

From Orchinnikov[125], by courtesy of Butterworths

complex (45)[80] shows that the potassium ion is held in a polar core of the correct dimensions for binding the non-hydrated ion, while at the same time the complex presents a non-polar exterior to the surrounding medium. A comparison of (43) and (45) shows that although there are still six β-turns in (45) the ester carbonyls are oriented in an opposite direction to the orientation adopted in (43), and that the chirality of the amide groups and hydrogen bonds is also opposed in these two structures. It is also interesting to note that the conformational flexibility of the 36-membered ring is lost on complexation, since a plot of conformational energy versus the size of the polar core shows that the potential-energy curve rises steeply from its lowest value, which corresponds to the K^+ complex, as the change in the orientation of the carbonyl groups changes the size of the core.

^{13}C N.M.R. studies[87, 97] show that, in addition to the considerable downfield shift (3–5 p.p.m.) which occurs in the ester carbonyl resonances on complexation with K^+, the amide carbonyl resonances also shift downfield (0.5–2 p.p.m.), and it has been suggested[87], that these latter carbonyl oxygens may also be participating in weak ion–dipole interaction with the K^+ ion, a possibility which had been noted previously[91] as a result of manipulation of space-filling models.

The information gained from these conformational studies of ionophores, used in conjunction with that obtained from other studies[98–110] of the ionophoric behaviour of depsipeptides, has resulted in the emergence of a better understanding of both the mode of action of certain depsipeptide antibiotics, and of ion transport mechanisms and membrane functions. With the continuation of this biological interest and the development of new techniques, such as routine ^{13}C and ^{15}N n.m.r. spectroscopy, the conformational aspects of depsipeptides will continue to attract attention for some time to come.

8.4 SYNTHESIS

This area of depsipeptide chemistry, in contrast to the structural studies dealt with in the previous section, is one in which there has been very little progress. There are several reasons for this lack of progress, one of which is that new methodology in this field arises mainly from developments in peptide synthesis. While there have been benefits from the development of new protecting groups which may be removed under very mild conditions, some of the newer methods, such as 'solid-phase' synthesis, may only be applied in a modified form. Other reasons for the slow progress include the difficulties associated with the synthesis of some of the novel amino acids which occur in depsipeptides, and the fact that synthesis is, in many cases, no longer a necessary requirement for the confirmation of a proposed structure.

The general approach to the synthesis of a cyclic depsipeptide involves the initial formation of the ester bond, since its formation at a later stage, either for chain elongation or for ring closure, introduces problems. Limitations are thus imposed on the protecting groups which may be employed, since an ester bond must be treated more circumspectly than an amide linkage. Typical amino protecting groups are the t-butyloxycarbonyl group, which may be removed under mild conditions[111, 112], aralkyloxycarbonyl groups[113] such as the di-(*p*-methoxyphenyl)-methyloxycarbonyl group, which is removed 66 000 times faster than the t-butyloxycarbonyl groups, and the *o*-nitrophenylsulphenyl group. Groups which may be used for carboxyl protection include the t-butyl, the 2-(*p*-nitrophenylthio)ethyl[114] and the

4-(methylthio)phenyl group[115]. This latter group can be converted to the 4-(methylsulphonyl)phenyl group, which may be used as an activating group for peptide bond formation. Generally peptide bond formation may be accomplished using any of the standard coupling procedures. Ester bond formation, on the other hand, requires the use of the more strongly activating procedures, such as the acid anhydride or chloride method, although numerous di- and tri-depsipeptides have been prepared using other methods for the formation of the ester linkage[116–124]. When the synthesis of the linear depsipeptide has been achieved, ring closure, via peptide bond formation, employing high dilution techniques gives the cyclic depsipeptide. However, the yield in the cyclisation is very dependent on factors such as the configuration of the amino and hydroxy acids, and on whether or not the nitrogen

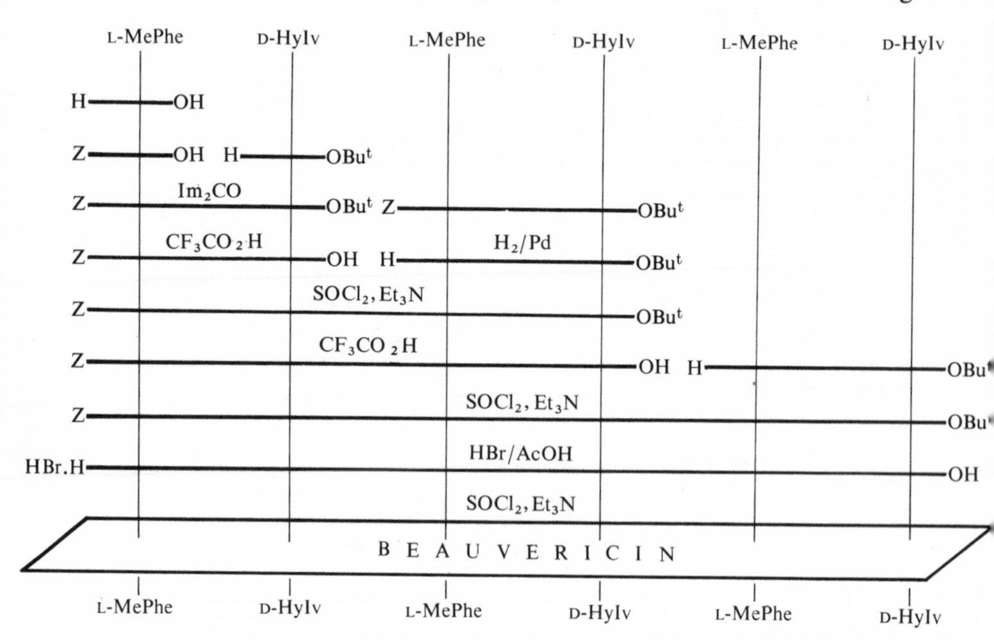

Figure 8.3 From Ovchinnikov, Ivanov and Mikhaleva[85], reproduced by courtesy of Pergamon Press

participating in the amide bond formation is alkylated[125]. Other approaches which have been utilised for depsipeptide syntheses include hydroxyacyl incorporation reactions, and, in the case of tetra- hexa depsipeptides, twinning[125].

The recently reported syntheses of depsipeptides have been, in the main, directed towards achieving a better understanding of structure–activity relationships. A review of depsipeptide synthesis[125] contains tables of the properties of numerous valinomycin and enniatin analogues which differ in ring size and in the nature and configuration of the amino and hydroxy acid residues. The syntheses of some of these valinomycin analogues, utilising the benzyloxycarbonyl group for amino protection and t-butyl esters for carboxyl group protection have been reported[126–128]. Ester bonds were

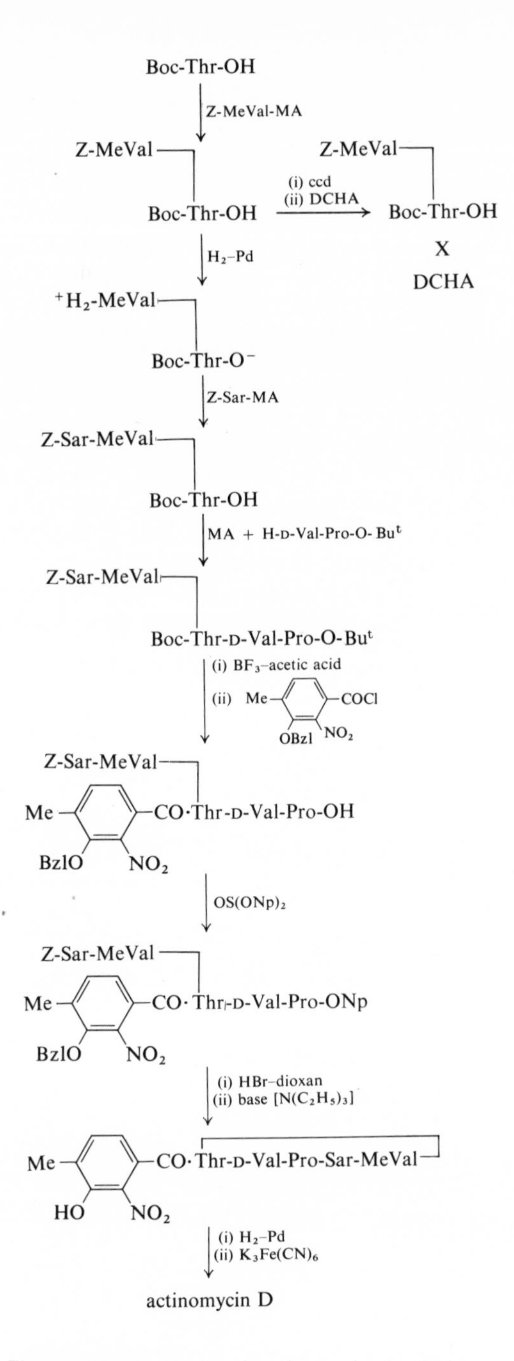

Figure 8.4 Synthesis of actinomycin D (MA, mixed anhydrido; ccd, counter current distribution; DCHA, dicyclohexylamine
(From Meienhofer[134], reproduced by courtesy of Pergamon Press)

formed by the mixed anhydride method, using benzene sulphonyl chloride, and the amide bonds formed from acid chlorides and DCCI* in the presence of *N*-hydroxysuccinimide[128]. Enniatin analogues and beauvericin have been synthesised via the route shown in Figure 8.3[85].

Syntheses of the actinomycin group of depsipeptides, and of several actinocyl derivatives, have been reported[129–138]. These latter compounds were synthesised either as model compounds for studying the binding of actinomycins to DNA, or in an attempt to obtain compounds of therapeutic value, since, apart from actinomycin D and C_3, none of the natural or synthetic actinomycins, or their derivatives, have chemotherapeutic applications. The synthesis of actinomycin D involving cyclisation of the pentapeptide lactone ring at the proline–sarcosine amide bond, as shown in Figure 8.4 has been accomplished[134].

The yield in the cyclisation step, which was carried out at high dilution in pyridine was 24–26%. Pseudoactinomycin D[135] (46) *aniso*-actinomycin C_2 (47) and *aniso*-actinomycin iso-C_2[136] (48) and four actinomycin D analogues[137] (49) have been synthesised, and it has been shown[138] that reaction of the *β*-pentadepsipeptide *O*-(benzyloxycarbonylsarcosyl-L-*N*-methylvalyl)-L-threonyl-D-valyl-L-proline with 2-nitro-3-benzyloxy-4-methylbenzoyl chloride results in oligomers of the type (50). Only the

COX, COY, O, N, NH$_2$, Me, Me

(46) X = Y = L-Thr-D-Val-L-Pro-Sar-L-MeVal (lactone ring to Thr)

COX, COY, N, O, NH$_2$, R, R

(47) X = L-Thr-D-Val-L-Pro-Sar-L-MeVal (lactone)
Y = L-Thr-D-*a*Ile-L-Pro-Sar-L-MeVal (lactone)
R = Me

(48) X = L-Thr-D-Val-L-Pro-Sar-L-MeVal (lactone)
Y = L-Thr-D-*a*Ile-L-Pro-Sar-L-MeVal (lactone)
R = Me

(49) X = Y = L-Thr-D-Val-L-Pro-Sar-L-MeVal (lactone)
R = H, OMe, Et, Bu^t

L-Pro—OH
D-Val
L-Thr
[L-Pro
D-Val
L-Thr-L-MeVal-Sar-Z]$_n$
L-Pro
D-Val
L-Thr-L-MeVal-Sar-Z
CO
NO$_2$, OBzl, Me

n = 0–3

(50)

*DCCI = *N*,*N*′-dicyclohexylcarbodi-imide

desired monomer was obtained if the addition of base was carefully controlled, but the best yields (80%) of the monomer were obtained from the symmetrical anhydride generated by the reaction of DCCI with 2-nitro-3-benzyloxy-4-methylbenzoic acid.

As has been previously mentioned, solid-phase synthesis has only been applied to depsipeptides by using modified procedures. Thus, the synthesis of valinomycin was achieved[139] by preforming all the ester bonds in solution, so that instead of eleven alternating amide- and ester-forming steps, only five peptide bond-forming steps were required, using units of the type $H_2N{\cdot}CHR^1{\cdot}CO_2{\cdot}CHR^2{\cdot}CO_2H$. D-Val-L-Lac was used as the *C*-terminal fragment, and was bound as an *N*-Boc* derivative to a chloromethylated styrene-divinylbenzene resin. The dodecadepsipeptide obtained after cleavage from the resin was cyclised in 51% yield using thionyl chloride to form the acid chloride and high dilution under basic conditions for the cyclisation step. The overall yield was 33%. Similar yields were obtained[140] by two-fold fragment-condensation of H-D-Val-L-Lac-L-Val-D-Hyiv-*O*-Ⓟ with Boc-D-Val-L-Lac-L-Val-D-Hyiv-OH and by stepwise *N*-terminal lengthening of H-L-Val-D-Hyiv-*O*-Ⓟ with Boc-aminoacyl-hydroxy-acids. Again, cyclisation of the linear dodecadepsipeptide obtained after cleavage from the support was achieved by the acid chloride method.

It has been stated[141] that 'only three significant papers concerned with the synthesis of depsipeptides have appeared in the last two years'. The period referred to was 1968–1970 and a similar statement can be justifiably made concerning the period covered in this review. Although the present methodology is adequate for the synthesis of most cyclic depsipeptides, the procedures are often tedious and time consuming, and the syntheses so far achieved have been of relatively simple depsipeptides. It is obvious that considerable progress will have to be made if syntheses of large cyclodepsipeptides containing irregular arrays of novel amino and hydroxy acids are to become commonplace.

8.5 BIOSYNTHESIS

The only statement which can be made with confidence regarding this subject is that the mechanism involved in the biogenesis of depsipeptides differs from that for cellular proteins[2,142,143]. One obvious difference is the fact that inhibitors of protein synthesis do not, in general affect depsipeptide production. Another point of interest is the presence in cyclic depsipeptides of imino acids, *N*-methylamino acids and D-amino acids.

If the biosynthetic route to depsipeptides is similar to that for peptide antibiotics, such as gramicidin S or tyrocidine, then a multi-enzyme complex, produced through a DNA-derived message on ribosomes, serves as a template for the proper sequencing of the amino and hydroxy acids. The lower specificity of these antibiotic synthesising systems, compared to protein synthesising

*Boc = t-butyloxycarbonyl

systems, is demonstrated by the ready replacement of one amino acid in a sequence by another of a similar nature, e.g. D-*allo*isoleucine for D-val in the actinomycins. This is presumed to be due to the fact that aminoacyl-tRNA synthetases, specific tRNAs or mRNA are not involved in the biosynthesis of these compounds. It has also been pointed out that there are some similarities between peptide antibiotic biosynthesis and fatty acid elongation, which is catalysed by polyenzyme systems[144]. In this context, the presence of fatty acid chains in several depsipeptides, e.g. serratamolide, the enduracidins and the S-250 complex, may be of some significance.

One structural aspect of depsipeptides which has received much attention is the presence of D-amino acid residues in the majority of depsipeptides. In many cases they have been shown to be derived from L-amino acids. Suggestions that these D-amino acid residues are formed from L-amino acids after incorporation into stereochemically labile intermediates have been forwarded[145,146], although the mechanism remains obscure[147]. The 'rule of α-epimerisation' was proposed[148] following the observation that D-*allo*isoleucine and L-isoleucine were present in many depsipeptides, but that L-*allo*isoleucine and D-isoleucine were not. The presence of D-isoleucine in some of the monamycin congeners[20], and of *N*-methyl-γ-methyl-L-*allo*isoleucine in etamycin, has been explained in terms of a mechanism involving dehydrogenation, isomerisation and rehydrogenation[149,150]. It would seem likely that the presence of D-amino acid residues in cyclodepsipeptides, especially the smaller cyclodepsipeptides, fulfils some conformational requirement for ring closure, but it has been suggested[151] that in the case of angolide, cyclo(L-Hyiv-L-Ile-L-Hyiv-D-aIle), produced by *Pithomyces sacchari*, the biosynthesis proceeds via an all L-cyclotetradepsipeptide precursor, the single D-residue being introduced by random inversion of one or the other of the two equivalent L-amino acid residues.

Studies on the biosynthesis of actinomycin D[152] have led to the isolation of an enzyme which probably activates L-valine prior to its incorporation into actinomycin D. It would be of interest to determine whether this enzyme is also responsible for the formation of D-valine, of which L-valine is the precursor, since an enzyme which both activates and racemises L-phenylalanine has been isolated from *B. brevis* stains that produce the peptide antibiotics gramicidin S and tyrocidine[153,154]. Other studies[155,156] on actinomycin D indicate that 3-hydroxy-4-methylanthranilate is involved in the biosynthesis of the actinomycin chromophore which is derived from tryptophan, and that, contrary to previous reports, chloroamphenical does not really affect the incorporation of proline into actinomycin[157]. A monolactone of actinomycin D has been isolated[158] from ethyl acetate extracts of media fermented by *S. antibioticus* 3720, although whether it is involved in the biosynthetic route to actinomycin D is doubtful.

Work on the biosynthesis of serratamolide[159,160] has shown that [^{14}C]serine, [^{14}C]acetate and [^{14}C]octanoate were incorporated by *Serratia marcescens*, but that [^{14}C]decanoate, although metabolised was not incorporated. [^{14}C]Serratamic acid appears to be degraded before being incorporated as [^{14}C]serine, and it also appears that the streptomycin-mediated inhibition of cyclic depsipeptide biosynthesis is not related to the effect of this antibiotic on protein biosynthesis[161,162].

Although recent progress in the field of peptide antibiotic biosynthesis may well prove to be of value in the elucidation of the biogenetic mechanisms for depsipeptide synthesis, problems such as the mechanisms of activation, sequentialisation and cyclisation have to be solved if a clear understanding of depsipeptide biosynthesis is to be reached.

8.6 CONCLUSION

The period covered by this review has been one in which much useful information has emerged as a result of pertinacious research, rather than as a consequence of any dramatic discovery. The most obvious advances have occurred in the field of structural elucidation, and a greater insight into structure–activity relationships has thus been obtained. Another area in which considerable progress has been made is the study of the functioning of various membrane systems, and this appears to be an aspect of depsipeptide research on which attention will continue to be focused. Many questions concerning the biogenesis of depsipeptides, their physiological role in the producing micro-organism and their mode of action remain unanswered. However, if these fundamental problems were to receive the attention that has been concentrated on structural and, in particular, chemotherapeutic aspects of depsipeptides, there is no doubt that a better understanding would be forthcoming.

ACKNOWLEDGEMENTS

I wish to thank Drs J. S. Davies, W. A. Thomas and M. J. Hall for many helpful discussions.

References

1. Shemyakin, M. M. and Ovchinnikov, Yu. A. (1967). The Chemistry of Natural Cyclodepsipeptides in *'Recent Developments in the Chemistry of Natural Carbon Compounds'*, 1. (R. Bognar, V. Bruckner, G. Fodor, Cs. Szantay, editors). (Budapest: Publishing House of the Hungarian Academy of Sciences)
2. Taylor, A. (1970). *Advan. Appl. Microbiol.*, **12,** 189
3. Davies, J. S. (1970). *Amino-acids, Peptides and Proteins,* Vol. 2, 192. (G. T. Young, editor). (London: The Chemical Society). Davies, J. S. (1971). *Amino-acids, Peptides and Proteins,* Vol. 3, 276. (G. T. Young, editor). (London: The Chemical Society)
4. Sengupta, S., Banerjee, A. B., Majumder, S. K. and Bose, S. K. (1970). *J. Sci. Ind. Res.*, **29,** 451
5. Bocker, H. and Thrum, H. (1970). *Ger. Offen.* 2, 039, 569
6. Kuznetsov, V. D. and Vikhrova, N. M. (1969). *Prikl. Biokhim. Mikrobiol.*, **5,** 549
7. Vinogradova, K. A., Poltorak, V. A., Petrova, L. I. and Silaev, A. B. (1970). *Antibiotiki,* **15,** 718
8. Petrova, L. I. and Padron, E. (1970). *Biol. Nauki.*, 75
9. Smirnova, G. M., Blinova, I. N., Koloditskaya, T. A. and Khokhlov, A. S. (1970). *Antibiotiki,* **15,** 387
10. Yagi, S., Kitai, S. and Kimura, T. (1971). *Appl. Microbiol.*, **22,** 153
11. Ratledge, C. and Hall, M. J. (1971). *J. Bacteriol.*, **108,** 314
12. Kuzhetsova, V. S., Orlova, T. I. and Silaev, A. B. (1971). *Antibiotiki,* **16,** 18

13. Katz, E., Kawai, Y. and Shoji, J. (1971). *Biochem. Biophys. Res. Commun.*, **43,** 1035
14. Arif, A. J., Singh, C., Bhaduri, A. P., Gupta, C. M., Khan, A. W. and Dhar, M. M. (1970). *Indian J. Biochem.*, **7,** 193
15. Suzuki, A., Taguchi, H. and Tamura, S. (1970). *Agri. Biol. Chem.*, **34,** 813
16. Mizuno, K., Asai, M., Horii, S., Hori, M., Iwasaki, H. and Ueyanagi, J. (1970). *Antimicrob. Ag. Chemother.*, 6
17. Meyers, E., Parker, W. L., Weisenborn, F. L., Pansy, F. E. and Principe, P. A. (1970). *Ger. Offen.* 2 035 655
18. Bermingham, M. A. C., Deol, B. S. and Still, J. L. (1970). *Biochem. J.*, 116, 759
19. Hall, M. J. and Hassall, C. H. (1970). *Appl. Microbiol.*, **19,** 109
20. Bevan, K., Davies, J. S., Hassall, C. H., Morton, R. B. and Phillips, D. A. S. (1971). *J. Chem. Soc. C*, **1971,** 514
21. Ovchinnikov, Yu. A., Ivanov, V. T., Antonov, V. K., Shkrob, A. M., Mikhaleva, I. I., Evstratov, A. V., Malenkov, G. G., Melnik, E. I. and Shemyakin, M. M. (1968). *Peptides*, 56. (E. Bricas, editor). (Amsterdam: North Holland Publishing Co.)
22. Hassall, C. H. and Thomas, W. A. (1971). *Chem. in Britain*, **7,** 145
23. Katz, E. (1969). *Progr. Antimicrob. Anticancer Chemother., Proc. 6th Int. Congr. Chemother.*, Vol. 2, 1138. (Univ. Tokyo Press, 1970)
24. Horii, S. and Kameda, Y. (1968). *J. Antibiot. (Tokyo)*, **21,** 665
25. Bodanszky, M., Marconi, G. and Bodanszky, A. (1969). *J. Antibiot. (Tokyo)*, **22,** 40
26. Bodanszky, M., Izdebski, J. and Muramatsu, I. (1969). *J. Amer. Chem. Soc.*, **91,** 2351
27. Thomas, D. W., Lederer, A., Bodanszky, M., Izdebski, J. and Muramatsu, I. (1968). *Nature (London)*, **220,** 580
28. Silaev, A. B., Katrukha, G. S., Trifonova, Z. P., Li, R. I. and Melent'eva, T. M. (1971). *Khim. Prir. Soedin.*, **7,** 130
29. Sheehan, J. C., Mania, D., Nakamura, S., Stock, J. and Maeda, K. (1968). *J. Amer. Chem. Soc.*, **90,** 462
30. Smirnova, I. G., Silaev, A. B. and Katrukha, G. S. (1971). *Khim. Prir. Soedin.*, **7,** 544
31. Shoji, J., Kojuki, S., Mayama, M. and Shimoaka, N. (1970). *J. Antibiot. (Tokyo)*, **23,** 429
32. Shoji, J. and Sakazaki, R. (1970). *J. Antibiot. (Tokyo)*, **23,** 432
33. Shoji, J. and Sakazaki, R. (1970). *J. Antibiot. (Tokyo)*, **23,** 519
34. Shoji, J. and Sakazaki, R. (1970). *J. Antibiot. (Tokyo)*, **23,** 418
35. Hassall, C. H., Ogihara, Y. and Thomas, W. A. (1971). *J. Chem. Soc. C*, **1971,** 522
36. Hassall, C. H., Morton, R. B., Ogihara, Y. and Phillips, D. A. S. (1971). *J. Chem. Soc. C*, **1971,** 526
37. Shemyakin, M. M. (1968). *Pure Appl. Chem.*, **17,** 313
38. Lederer, E. (1968). *Pure Appl. Chem.*, **17,** 489
39. Bogdanova, I. A., Kiryushkin, A. A., Rozynov, B. and Burikov, V. M. (1969). *Zh. Obshch. Khim.*, **39,** 891
40. Compernolle, F., Vanderhaeghe, H. and Janssen, G. (1972). *Org. Mass Spectrometry*, **6,** 151
41. Suzuki, A., Takahashi, N. and Tamura, S. (1970). *Org. Mass Spectrometry*, **4.** (Suppl.), 175
42. Terlain, B. and Thomas, J-P. (1971). *Bull. Soc. Chim. France*, 2357
43. Terlain, B. and Thomas, J-P. (1971). *Bull. Soc. Chim. France*, 2363
44. Vanderhaege, H., Janssen, G. and Compernolle, F. (1971). *Tetrahedron Letters*, 2687
45. Thomas, D. W. and Ito, T. (1969). *Tetrahedron.*, **25,** 1985
46. Ovchinnikov, Yu. A., Ivanov, V. T., Kostetskii, P. V. and Shemyakin, M. M. (1968). *Khim. Prir. Soedin.*, **4,** 236
47. Hiramoto, M., Okada, K. and Nagai, S. (1970). *Tetrahedron Letters*, 1087
48. Hiramoto, M., Okada, K., Nagai, S. and Kawamoto, H. (1969). *Biochem. Biophys. Res. Commun.*, **35,** 702
49. Hiramoto, M., Okada, K., Nagai, S. and Kawamoto, H. (1971). *Chem. Pharm. Bull.*, **19,** 1308
50. Godtfredsen, W. O., Vangedal, S. and Thomas, D. W. (1970). *Tetrahedron*, **26,** 4931
51. Kiryushkin, A. A., Fales, H. M., Axenrod, T., Gilbert, E. J. and Milne, G. W. A. (1971). *Org. Mass Spectrometry*, **5,** 19
52. Hignite, C. and Biemann, K. (1969). *Org. Mass Spectrometry*, **2,** 1215
53. Kiryushkin, A. A., Rozynov, B. V. and Ovchinnikov, Yu. A. (1968). *Khim. Prir. Soedin.*, **4,** 182

54. Hamill, R. L., Higgens, C. E., Boaz, H. E. and Gorman, M. (1969). *Tetrahedron Letters,* **1969,** 4255
55. Mauger, A. B. (1971). *Chem. Commun.,* **1971,** 39
56. Schilling, G., Berti, D. and Kluepfel, D. (1970). *J. Antibiot. (Tokyo),* **23,** 81
57. Endo, T. and Yonehara, H. (1970). *J. Antibiot. (Tokyo),* **23,** 91
58. Bodanszky, M., Scozzie, J. A. and Muramatsu, I. (1969). *J. Amer. Chem. Soc.,* **91,** 4934
59. Anderson, B., Hodgkin, D. C. and Viswamitra, M. A. (1970). *Nature (London),* **225,** 233
60. Bodanszky, M., Scozzie, J. A. and Muramatsu, I. (1970). *J. Antibiot. (Tokyo),* **23,** 9
61. Andreeva, L. I., Ivanova, T. M., Efremov, E. P., Antonov, V. K. and Shemyakin, M. M. (1970). *Zh. Obshch. Khim.,* **40,** 475
62. Popov, E. M. and Pletnev, V. Z. (1971). *Biofizika,* **16,** 407
63. Pletnev, V. Z. and Popov, E. M. (1970). *Izv. Akad. Nauk SSSR, Ser. Khim.,* **5,** 991
64. Konnert, J. and Karle, I. L. (1969). *J. Amer. Chem. Soc.,* **91,** 4888
65. Hassall, C. H., Moschidis, M. C. and Thomas, W. A. (1971). *J. Chem. Soc. B,* **1971,** 1757
66. Waring, M. (1970). *J. Molec. Biol.,* **54,** 247
67. Victor, T. A., Hruska, F. E., Bell, C. L. and Danyluk, S. S. (1969). *Tetrahedron Letters,* **1969,** 4721
68. Conti, F. and De-Santis, P. (1970). *Nature (London),* **227,** 1239
69. Arison, B. H. and Hoogstein, K. (1970). *Biochemistry,* **9,** 3976
70. Lackner, H. (1971). *Tetrahedron Letters,* **1971,** 2221
71. Lackner, H. (1971). *Chem. Ber.,* **104,** 3653
72. De-Santis, P., Rizzo, R. and Ughetto, G. (1972). *Biopolymers,* **11,** 279
73. De-Santis, P., Rizzo, R. and Ughetto, G. (1971). *Tetrahedron Letters,* 4309
74. Lackner, H. (1970). *Tetrahedron Letters,* 3189
75. Lackner, H. (1970). *Tetrahedron Letters,* **1970,** 2807
76. Ascoli, F., De-Santis, P. and Savino, M. (1970). *Nature (London),* **227,** 1237
77. Sobell, H. M., Jain, S. C., Sakora, T. D. and Nordman, C. E. (1971). *Nature New Biology,* **231,** 200
78. Kopple, K. D., Go, A., Logan, R. H. and Savrda, J. (1972). *J. Amer. Chem. Soc.,* **94,** 973
79. Urry, D. W. and Ohnishi, M. (1970). *Spectroscopic Approaches to Biomolecular Conformation,* 263. (D. W. Urry, editor). (Chicago: American Medical Association Press)
80. Ivanov, V. T. and Ovchinnikov, Yu. A. (1969). *Conform. Anal., Pap. Int. Symp.* (Pub. 1971), 111. (G. Chiurdoglu, editor). (New York: Academic Press)
81. Popov, E. M., Pletnev, V. Z., Evstratov, A. V., Ivanov, V. T. and Ovchinnikov, Yu. A. (1970). *Khim. Prir. Soedin.,* **6,** 616
82. Shemyakin, M. M., Antonov, V. K., Bergelson, L. D., Ivanov, V. T., Malenkov, G. G., Ovchinnikov, Yu. A. and Shkrob, A. M. (1968). *Molec. Basis Membrane Funct., Symp. 1968.* (Pub. 1969). (D. C. Tosteson, editor). (Englewood, N. J.: Prentice-Hall, Inc., Pub. 1969)
83. Ovchinnikov, Yu. A., Ivanov, V. T., Evstratov, A. V., Bystrov, V. F., Abdullaev, N. D., Popov, E. M., Lipkind, G. M., Arkhipova, S. F., Efremov, E. S. and Shemyakin, M. M. (1969). *Biochem. Biophys. Res. Commun.,* **37,** 668
84. Shemyakin, M. M., Ovchinnikov, Yu. A., Ivanov, V. T., Antonov, V. K., Vinogradova, E. I. Shkrob, A. M., Malenkov, G. G., Evstratov, A. V., Laine, I. A., Melnik, E. I. and Ryabova, I. D. (1969). *J. Membrane Biol.,* **1,** 402
85. Ovchinnikov, Yu. A., Ivanov, V. T. and Mikhaleva, I. I. (1971). *Tetrahedron Letters,* 159
86. Dobler, M., Dunitz, J. D. and Krajewski, J. (1969). *J. Molec. Biol.,* **42,** 603
87. Bystrov, V. F., Ivanov, V. T., Koz'min, S. A., Mikhaleva, I. I., Khalilulina, K. Kh., Ovchinnikov, Yu. A., Fedin, E. I. and Petrovskii, P. V. (1972). *FEBS Letters,* **21,** 34
88. Ivanov, V. T., Laine, I. A., Abdullaev, N. D., Pletnev, V. Z., Lipkind, G. M., Arkhipora, S. F., Seryavina, L. B., Meshcheryakova, E. N., Popov, E. M., Bystrov, V. F. and Ovchinnikov, Yu.A. (1971). *Khim. Prir. Soedin.,* **7,** 221
89. Pinkerton, M., Steinway, L. K. and Dawkins, P. (1969). *Biochem. Biophys. Res. Commun.,* **35,** 512
90. Ivanov, V. T., Laine, I. A., Abdullaev, N. D., Senyavina, L. B., Popov, E. M., Ovchinnikov, Yu., A. and Shemyakin, M. M. (1969). *Biochem. Biophys. Res. Commun.,* **34,** 803
91. Pressman, B. C. and Haynes, D. H. (1969). *Molec. Basis Membrane Funct. Symp. 1968.* (Pub. 1969), 221. (D. C. Tosteson, editor). (Englewood, N. J.: Prentice-Hall, Inc.)
92. Ohnishi, M. and Urry, D. W. (1969). *Biochem. Biophys. Res. Commun.,* **36,** 194
93. Haynes, D. H., Kowalsky, A. and Pressman, B. C. (1969). *J. Biol. Chem.,* **244,** 502

94. Ohnishi, M. and Urry, D. W. (1970). *Science,* **168,** 1091
95. Izatt, R. M., Nelson, D. P., Rytting, J. H., Haymore, B. L. and Christiansen, J. J. (1971). *J. Amer. Chem. Soc.,* **93,** 1619
96. Mayers, D. F. and Urry, D. W. (1972). *J. Amer. Chem. Soc.,* **94,** 77
97. Ohnishi, M. Fedarko, M-C., Baldeschwielder, J. R. and Johnson, L. F. (1972). *Biochem. Biophys. Res. Commun.,* **46,** 312
98. Ashton, R. and Steinrauf, L. K. (1970). *J. Molec. Biol.,* **49,** 547
99. Pressman, B. C. (1969). *Proc. Int. Congr. Pharmacol., 4th, 1969.* (Pub. 1970), 4, 383. (R. Eigenmann, editor). (Basel, Switz.: Schwabe)
100. Pressman, B. C. (1969). *Antimicrob. Ag. Chemother.,* **1969.** (Pub. 1970). 28. (G. L. Hobby, editor). (Bethesda, Md.: Amer. Soc. for Microbiol.)
101. Dorschner, E. and Lardy, H. (1968). *Antimicrob. Ag. Chemother.* **1968** (Pub. 1969). 11. (G. L. Hobby, editor). (Bethesda, Md.: Amer. Soc. for Microbiol.)
102. Andreev, I. M., Malenkov, G. G., Shkrob, A. M. and Shemyakin, M. M. (1971). *Molec. Biol.,* **5,** 614
103. Moeschler, H. J., Weder, H. G. and Schwyzer, R. (1971). *Helv. Chim. Acta,* **54,** 1437
104. Eyal, E. and Rechnitz, G. A. (1971). *Analyt. Chem.,* **43,** 1090
105. Haynes, D. H. (1972). *FEBS Letters,* **20,** 221
106. Pressman, B. and Haynes, D. H. (1968). *Molec. Basis Membrane Funct.,* Symp., 1968. (Pub. 1969). 221. (D. C. Tosteson, editor). (Englewood, N. J.: Prentice Hall Inc.)
107. Länger, P. and Stark, G. (1970). *Biochem. Biophys. Acta,* **211,** 458
108. Morf, W. E. and Simon, W. (1971). *Helv. Chim. Acta,* **54,** 2683
109. Frumkin, A. M., Gugeshashvili, M. I. and Boguslavskii, L. I. (1971). *Dokl. Akad. Nauk SSR,* **198,** 1452
110. Feinstein, M. B. and Felsenfeld, H. (1971). *Proc. Nat. Acad. Sci.,* **68,** 2037
111. Gray, C. T. and Khoujah, A. M. (1969). *Tetrahedron Letters,* **1969,** 2647
112. Hiskey, R. G., Beacham, L. M., Matl, V. G., Smith, J. N., Williams, Jr., E. B., Thomas, A. M. and Wolters, E. T. (1971). *J. Org. Chem.,* **36,** 488
113. Sieber, P. and Iselin, B. (1968). *Helv. Chim. Acta,* **51,** 614
114. Amaral, M. J. S. A. (1969). *J. Chem. Soc. C,* **1969,** 2495
115. Johnson, B. J. (1969). *J. Org. Chem.,* **34,** 1178
116. Stewart, F. H. C. (1968). *Aust. J. Chem.,* **21,** 1639
117. Krit, N. A., Ravdel, G. A. and Shchukina, L. A. (1968). *Zh. Obshch. Khim.,* **38,** 1015
118. Wasielewski, C. (1968). *Rocz. Chem.,* **42,** 1479
119. Wasielewski, C. (1969). *Rocz. Chem.,* **43,** 1419
120. Stewart, F. H. C. (1969). *Aust. J. Chem.,* **22,** 1291
121. Andreeva, J. I., Lyakisheva, A. G., Antonov, V. K. and Shemyakin, M. M. (1969). *Zh. Obshch. Khim.,* **39,** 2774
122. Wesielewski, C. and Hoffmann, M. (1971). *Rocz. Chem.,* **45,** 995
123. Lloyd, K. and Young, G. T. (1971). *J. Chem. Soc. C,* 2890
124. Dutta, A. S. and Morley, J. S. (1971). *J. Chem. Soc. C,* 2896
125. Ovchinnikov, Yu. A. (1971). *XXIIIrd. Int. Congr. Pure Appl. Chem.* (Boston, USA, 1971). Vol. 2, 121. (London: Butterworths)
126. Fonina, L. A., Sanasaryan, A. A. and Vinogradova, E. I. (1971). *Khim. Prir. Soedin.,* **7,** 69
127. Ivanov, V. T., Laine, I. A., Ryabova, I. D. and Ovchinnikov, Yu. A. (1970). *Khim. Prir. Soedin,* **6,** 744
128. Sanasaryan, A. A., Fonina, L. A., Shvetsov, Yu. B. and Vinogradova, E. I. (1971). *Khim. Prir. Soedin.,* **7,** 81
129. Glibin, E. N., Korshunova, Z. I., Guizburg, O. F. and Larionov, L. F. (1969). *Khim. Geterosikl Soedin.,* **1969,** 602
130. Sinitsyu, V. G., Glibin, E. N. and Guizburg, O. F. (1970). *Zh. Org. Khim.,* **6,** 500
131. Glibin, E. N., Sinitsyn, V. G. and Guizburg, O. F. (1970). *Zh. Org. Khim.,* **6,** 1020
132. Korshunova, Z. I., Zakhs, E. R. and Guizburg, O. F. (1970). *Zh. Org. Khim.,* **6,** 504
133. Seela, F. (1971). *Z. Naturforschung,* **26b,** 875
134. Meienhofer, J. (1970). *J. Amer. Chem. Soc.,* **42,** 3771
135. Brockmann, H. and Schulze, E. (1971). *Tetrahedron Letters,* **1971,** 1489
136. Lackner, H. (1970). *Chem. Ber.,* **103,** 2476
137. Brockmann, H. and Seela, F. (1971). *Chem. Ber.,* **104,** 2751
138. Meienhofer, J., Cotton, R. and Atherton, E. (1971). *J. Org. Chem.,* **36,** 3746

139. Gisin, B. F., Tosteson, D. C. and Merrifield, R. B. (1969). *J. Amer. Chem. Soc.*, **91,** 2691
140. Losse, G. and Klengel, H. (1971). *Tetrahedron,* **27,** 1423
141. Pettit, G. R. (1971). *Synthetic Peptides,* Vol. 2, 171. (New York: Van Nostrand Reinhold Company)
142. Katz, E. (1971). *Pure Appl. Chem.,* **28,** 551
143. Perlman, D. and Bodanszky, M. (1971). *Ann. Rev. Biochem.,* **40,** 449. (E. E. Snell, editor)
144. Lipmann, F. (1971). *Science,* **173,** 875
145. Manger, A. B. (1968). *Experientia,* **24,** 1068
146. Bycroft, B. W. (1969). *Nature (London),* **224,** 595
147. Katz. E. (1969). *Progr. Antimicrob. Anticancer Chemother.,* Proc. 6th Int. Congr. Chemother, 1969. (Pub. 1970). Vol. 2, 1138. (Uni. Tokyo Press)
148. Bodanszky, M. and Perlman, D. (1968). *Nature (London),* **218,** 291
149. Walker, J. E., Bodanszky, M. and Perlman, D. (1970). *J. Antibiot. (Tokyo),* **23,** 255
150. Walker, J. E. and Perlman, D. (1971). *Biotech. Bioeng.,* **23,** 371
151. Okotore, R. O. and Russell, D. W. (1971). *Experientia,* **27,** 380
152. Walker, J. E., Otani, S. and Perlman, D. (1972). *FEBS Letters,* **20,** 162
153. Itoh, H., Yamada, M., Tomino, S. and Kwahashi, K. (1968). *J. Biochem. (Tokyo),* **64,** 259
154. Roskoski, R., Gevers, W., Kleinkauf, H. and Lipmann, F. (1970). *Biochemistry,* **9,** 4839
155. Orlova, T. I., Nefelova, M. V., Kulida, L. I. and Silaev, A. B. (1970). *Biokhimiya,* **35,** 648
156. Golub, E. E. (1970). *Diss. Abstr. Int. B,* **31,** 1699
157. Nefelova, M. V., Karpov, V. L. and Khasigov, P. Z. (1970). *Biokhimiya,* **35,** 585
158. Perlman, K. L., Walker, J. and Perlman, D. (1971). *J. Antibiot. (Tokyo),* **24,** 135
159. Bermingham, M. A. C., Deol, B. S. and Still, J. L. (1971). *J. Gen. Microbiol.,* **67,** 319
160. Bermingham, M. A. C., Deol, B. S. and Still, J. L. (1972). *Biochem. J.,* **127,** 45P
161. Bermingham, M. A. C., Deol, B. S. and Still, J. L. (1970). *Biochem. J.,* **117,** 29P
162. Bermingham, M. A. C., Deol, B. S. and Still, J. L. (1970). *Biochem. J.,* **119,** 861

9
Penicillins and Cephalosporins

R. D. G. COOPER

Lilly Research Laboratories, Indianapolis, Indiana

9.1 INTRODUCTION

Increased research into sulphur chemistry is general and the availability of new reagents to accomplish novel transformations has caused the number of publications concerning the chemistry of the β-lactam antibiotics, namely penicillin and cephalosporin, to increase considerably. An expanding number of research groups is recognising the challenge and stimulation of the intimate assemblage of functional groups that constitute these molecules.

The publication in 1962 of the rearrangement of a penicillin sulphoxide to a deacetoxycephalosporin—the first interconversion of the two molecular systems brought about a renaissance of β-lactam chemistry. Further impetus has been added recently by the isolation[2] of a new type of cephalosporin with the intriguing structural relationship of a 7-methoxyl substitution in the β-lactam ring.

In recent months, reviews have appeared concerning aspects of cephalosporin chemistry[3, 4] cephalosporin synthesis[5] and structure–activity relationships of the β-lactam antibiotics[6]. In addition, a 'monograph' comprising a comprehensive review of all chemical and biochemical aspects of the cephalosporin antibiotics prior to 1971 has been published[7].

9.2 PENICILLIN CHEMISTRY

The preconceived idea, long in vogue, that the penicillin molecule was too unstable to permit many chemical transformations, was somewhat dispelled by the reported rearrangement of penicillin (1) to anhydropenicillin (2)[8]. This was the first reported instance of a molecular reorganisation of the thiazolidine ring which retained the β-lactam function intact. Recently, in another variation the *N*-tritylpenicillin ester (3) was reacted with sodium hydride/methyl iodide, and the azetidinone (4) was produced[9]. Conversion to azetidinones containing conventional penicillin side chains, e.g. (5), was accomplished by standard procedures. Unfortunately, these derivatives had

H H H RCON S N O H CO_2H (1) $\xrightarrow[\text{NEt}_3]{\text{ClCO}_2\text{Et}}$ H H H RCON S O N O (2)

R = $PhOCH_2$—

H H H Ph_3CN S N O H CO_2R^2 (3) ⟶ H H H Ph_3CN SMe N O CO_2R^2 (4) ⟹ H H H R^1CON SMe N O CO_2H (5)

R^1 = $PhOCH_2$—
R^2 = (*p*)$MeOPhCH_2$—

no significant antibacterial activity. When the phthalimidopenicillin derivative (6) was treated with triethylamine[10], a mixture of the two 7-epimers (7) and (8) resulted, their ratio depending on solvent and base used. Another product (9) isolated in this rearrangement, was obtained by cleavage of the β-lactam ring. Rearrangement of penicillin (10) gave the 7β-isomer (11) exclusively. But, perhaps the most startling and potentially useful cleavage

H H R=N S N O H $COCH_2Cl$ $\xrightarrow{\text{NEt}_3}$ H H R=N S N O O + H H R=N S N O O (8)

(6) R = [phthaloyl, O O]

(7) R = [phthaloyl, O O]

H H R=N S N H HO_2C O (9)

(10) R = $PhOCH_2$—; H

(11) R = $PhOCH_2$—; H

of the penicillin thiazolidine ring reported involved the electrophilic cleavage of the S—C-5 bond with chlorine[11]. Penicillin (12), on reaction with chlorine, gave excellent yields of both sulphenyl chlorides (13) and (14) or olefins (15) and (16) depending on the reaction conditions. In each case a 4:1 *trans/cis* mixture of the β-lactam isomers was obtained. Furthermore, (13) could be easily transformed into (15) with triethylamine (also (14) into (16)). The potential utility of these derivatives was amply illustrated when (13), after treatment with anhydrous stannous chloride, recyclised to 5-epi-penicillin (17).

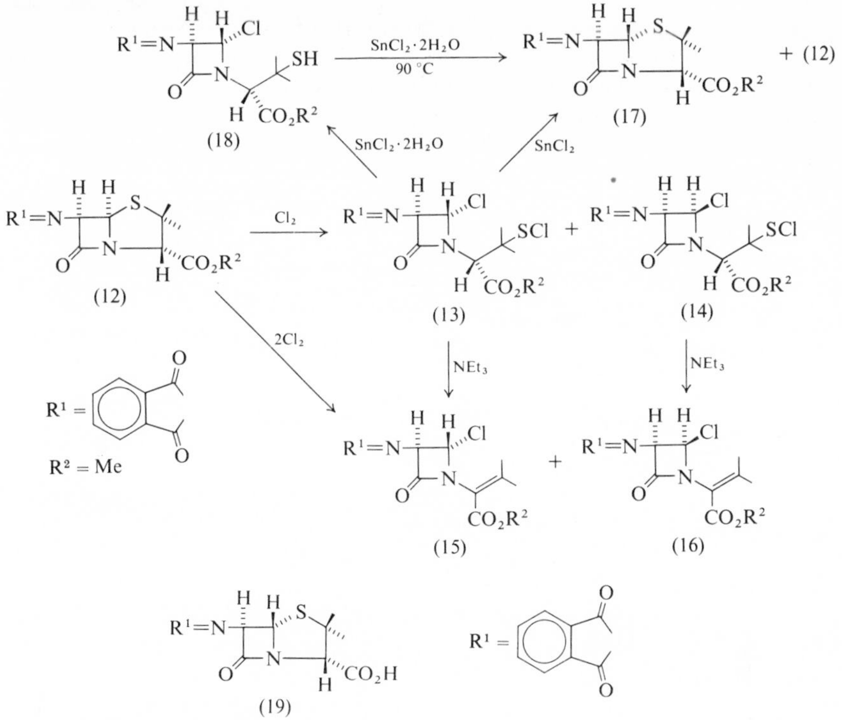

$SnCl_2 \cdot 2H_2O$, 90 °C; + (12); (18); (17); $SnCl_2 \cdot 2H_2O$; $SnCl_2$; Cl_2; (12); (13); (14); $2Cl_2$; NEt_3; (15); (16); (19)

R^1 = phthalimido; R^2 = Me

Use of stannous chloride dihydrate gave a 5:1 mixture of (17) and (12). The thiol intermediate (18) could be isolated when shorter reaction times were employed. Cyclisation of (12) with anhydrous $SnCl_2$ gave only the 5-epi isomer (17).

This represented the first synthesis of this stereoisomer of the penicillin molecule. However, the 6-phthalimido-5-epipenicillanic acid (19) was microbiologically inactive.

Considerable effort has been expended in investigating the mechanism of the base-catalysed epimerisation of substituents at C—6 in the penicillins. This epimerisation was first noted in the phthalimidopenicillin ester (12) when NaH/DMF caused formation of the 6-epi compound (20)[12], but no reverse reaction occurred although deuterium exchange at the 6β position was observed. The original work proposed a β-elimination mechanism to thiolanion. However, this now seems to be discounted in favour of a carbanion Elcb mechanism[13, 14]. This epimerisation was observed in penicillins substituted at C-6 by a variety of groups, e.g. phthalimido[13], trimethylammonium[15], bromo[15], acylalkylamido[16], and Schiff's bases[17], but not penicillins with an acylamido side chain, suggesting that a secondary amide-type substitution has too great an activation energy for the process. Generally, at equilibrium there were no detectable amounts of the 6β-isomer; however, use of 2-hydroxy-1-naphthylamino Schiff's base (21) gave an equilibrium containing 39% of the β-isomer[17, 18]. The results were consistent with a steric compression problem between a 6β side chain and the 2β methyl group. The stereochemistry of the epimerised derivatives, demonstrating the lack of involvement of the 5 position, i.e. the product was not a 5α,6β-derivative, was determined when the same 6α-chloroderivative (23) was obtained from both the 6β-compound and its epimerised derivative (22)[18]. Thiazepine (24), a by-product of the epimerisation, was formed preferentially when a weak base was used and was postulated as proceeding through the enethiolate.

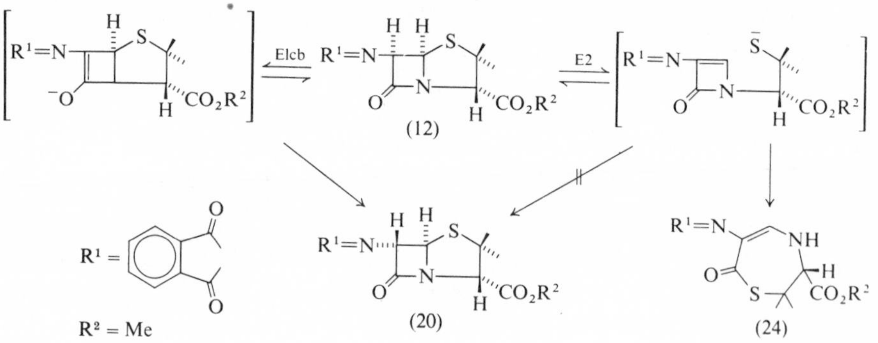

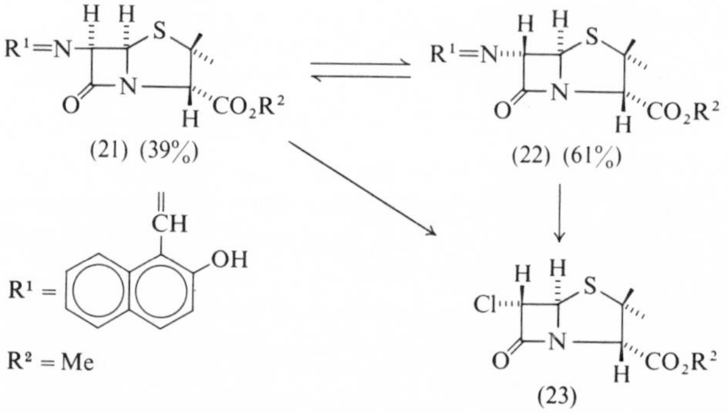

Epimerisation of the side chain proceeded with relative ease in a penicillin sulphoxide, e.g. phthalimidopenicillin sulphoxide epimerised rapidly in the presence of an amine base[19], and penicillin sulphoxides containing secondary amide side chains, e.g. (25), could now be epimerised under essentially non-basic conditions employing trimethylsilylacetamide to yield a 1:4 equilibrium of 6β:6α [20]. When the silyl epimerisation procedure was applied to an acylamidopenicillin, e.g. (27), a base was required also[21], demonstrating the lowering of the activation energy for the epimerisation by the sulphoxide group.

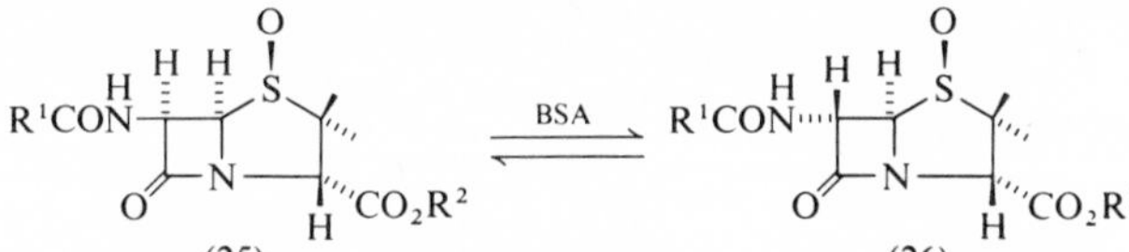

(25) ⇌ (26) [BSA]

R^1 = $PhOCH_2$—
R^2 = CCl_3CH_2—

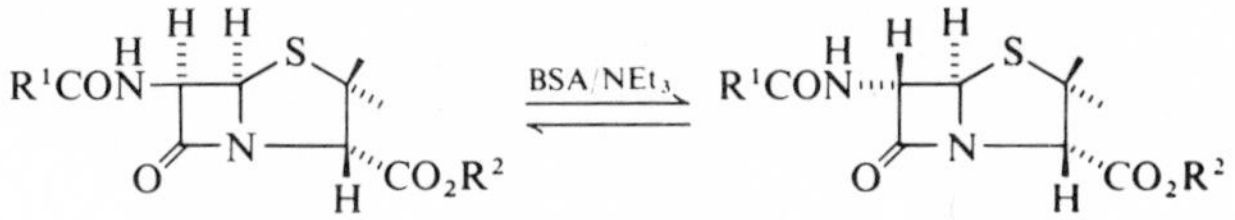

(27) ⇌ (28) [BSA/NEt_3]

R^1 = $PhOCH_2$—
R^2 = $PhCH_2$—

6-Diazopenicillin (29), first reported some years ago as being prepared in very small yields[22], has now been synthesised in considerably higher yield[23]; and although catalytic reduction gave (31), reaction with triphenylphosphine in wet ether gave a stable complex of triphenylphosphine oxide and the 6-hydrazonopenicillin (32), which could be acylated directly to the 6-acylhydrazono derivative (33). Formation of 6β-acylhydrazinopenicillin (34) was achieved by a stereoselective sodium borohydride reduction. The hydrazinopenicillins, e.g. (35), were found to be more microbiologically active than the hydrazonoderivatives against *Staphylococcus aureus* but

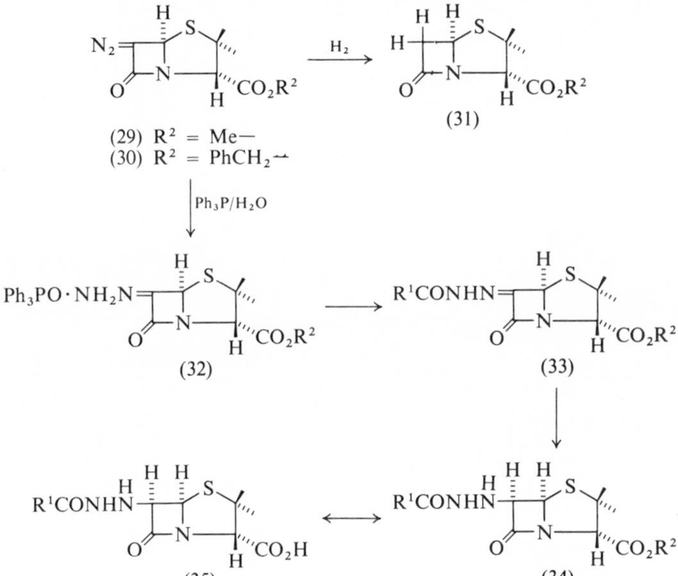

(29) R^2 = Me—
(30) R^2 = $PhCH_2$—

(29) →[H_2] (31)

(29) ↓[Ph_3P/H_2O] (32)

(32) → (33) → (34) ⟷ (35)

$Ph_3PO \cdot NH_2N$= (32)

R^1CONHN= (33)

R^1 = R^2 = $PhCH_2$—

(36) $\xrightarrow{Pb(OAc)_2}$ (37)

(36) → (39)

(39) $\xrightarrow{Ac_2O}$ (38) + (37)

(39) → (40)

$R^1 = PhOCH_2-$
$R^2 = CCl_3CH_2-$

(40) → (41) + (42)

(41) → (44)

(42) → (45)

(40) $\xrightarrow{R^3SH}$ (43)

(43) → (46)

(40) → (47)

$R^1 = PhOCH_2-$
$R^2 = CCl_3CH_2-$

considerably less effective than the parent penicillins. However, they did possess one surprising virtue: they were resistant to *B. cereus* penase, a β-lactam-destroying enzyme.

9.3 CEPHALOSPORIN CHEMISTRY

In an attempt to obtain enhanced microbiological activity, the major chemical investigation on cephalosporin has concerned the modification of the structure, usually at the C-7 and the C-3 methylene positions. These chemical manipulations will not be discussed here. An alternate approach in this search utilised the reactivity of the methylene group adjacent to the sulphur atom. Oxidation of (36) with lead tetra-acetate gave the 2α-acetoxy derivative (37)[24], which also could be obtained together with (38) by a Pummerer rearrangement of the sulphoxide (39) with acetic anhydride[25]. The sulphoxide activated the 2-position such that a Mannich reaction with formaldehyde gave the 2-methylene derivative (40)[26] in good yield. Reduction of (40) provided a mixture of the isomeric 2-methyl substituted compounds (41) and (42). Michael addition of thiol (R^3SH) to (40) gave the adduct (43). The acids (44), (45), (46) and (47) showed good microbiological activity but were not superior to the unsubstituted compound.

9.3.1 6-Methylpenicillin and 7-methylcephalosporin synthesis

The considerable interest in this β-lactam substitution originated from the suggestion that a 6-methyl group would enhance the microbiological activity of a penicillin[27]. This theory was based on the analogy between a penam structure and the D-alanyl-D-alanine terminal portion of the cell wall peptide, the part involved in the final cross-linking process in the cell wall synthesis. A 6-methylpenam would represent a better structural analogue. The first reported attempt to synthesise 6-methylpenicillin was unsuccessful[28], and involved the condensation of a substituted oxazalone with the thiazoline (48), prepared by treatment of the penicillin ester (49) with trifluoroacetic acid[29]. A crystalline product, (50), was obtained, but no information on the product stereochemistry was available. All attempts to close the β-lactam ring failed. The first stereospecific functionalisation of this position involved an interesting Stevens rearrangement[30]. The *NN*-dimethylpenicillin (51) was quaternised with allyl bromide, then treated with sodium

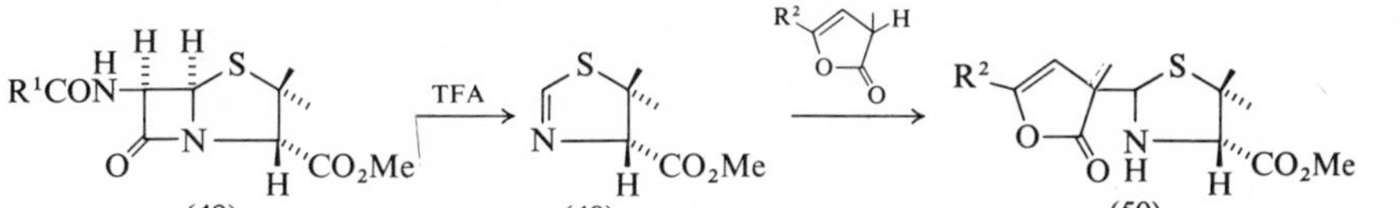

(49) TFA → (48) → (50)

$R^1 = PhCH_2-$

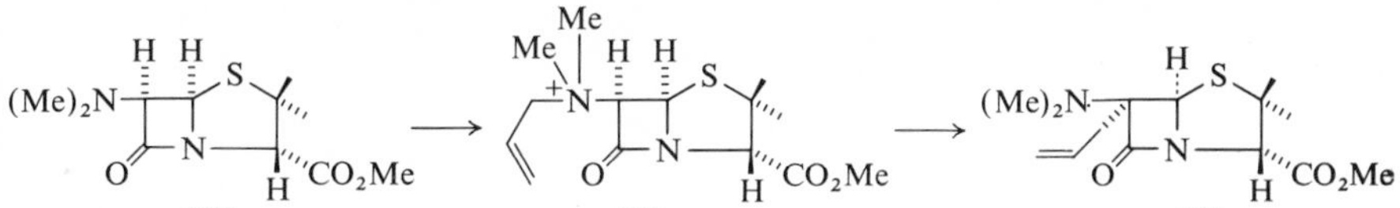

(51) → (52) → (53)

hydride. The amine (53) was isolated in 75% yield. Proof of structure using n.m.r. indicated the 6α-allyl substitution. This approach, however, has not yielded the 6-methyl substituted derivatives. The 6α-methylpenicillins have finally been synthesised successfully and found to possess greatly diminished biological activity[31]. The synthesis involved alkylation with NaH/MeI on the benzylidene derivative (54), and a 9:1 mixture of the 6α- and 6β-methyl-isomers (55) and (56) resulted. Acid hydrolysis of (55), followed by acylation, gave the penicillin ester (57). By using a cleavable ester group, the 6α-methyl-penicillin (58) was obtained. This procedure was also applied to the cephalosporins to prepare 7α-methylcephalothin (59).

H H, R^1=N, S, N, O, H, CO_2R^2
(54)

→

Me H, R^1=N, S, N, O, H, CO_2R^2
(55)

R^1 = p-NO_2PhCH=
R^2 = Me and p-$MeOPhCH_2$—

↓

Me H, H, $PhCH_2CON$, S, N, O, H, CO_2H
(58)

S, CH_2CON, H, Me H, S, N, O, OAc, CO_2H
(59)

9.3.2 6-Methoxypenicillin and 7-methoxycephalosporin synthesis

A recent report[2] disclosed the isolation and identification of three new cephalosporins (60), (61) and (62) from two species of Streptomyces, taxonomically different organisms from the species *Cephalosporium acremonium* from which cephalosporin C was isolated. Besides this being the first different organism to produce a cephalosporin, the cephalosporins (60), (61) and

R^1 H, H, $HO_2CCH(CH_2)_3CON$, NH_2, S, N, O, CH_2OCOR^2, CO_2H

(60) R^1 = H; R^2 = NH_2
(61) R^1 = OMe; R^2 = Me
(62) R^1 = OMe; R^2 = NH_2
(63) R^1 = OMe; R^2 = C(OMe)=C–C_6H_4–OSO_3H
(64) R^1 = OMe; R^2 = C(OMe)=C–C_6H_4–OH

H H, H, $HO_2CCH(CH_2)_3CON$, NH_2, S, N, O, H, CO_2H
(65)

(62) represented the first naturally-occurring structural variations on the cephalosporin C nucleus. The structural variation possessed by these new antibiotics, (61) and (62), is the 7-methoxyl substituent with presumably the same amido side-chain stereochemistry at C-7 as found in cephalosporin C. Coincidentally, another research group also has isolated independently a series of new cephalosporins, (62), (63) and (64), from several different species of Streptomyces[32]. Penicillin N (65) is also found in conjunction with these new cephalosporins just as in the *Cephalosporium acremonium* case. The structure of (60) was established by synthesis from cephalosporin C[33]. The stimulation caused by these new cephalosporins originated from their greater activity against gram negative organisms and their greater stability to the cephalosporin destroying enzyme (cephase) producing organisms. Liberation of the nucleus from a 7-methoxy-substituted cephem would generate an α-hydroxyamine, a species expected to have considerable instability especially under the chemical conditions normally used for removal of the α-aminoadipoyl moiety in ceph C (e.g. PCl_5, pyridine and methanol). This problem was solved[34] by a direct acylation (with RCOCl) of the silylated amido function of (66) to give imide (67). Liberation of the ω-amino group caused a spontaneous cyclisation and cleavage of the α-aminoadipoyl fragment as a cyclic lactam, leaving the desired 7-methoxy derivatives (68) (see scheme below) from which the acids (69) were readily obtained.

$R^1OCOCH(CH_2)_3CON(R^3)$–, NHR^2; MeO, H, S, N, O, CH_2OCONH_2, CO_2R^1
(62) → (66) → (67)

MeO, H, $R^4CON(H)$–, S, N, O, CH_2OCONH_2, CO_2R^1
(68)

(69) ⟵ (68)

(62) $R^1 = R^2$; $R^3 = H$
(66) $R^1 = Ph_2CH$—; $R^2 = —OCOCH_2CCl_3$; $R^3 = H$
(67) $R^1 = Ph_2CH$—; $R^2 = —OCOCH_2CCl_3$; $R^3 = R^4CO$; R^4 = 2-thienyl-CH_2—
(68) $R^1 = Ph_2CH$—; R^4 = 2-thienyl-CH_2—
(69) $R^1 = H$; R^4 = 2-thienyl-CH_2—

An alternative route, developed for the introduction of a methoxy group into a penicillin or cephalosporin, has the added versatility of allowing the introduction of groups other than methoxy although this is the only substituent so far reported as being introduced[35]. The key step in this sequence was the reaction of the previously reported 6-diazopenicillin[22] (30) with bromine azide to give the two azido-bromide isomers (70). Treatment of the isomeric mixture with AgF_4/MeOH gave only one methoxy azide, the 6α-methoxy isomer (71). Hydrogenation of (71) gave the nucleus (72), stable under the careful conditions used. Subsequent acylation (e.g. phenylacetyl chloride) and removal of the ester protecting group gave the 6-methoxypenicillin derivative (73). Reaction of *N*-bromoacetamide with (30) in methanol, resulted in the 6β-bromo isomer (74) which could be equilibrated with LiBr in DMF to give an isomeric mixture of methoxy bromides or, alternatively, reacted with NaN_3/DMF. The resulting azide (75) was then transformed by the procedure used previously to give the 6β-methoxypenicillin derivatives (76). The same procedure was also used for the synthesis of 7-methoxy cephems (77) from 7-diazocephem (78).

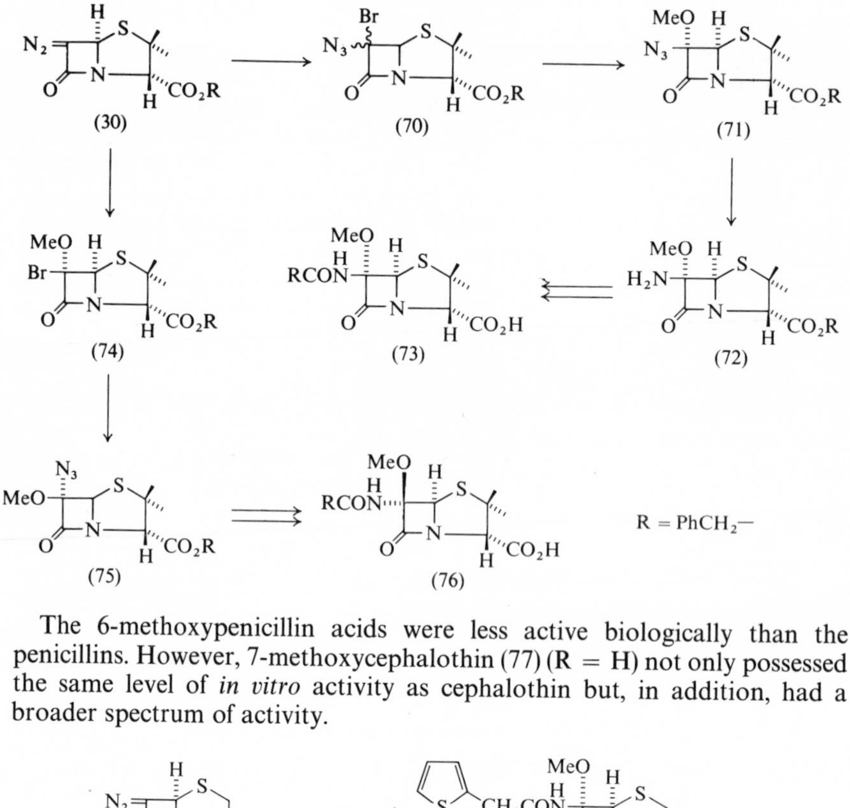

(30) (70) (71) (74) (73) (72) (75) (76)

R = $PhCH_2$—

The 6-methoxypenicillin acids were less active biologically than the penicillins. However, 7-methoxycephalothin (77) (R = H) not only possessed the same level of *in vitro* activity as cephalothin but, in addition, had a broader spectrum of activity.

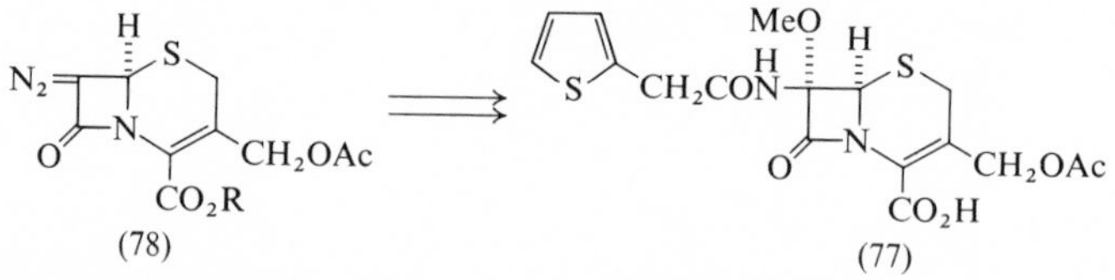

(78) (77)

9.4 PENICILLIN SULPHOXIDE CHEMISTRY

Until 1962 penicillin sulphoxide was a relatively uninteresting derivative of penicillin. Then a pioneering publication appeared[1] describing a novel rearrangement of the sulphoxide (27) (R^2 = Me) to the acetoxymethyl penicillin (78) and the two deacetoxycephalosporin derivatives (79) and (36) (R^1 = Me), albeit in low yields. Recognition that this reaction had potential for use in a synthetic route to cephalexin (80) (an orally adsorbed antibiotic)[36] initiated much research into its development and mechanism. A synthesis of (80) was recently reported[37] in which the rearrangement of (27) (R^2 = CCl_3CH_2) to (36) (R^2 = CCl_3CH_2) was accomplished in 60% yield using acetic anhydride in DMF. Further development has resulted in >75% yield for the rearrangement using methylsulphonic acid in DMAC/benzene[38].

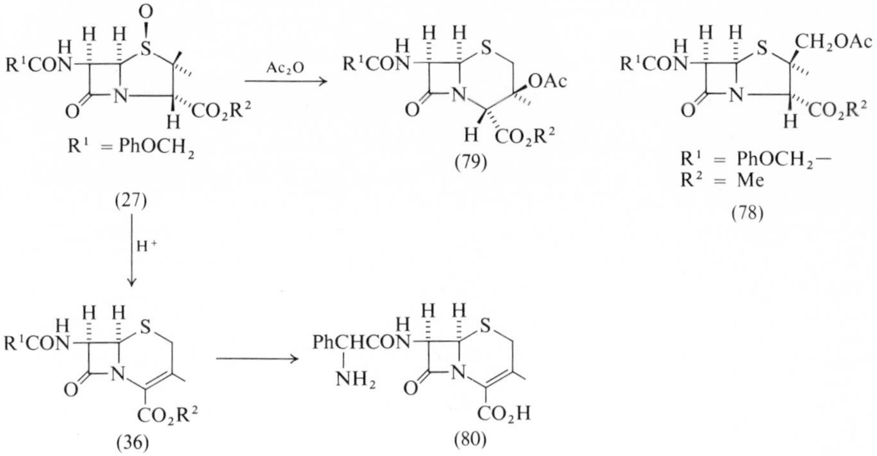

R^1 = $PhOCH_2$

(27)

(79)

R^1 = $PhOCH_2$—
R^2 = Me

(78)

(36)

(80)

The stereochemistry of the sulphoxide bond in a penicillin sulphoxide has been intensively investigated, mainly by n.m.r. procedures[39, 40, 41]. The sulphoxide obtained by oxidation of a penicillin containing a secondary amido side chain, e.g. (27), was found to be (S). This configuration was also confirmed by x-ray crystallography[39]. Phthalimidopenicillin derivatives, e.g. (12), gave (R)-sulphoxide (81)[40]. The difference in product stereochemistry was due to reagent approach control (with the oxidising agent being involved in hydrogen bonding to the amido NH) *v.* steric control. Use of either iodobenzene dichloride[41] or ozone[42] enabled both sulphoxide isomers of (27) to be obtained. The facile epimerisation of the (R)-sulphoxide

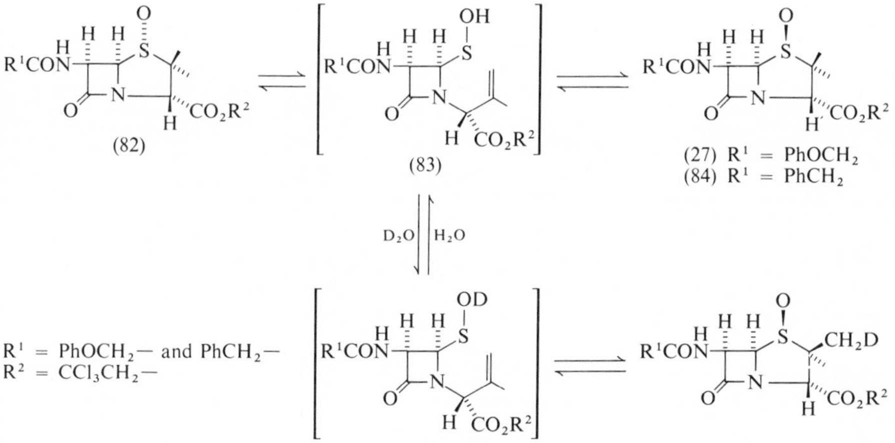

(82)

(83)

(27) R^1 = $PhOCH_2$
(84) R^1 = $PhCH_2$

R^1 = $PhOCH_2$— and $PhCH_2$—
R^2 = CCl_3CH_2—

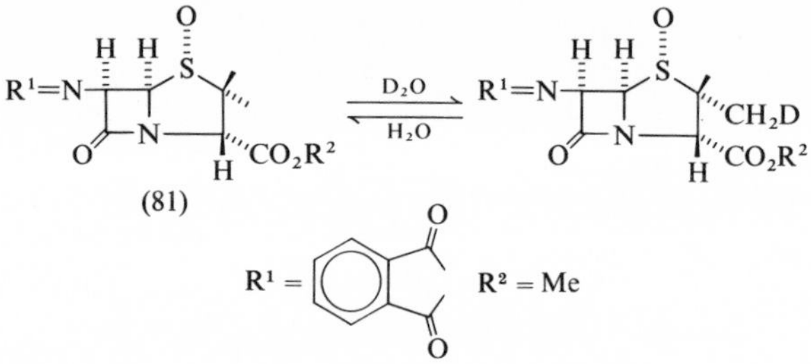

(81)

R^1 = (phthalimido) R^2 = Me

(82) to the (S) was shown[43] to occur via a sulphenic acid intermediate (83) by using Bu^{t}OD as solvent for the isomerisation and obtaining deuterium incorporation in the 2β-methyl group. This sulphenic acid had been proposed[1] to be involved in the rearrangement pathway of (27) (R^2 = Me) to the deacetoxycepham (36) (R^2 = Me). Proof of its thermal generation from (27) was provided by refluxing (27) ($R^2 = CCl_3CH_2$) in dioxane containing D_2O[44]. Deuterium was incorporated into the 2β-methyl group. The phthalimido-sulphoxide (81) incorporated deuterium into the 2α-methyl group. The incorporation was stereospecific and was considered to involve a six-electron sigmatropic reversible process.

The facile thermal generation of the sulphenic acid has initiated study of its chemical reactivity. Reduction with trimethylphosphite was reported[45] to give an intermediate thiol (84) which underwent subsequent intramolecular condensation with the amido side chain to give high yields of a thiazolidine azetidinone (84). In the presence of acetic anhydride, the major product became the thioacetyl compound (86)[46]. Use of a phthalimido side chain to

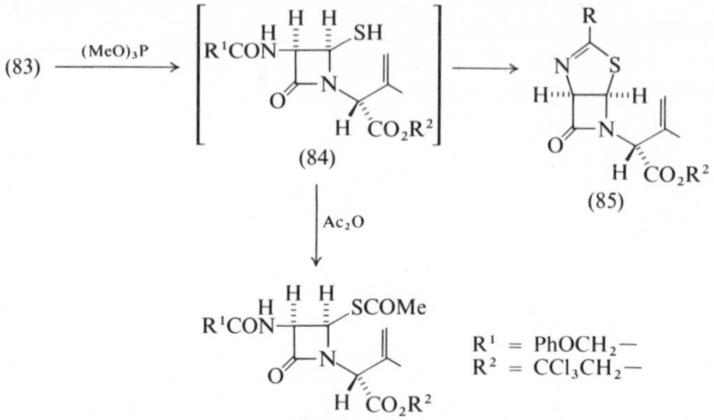

(84) (85) (86)

$R^1 = PhOCH_2$—
$R^2 = CCl_3CH_2$—

prevent the intramolecular reaction of the thiol caused the isolation of a new series of products[25], the major one being the alkylated thiol ether (87). Two minor products (88) and (89) indicated the probable intermediacy of (90) in the reduction of a sulphenic acid with trialkylphosphite.

Other intramolecular reactions of the sulphenic acid which amply demonstrated its nucleophilic properties were the reaction of the sulphoxide alcohol (91) to give (92)[25] and the diazoketone-sulphoxide (93) to give (94)[25] (as a mixture of sulphoxide isomers).

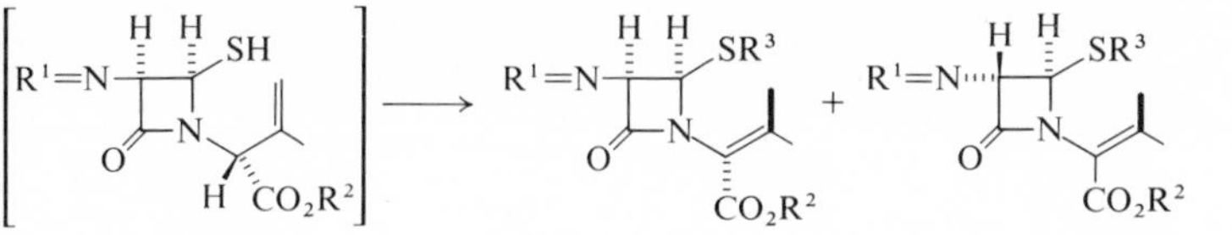

(87) R^3 = Me
(88) R^3 = $P(O)(OMe)_2$
(89) R^3 = $P(O)(OMe)_2$ 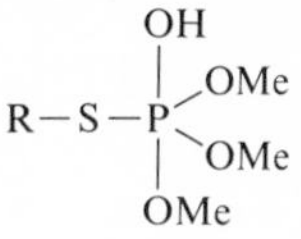
(90)

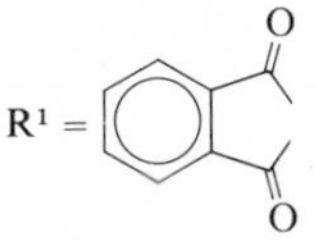

R^2 = Me

Examples of intermolecular trapping of the sulphenic acid have been reported[47]. Addition to norbornadiene gave the adduct (95)[47]; whereas, dihydropyran reacted by nucleophilic displacement on sulphur to yield (96)[47]. Other vinyl ethers, e.g. isobutyl vinyl ether gave the expected adduct[48] (97) mainly as the *trans*-substituted derivative. This derivative was somewhat

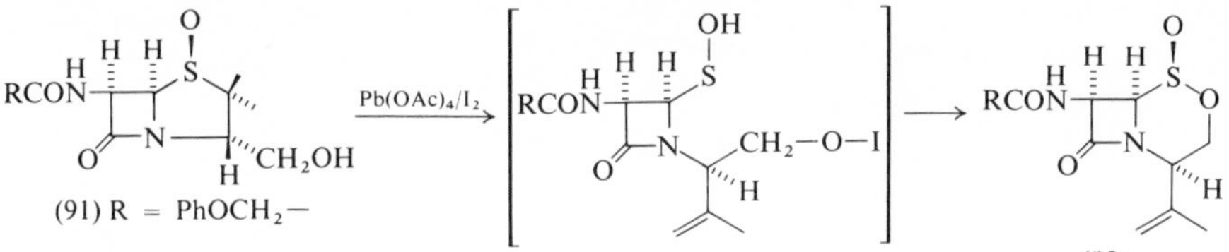

(91) R = $PhOCH_2$—

(92)

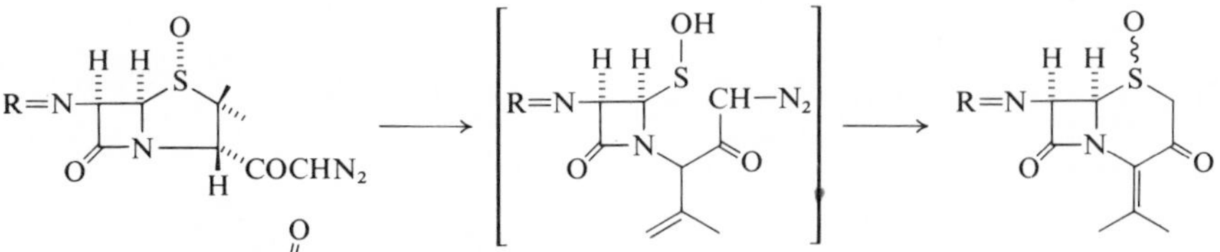

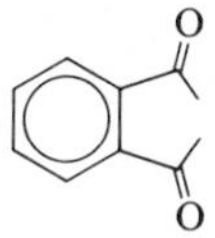

(93) R =

(94)

unstable and, on heating in the presence of a catalytic amount of acid, gave the dihydrothiazine (98). With 1,1-diethoxyethane the vinyl ether was not produced, but *in situ* hydrolysis produced the ester (99). Further examples of a nucleophilic displacement on the sulphenic acid were provided by its reactivity with thiols[49], e.g. 2-methylpropane-1-thiol, when (100) was the product. Reaction of (100) with triethylphosphite yielded the *S*-ethyl compound (101), presumably through the pentavalent phosphorus intermediate (102).

A variation of the acid-catalysed reaction of the sulphenic acid has appeared. Whereas, use of a sulphonic acid gave the cephem[1] (36) (R^2 = CCl_3CH_2), sulphuric acid gave the 3β-hydroxycepham (103)[50]. Another distinct and surprising difference in the reaction pathway caused by this reagent change was the reaction of the penicillin sulphoxide acid (104). With a sulphonic acid it had been shown previously that the decarboxylated cephem (105) was the main product[1]; with sulphuric acid the major product

(83) →

H H O
H | |
R^1CON— S
N CO_2R^2
O

(95)

R^1 = $PhCH_2$—
R^2 = CCl_3CH_2—

OBu^{iso}

EtO OEt

H H
H
R^1CON— S
N CO_2R^2
O

(96)

H H O
H
R^1CON— S OEt
N CO_2R^2
O

(99)

H H
H
R^1CON— S OBu^{iso}
N CO_2R^2
O

(97)

H^+

R^1CON S
N CO_2R^2
O

(98)

(83) $\xrightarrow{Bu^{iso}\text{-SH}}$

H H
H
R^1CON— S—SBu^{iso}
N CO_2R^2
O

(100)

$\xrightarrow{(EtO)_3P}$

SBu^{iso}
H H |
H
R^1CON— S—P(OEt)$_3$
N CO_2R^2
O

(102)

R^1 = $PhCH_2$—
R^2 = CCl_3CH_2—

H H
H
R^1CON— SEt
N CO_2R^2
O

(101)

was the 3β-hydroxycepham acid (106) [51]. A detailed argument concerning the mechanism of the acid-catalysed ring expansion has been published[52].

9.5 PENICILLIN AND CEPHALOSPORIN SYNTHESIS

To date there have been three total syntheses of cephalosporin and two partial syntheses (from penicillin) described.

The Squibb synthesis[53] has been completed in 20% yield by successful hydrolysis of the previously synthesised lactone (107) to the (±)-hydroxy acid (108). A review of the French procedure[5] described the resolution of

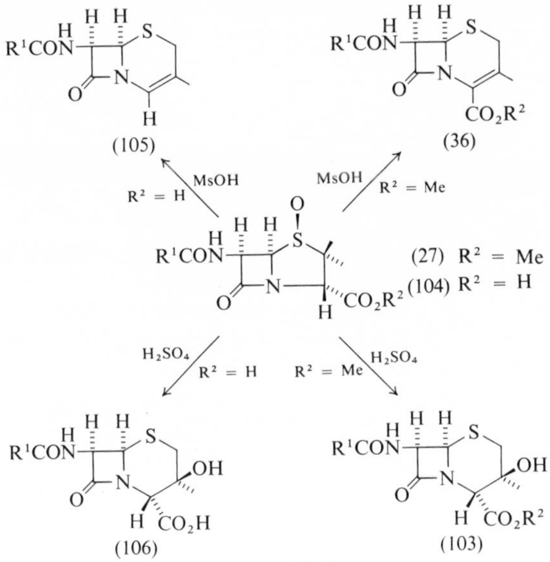

(105) (36)

$R^2 = H$ MsOH MsOH $R^2 = Me$

(27) $R^2 = Me$
(104) $R^2 = H$

H_2SO_4 $R^2 = H$ $R^2 = Me$ H_2SO_4

(106) (103)

$R^1 = PhOCH_2-$

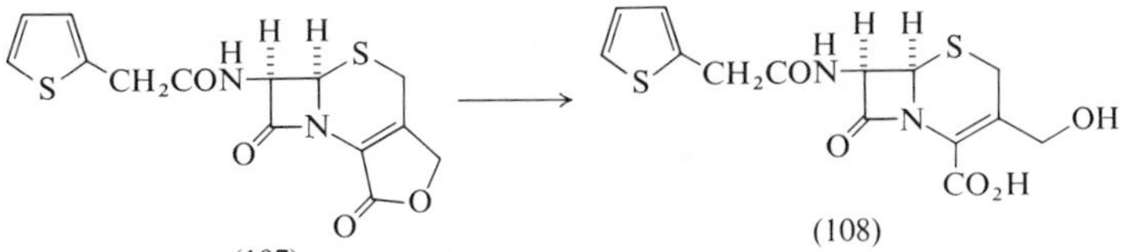

(107) (108)

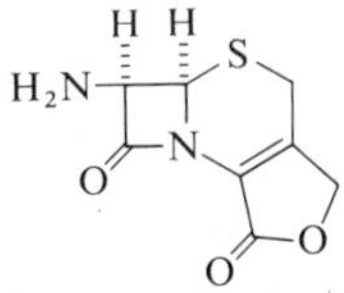

(109)

the (±)-amino-lactone (109).

The first reported synthesis[54] has been developed considerably and proven to be the most versatile for the preparation of structural modifications. The key thiazolidine-azetidinone (110) has been used to synthesise modified cephems otherwise not available[55]. Reaction of (110) with glyoxylic esters gave an equilibrium mixture containing the two isomeric carbinolamines (111) which, after conversion to their chlorides (112) and reaction with triphenylphosphine, gave a single, stable phosphorane (113). Unhindered or α-keto aldehydes readily reacted with (113) to give the double bond

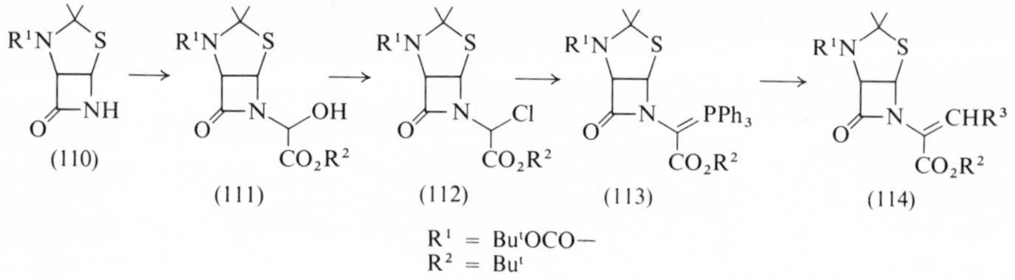

(110) (111) (112) (113) (114)

R^1 = Bu^tOCO—
R^2 = Bu^t

isomers (114). When R^3 = H, i.e. formaldehyde adduct, the Michael addition of a trityl mercaptan (or hydrogen sulphide) followed by removal of the t-butoxycarbonyl group (and trityl group if required) and oxidation with iodine gave the disulphide (116). Acylation of the amine group with phenoxyacetyl chloride gave (117) which had some *in vitro* antibacterial activity. The considerably more active dehydro derivative (118) was prepared by acylation of the mercaptan (119) with phenacyl bromide. Removal of the t-butyl

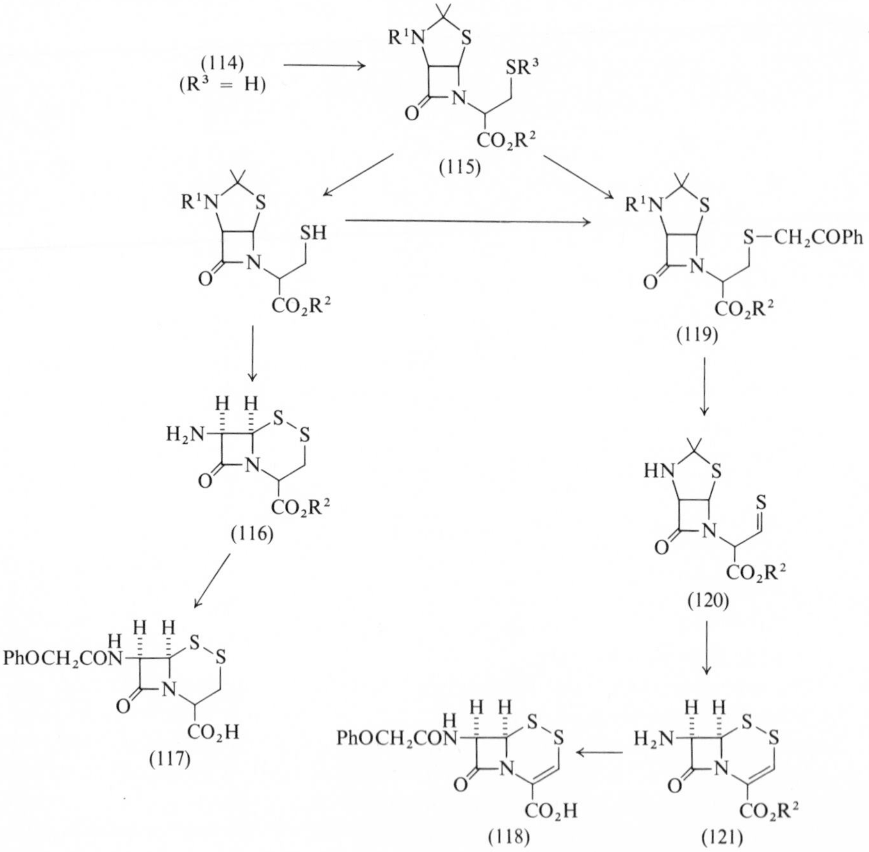

(114) (R^3 = H)

(115) (116) (117) (118) (119) (120) (121)

group and u.v. irradiation in the presence of pyridine promoted a Norrish type II cleavage to the thioaldehyde (120) which when oxidised with iodine and then acylated, furnished (118).

The phosphorane (113) also reacted with glyoxylates; e.g. $CHO{\cdot}CO_2Bu^t$ to give (122) as a mixture of *cis/trans* isomers. Reduction either catalytically or with cobalt tetracarbonyl hydride, separation of the isomeric dihydro derivatives (123) and (124), and subsequent treatment of (123) with trifluoroacetic acid/trifluoroacetic anhydride gave the thiolactone (125). Compounds possessing increased microbiological activity over (125) could be prepared by a variation of the previous scheme. Reaction of the phosphorane with α-keto aldehydes gave the double bond isomers of (126). Treatment of the *trans* isomer with trifluoroacetic acid caused ring closure and acylation gave the final antibiotic (127). Considerable variation in R^4 was reported but clear correlation of structure with activity could not be established.

(113) → (122) → (123) + (124)

(123) → (125)

(113) —$CHOCOCH_2R^4$→ (126) ⟹ (127)

R^1N, S, CO_2R^3, N, O, CO_2R^2 (122)

R^1N, S, CO_2R^3, N, O, H CO_2R^2 (123)

R^1N, S, CO_2R^3, N, O, H CO_2R^2 (124)

H H H, CF_3CON, S, O, N, O, H CO_2H (125)

R^1N, S, CH_2R^4, O, N, O, CO_2R^2 (126)

H H H, $PhCH_2CON$, S, R^4, N, O, CO_2H (127)

$R^1 = Bu^tOCO-$
$R^2 = Bu^t$ or CCl_3CH_2-
$R^3 = Bu^t$

Another use of the thiazoline-azetidinone involved the reaction of (128) with iodine to give disulphide (129)[56] which was then acylated with the desired side chain. Reduction of the disulphide with zinc in the presence of ethylene oxide gave alcohol (131) which was then protected as the oxycarbonyl derivative (132). The phosphorane (133), generated by the previously described route, underwent an internal reaction with the formyl group of (134) (prepared by DMSO/Ac_2O oxidation of the alcohol) to give 3-demethylcephalosporin (135). Both the nucleus (136) and D-phenylglycyl derivative (137) were synthesised.

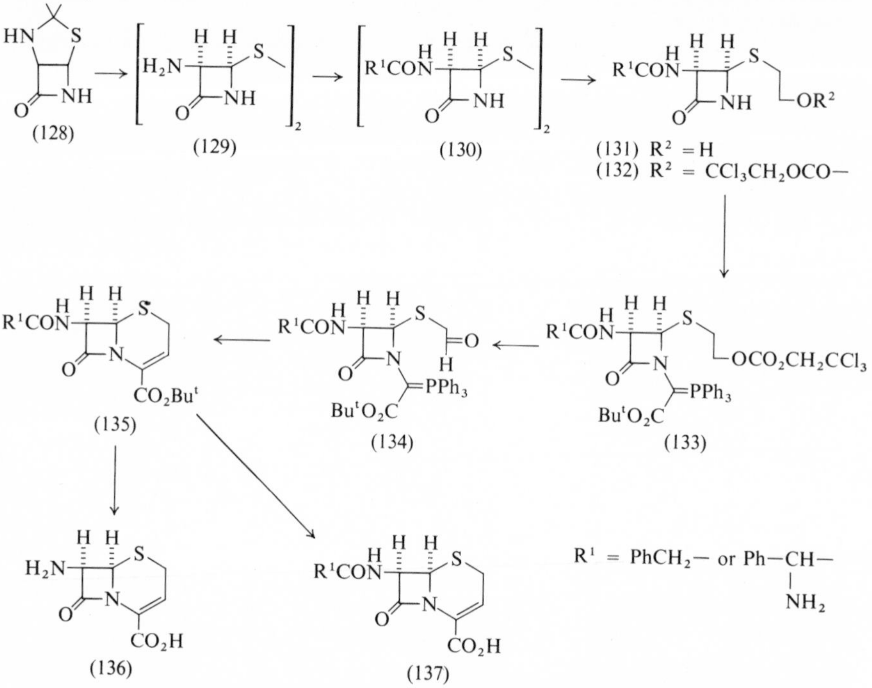

(128) → (129) → (130) → (131) $R^2 = H$; (132) $R^2 = CCl_3CH_2OCO-$ → (133) → (134) → (135) → (136), (137)

$R^1 = PhCH_2-$ or $Ph-CH(NH_2)-$

The success and versatility of this basic synthetic scheme have stimulated several alternate approaches to the synthesis of azetidinone intermediates of the same type as (128). The original workers devised an alternate route to (128) from penicillin[57]. Repeating the original cleavage[58] of the C_3–N bond via hydrolysis of the 3-isocyanate, they isolated in good yield the carbinolamine (140). The equilibrium between the carbinolamine (140) and the aldehyde (141) remained predominantly on the side of (140) in contrast to the phthalimidopenicillin where the aldehyde form predominated. The difference was due to the fact that an increase in steric requirements for the side chain increased the concentration of the aldehyde in the aldehyde/carbinolamine tautomerism. Reduction of (140) to the alcohol (142) and oxidation of (142) with lead tetra-acetate gave the two acetoxyazetidinones (143) and (144), together with (145) in 15% yield. Facile loss of the elements of acetic acid from (145) gave (146). Lead tetra-acetate oxidation of aldehyde (141), followed by loss of acetic acid, gave the *N*-formyl derivative (148) without the complications of the other side products. Deformylation to (145) was easily accomplished with ammonium hydroxide or rhodium chloride–triphenylphosphine complex. When this sequence was repeated on a penicillin having a t-butoxycarbonyl side chain and the side chain was removed from the deformylated product (146) ($R = Bu^tOCO$), thiazoline-azetidinone (128) was obtained. This was converted into (110) using phosgene and t-butanol. Use of the di-trichloroethyl protected compound (150)[59] gave on zinc reduction the azetidinone (151) in good yield. The intermediacy of a Schiff's base (152) was demonstrated by the synthesis of (152) from (150) using chromous chloride. Zinc/acetic acid reduction of (152) resulted in the thiazolidine (151).

An alternate simple procedure for the synthesis of (151) and similar-type

(1) → (138) → (139) → (140) ⇌ (141)

(141) → (147) → (148)

(141) → (142) → (143) + (144) + (145)

(145) → (146); (148) → (146)

(146) → (149) [R = Bu^tOCO—]

(149) → (128) → (110)

derivatives has been reported[60] using the thiazoline-azetidinone (153), which

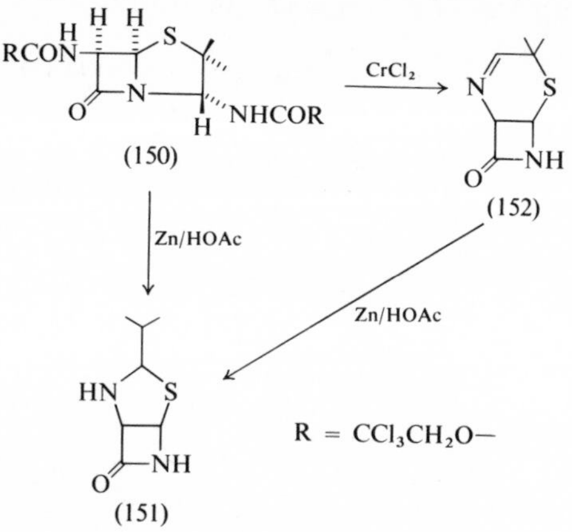

(150) $\xrightarrow{CrCl_2}$ (152); (150) $\xrightarrow{Zn/HOAc}$ (151); (152) $\xrightarrow{Zn/HOAc}$ (151)

R = CCl_3CH_2O—

had been synthesised in one step from penicillin sulphoxide (104) in excellent yield. Ozonolysis of (153) followed by methanolysis gave the azetidinone (155) which on reduction with Al/Hg yielded the thiazolidine (156). When the initial penicillin had the side chain R = $PhOCMe_2$—, (151) resulted. Alternatively, Al/Hg reduction and acylation of (153) gave the thiazolidine (158) which on subsequent ozonolysis and methanolysis yielded the

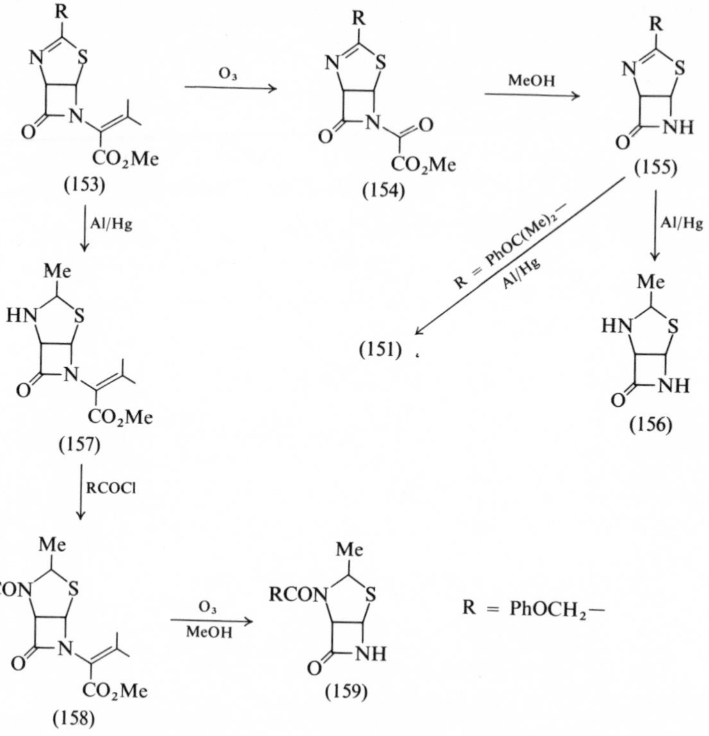

(153) $\xrightarrow{O_3}$ (154) $\xrightarrow{MeOH}$ (155); (155) $\xrightarrow{Al/Hg}$ (156); (155) $\xrightarrow{R = PhOC(Me)_2—,\ Al/Hg}$ (151); (153) $\xrightarrow{Al/Hg}$ (157) $\xrightarrow{RCOCl}$ (158) $\xrightarrow{O_3,\ MeOH}$ (159)

R = $PhOCH_2$—

azetidinone (159). All of these conversions reportedly proceeded in excellent yield.

Alternate procedures have been reported for the removal of the isopentenyl moiety to furnish the desired azetidinones[61]. The dipolar addition of diazomethane across the conjugated ester double bond gave quantitatively the two stereoisomeric products (160). These pyrazoline adducts were degraded to the azetidinone (161) by either potassium t-butoxide/t-butanol or zinc/acetic acid. This procedure, used successfully on a variety of substances, gave the respective azetidinones in excellent yield.

Oxidative liberation of azetidinones from the isopentenyl encumbrance involved reaction of the double bond with either osmium tetroxide or potassium permanganate[62]. Although this process appeared to be quite general (e.g. (86) → (162)) the yields were not so good as the previous two processes, being only of the order 24–56 %.

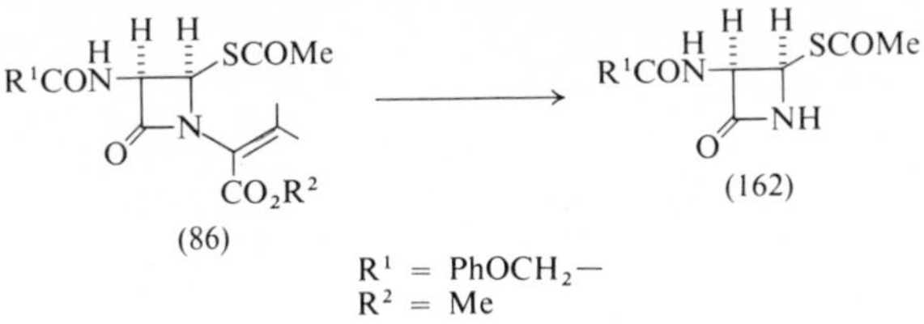

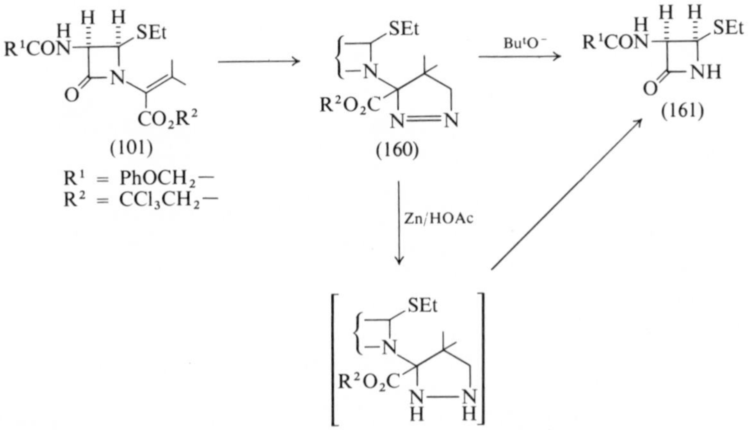

9.5.1 Partial synthesis from penicillin

This approach[63] utilised the reaction of penicillin sulphoxide with acetic anhydride to give the β-acetoxy methylsulphide (78) (R^1 = Me). Peracid oxidation of (78) gave the β-sulphoxide (163) which would not rearrange further. However, ozone oxidation of (78) furnished the two isomeric sulphoxides in a 1:1 ratio. The α-sulphoxide (164) rearranged to the cephem (165) and the 3β-hydroxycepham (166) via the β-sulphoxide-α-acetoxymethyl penam (167), thus completing the penam–cephem transformation.

Alternatively, photochemical epimerisation of (163) produced (164) in a mixture of the four possible acetoxymethyl sulphoxide isomers. Separation problems caused the previous procedure to be superior.

The ring expansion rearrangement of a penicillin sulphoxide has also been utilised for another cephalosporin synthesis. The remaining functionalisation problem of the allylic methyl group of the deacetoxycephem system was

Ac_2O

(78)

(163)

(164)

Δ

(166)

(165)

(167)

R^1 = Me

solved when the Δ^2-derivative (169)[64] (R^2 = *p*-MeOPhCH$_2$—) or the Δ^3-sulphoxide (39)[65] ($R^2 = CCl_3CH_2$—) were reacted with *N*-bromosuccinimide, followed by acetate displacement of the resulting allylic bromide. A new procedure was reported for the synthesis of the Δ^2-esters (169)[66] from the Δ^3-cephem acid ((36), (R^2 = H)) in which the acid chloride and an alcohol were reacted in the presence of a strong base. This reaction presumably proceeded through a ketene intermediate. A facile complete conversion of a Δ^2-cephem to a Δ^3-cephem was achieved[67] by oxidation of the Δ^2-sulphide (171) to the Δ^3-sulphoxide (172). Reduction of the sulphoxide to the Δ^3-sulphide (173) succeeded in high yield using acetyl chloride–sodium dithionite in DMF.

The thiazoline-azetidinone (85) previously reported as being prepared from penicillin sulphoxide (26) has been converted into a mixture containing both cephem and penam derivatives[68]. Treatment of (85) with *m*-chloroperbenzoic acid and trifluoroacetic acid gave the penicillin sulphoxide (26), the 2-hydroxymethyl penam derivatives (174) and (175) and the deacetoxycephems (39), (176) and (177).

A new synthesis of the penicillin-type structure has been developed from an isonitrile chemistry study[69]. Condensation of α-mercaptoisobutyraldehyde and ethyl β-aminocrotonate, followed by alkaline hydrolysis, gave the

R^1CON H H H S N O CO_2R^2 (36) → R^1CON H H H S N O CO_2H (168) —$R^2 = p$-MeOPhCH$_2$—→ R^1CON H H H S N O CO_2R^2 (169)

(36) —$R^2 = CCl_3CH_2$—→ R^1CON H H H O S N O CO_2R^2 (39)

(169) → R^1CON H H H S N O CH_2Br CO_2R^2 (170) → R^1CON H H H S N O CH_2OAc CO_2R^2 (171)

(39) ⇒ (172); (171) → (172)

R = PhOCH$_2$—

R^1CON H H H O S N O CH_2OAc CO_2R^2 (172) → R^1CON H H H S N O CH_2OAc CO_2R^2 (173)

R^1 N S O N H CO_2R^2 (85) → R^1CON H H H O S N O H CO_2R^2 (27)

\+

R^1CON H H H O S N O O H O (175) + R^1CON H H H O S CH_2OH N O H CO_2R^2 (174)

\+

R^1CON H H H O S N O CO_2R^2 (176) + R^1CON H H H O S OH N O H CO_2R^2 (177)

\+

R^1CON H H H O S N O CO_2R^2 (39)

R^1 = PhOCH$_2$—
R^2 = CCl_3CH_2—

thiazoline (178). Reaction of (178) with an isonitrile, e.g. EtNC, yielded the fused β-lactam amide (179), presumably through a bicyclic adduct (180). The mechanism indicated the ring closure to be stereospecific since the carboxamide group and the β-lactam ring were in a *cis* relationship. When applied to the thiazoline (181)[70], this procedure gave the 3-epipenicillin derivative (182).

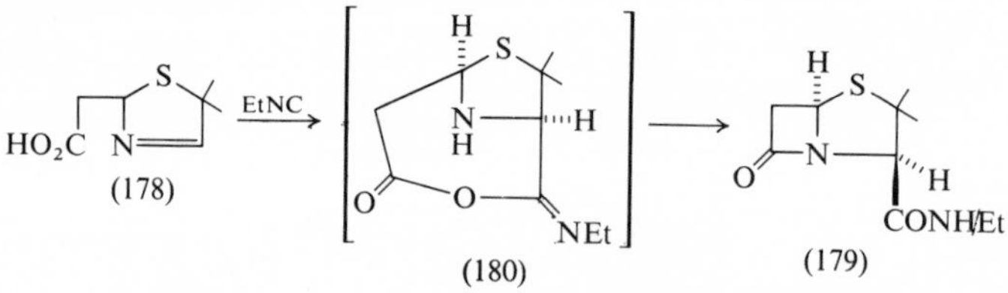

(178) (180) (179)

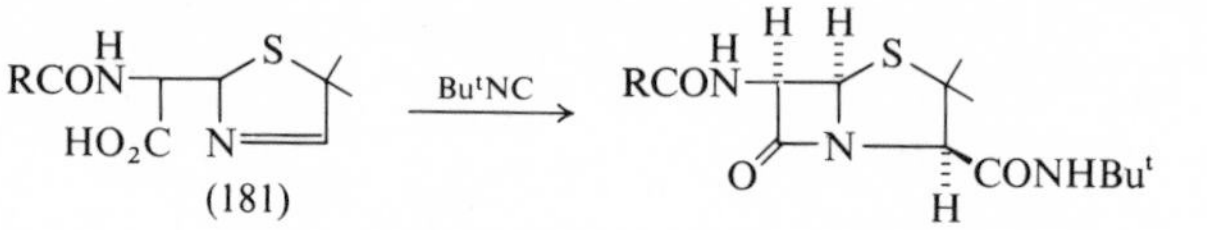

(181) (182)

9.6 β-LACTAM SYNTHESIS

This new momentum of recent years also has initiated numerous approaches into generalised syntheses of β-lactam systems. A ring expansion process via a cyclopropanone intermediate[71], e.g. carbinolamine of the type (183), produced the azetidinone (184) in good yield under a variety of conditions. Alternatively, a ring contraction process of cyclic keto amide (185) on treatment with alkaline periodate represented a new approach to azetidinones (186)[72]. However, the potential application of these processes to a penicillin synthesis faces the problem of stereochemical control in order to obtain only one of the eight possible isomers.

where X = Cl, OCOPh, N_2

(183) (184)

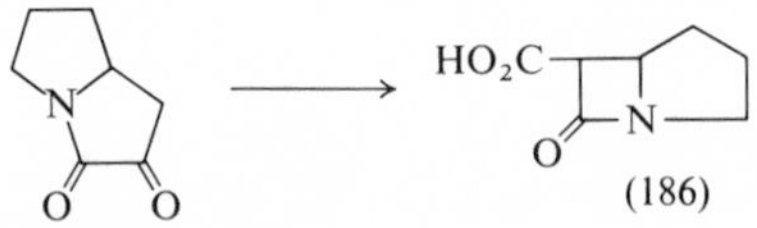

(185) (186)

Further work reported on the synthesis of penicillin-type β-lactam structures, involved the addition of a ketene to a thiazoline. Previously the reaction of azidoketene with the thiazoline (187) resulted in the 6-epi derivative (188) only[73], this being the major disadvantage of this route. Substitution of the thiazoline (189) gave the structural variation (190), whose stereochemistry as shown was indicated by n.m.r.[74]. Conversion to (191) was reported, but it had no biological activity. An initially more successful approach has been the addition of chloroketene to the olefin (192), yielding the azetidinone (193)[75]. Displacement of the chloro group with sodium azide gave the *cis*-azidoazetidinone (194).

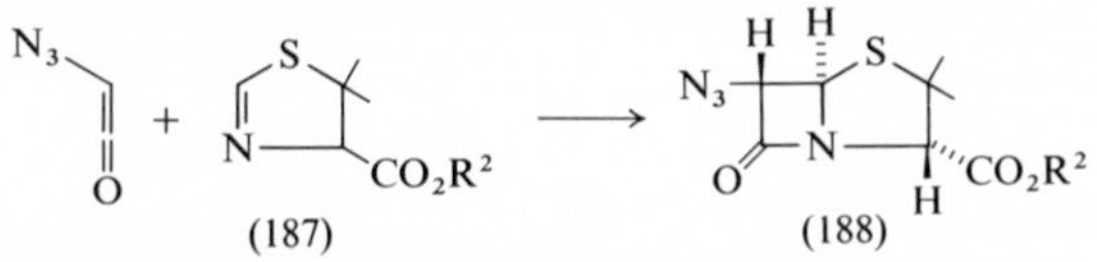

(187) (188)

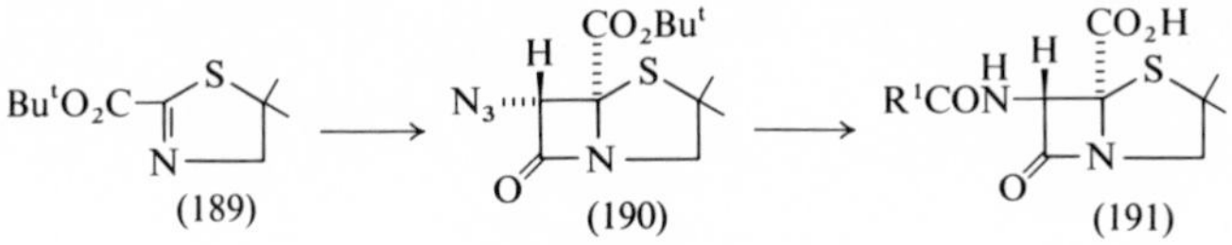

(189) (190) (191)

R^1 = $PhOCH_2$—
R^2 = Me

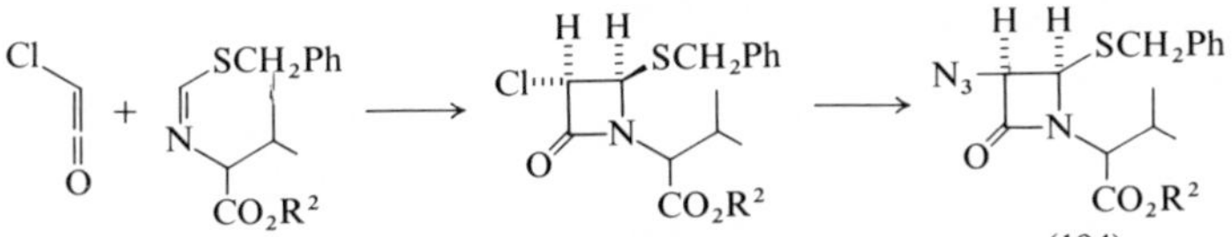

(192) (193) (194)

R^2 = Me

Perhaps the most original new approach has been the construction of the fused ring system by bond formation between two methylene carbon atoms of the potential β-lactam ring as first demonstrated by synthesis of a model system[76]. The ethyl malonic ester of piperidine (195) was converted into the diazo compound with tosyl azide. Photolysis of (196) gave the two isomeric β-lactam products (197) and (198) in 80% yield. The ratio of *cis*/*trans* ring

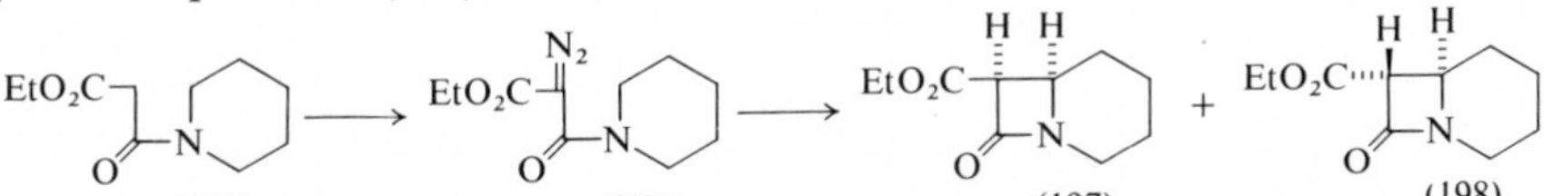

(195) (196) (197) (198)

fusion was 1:2. The utility of the β-lactam ring-forming reaction was demonstrated when the thiazoline (199) yielded 50% of the two β-lactam isomers (200)[76] and the D-piperolic acid derivative (201) gave (202) and (203)[77, 78].

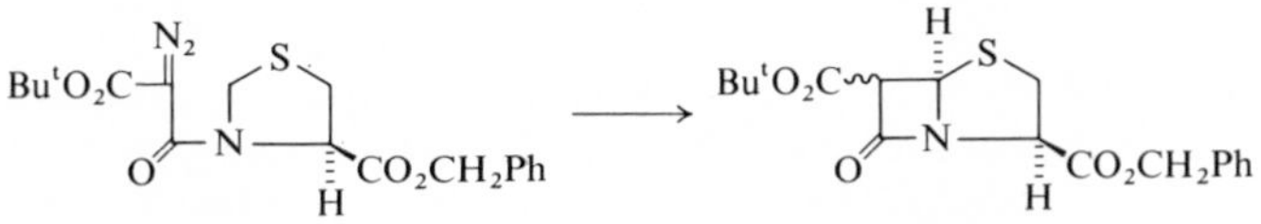

(199) (200)

Treatment of the acid (204) with t-butyl carbazate, removal of the t-butyl group, followed by reaction with sodium nitrite, gave the expected azide (206). Thermal rearrangement to the isocyanate was effected and, in the presence of t-butanol, gave the carbamate (207), from which the amine (208) was obtained. Although the acid (209) was found to have no activity against three different micro-organisms, the synthetic route seems to have considerable versatility.

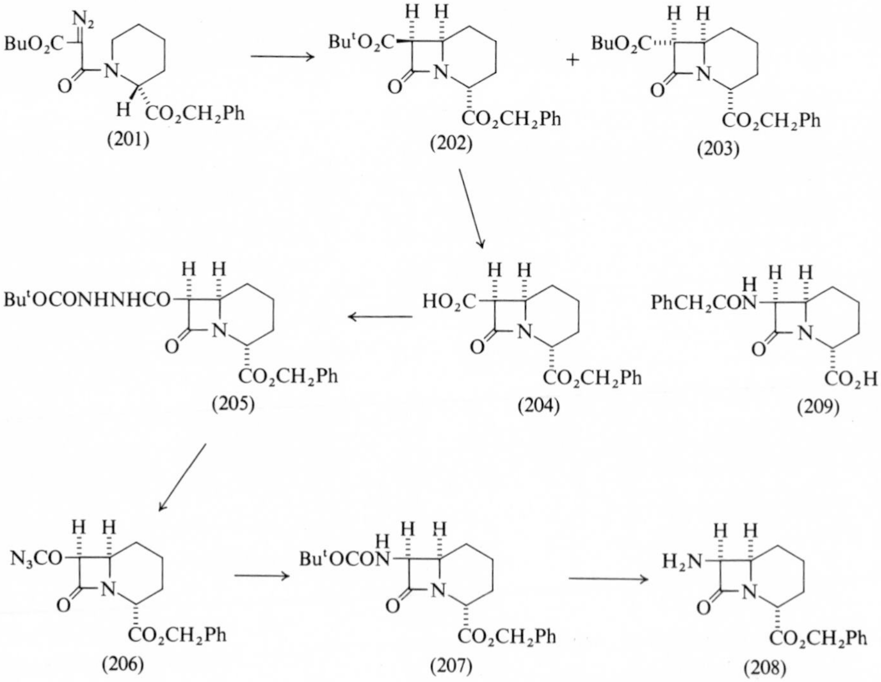

BuO_2C N_2 H CO_2CH_2Ph (201) → Bu^tO_2C H H CO_2CH_2Ph (202) + BuO_2C H H CO_2CH_2Ph (203)

HO_2C H H CO_2CH_2Ph (204) → $Bu^tOCONHNHCO$ H H CO_2CH_2Ph (205)

$PhCH_2CONH$ H H CO_2H (209)

N_3CO H H CO_2CH_2Ph (206) → Bu^tOCONH H H CO_2CH_2Ph (207) → H_2N H H CO_2CH_2Ph (208)

A somewhat related process[79] involving synthesis of the organo mercury intermediate (211) from phenylmercuric chloride and the thiazolidine (212) followed by heating (211) in bromobenzene yielded the 6α-bromopenicillinate (210) in low yield.

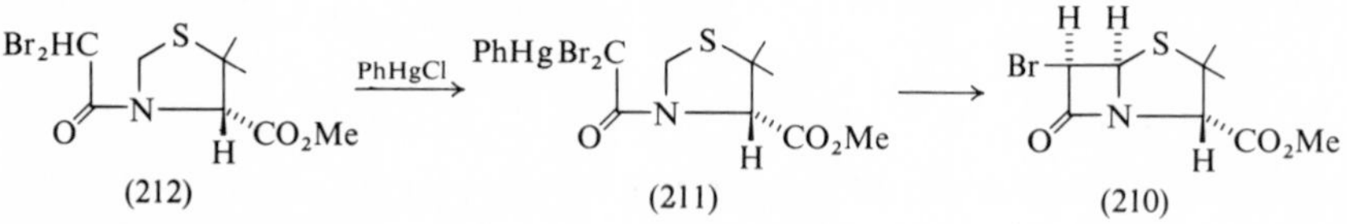

Br_2HC S H CO_2Me (212) —PhHgCl→ $PhHgBr_2C$ S H CO_2Me (211) → Br H H S H CO_2Me (210)

9.7 BIOSYNTHESIS

The biosynthetic pathway of the β-lactam antibiotics is still a mystery. The major problem is that conventional feeding techniques cannot be used since biosynthesis occurs inside the cell and the largest unit capable of permeating the cell wall is a monopeptide. Attempts to obtain a cell-free system for β-lactam production have been unsuccessful. The 'tripeptide theory', i.e. synthesis proceeds through an α-aminoadipoylcysteinyl-valine (213), received some support from the isolation[80] of the tripeptide from mycelia of *Penicillium chrysogenum.* Three related intracellular peptides, L-α-aminoadipoyl-L-cysteine-D-valine, α-AAA + cysteine + valine + glycine, and α-AAA + cysteine + γ-hydroxyvaline + glycine, were isolated from a *Cephalosporium* broken cell preparation[81]. However, all of these isolates may be only diversions from the biosynthetic pathway; and until a system is evolved which will directly incorporate these labelled precursors intact into the β-lactam compound, their isolation, though interesting, remains irrelevant to the solution of the problem. A variation has been proposed[68] in which the β-lactam was formed subsequent to the bicyclic system. Thus a tripeptide of type (214) represented the possible divergent point in the biosynthesis of the two distinct lactam fused-ring systems. The support for this postulate was that chemically it was possible to transform a derivative of type (85) into a mixture of both penicillin and cephalosporin derivatives. The application of ^{13}C n.m.r. in biosynthetic problems has been demonstrated[82] beautifully using the incorporation of ^{13}C NaOAc and ^{13}C valine into cephalosporin C.

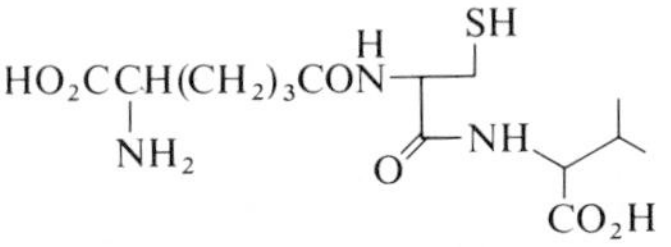

(213) 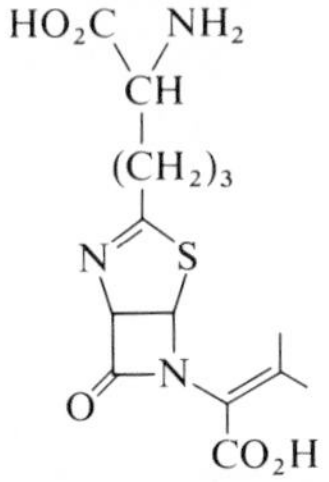 (214) 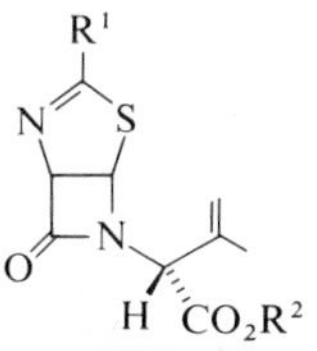 (85)

Note added in proof

Since the initial conception of this chapter, considerable further research in the area has been published. This added note will attempt to review briefly some of these advances.

Penicillin chemistry

Ring opening reactions

Extension of the earlier thiazolidine ring opening reaction with NaH/MeI has been applied to a variety of alkylating reagents[83]. Use of $\phi C{\equiv}CCH_2Br$ gave the azetidinone (1) from which, following the previously established route shown, the 3-phenylmethylcephem (2) was synthesised. This compound was reported to have good microbiological activity.

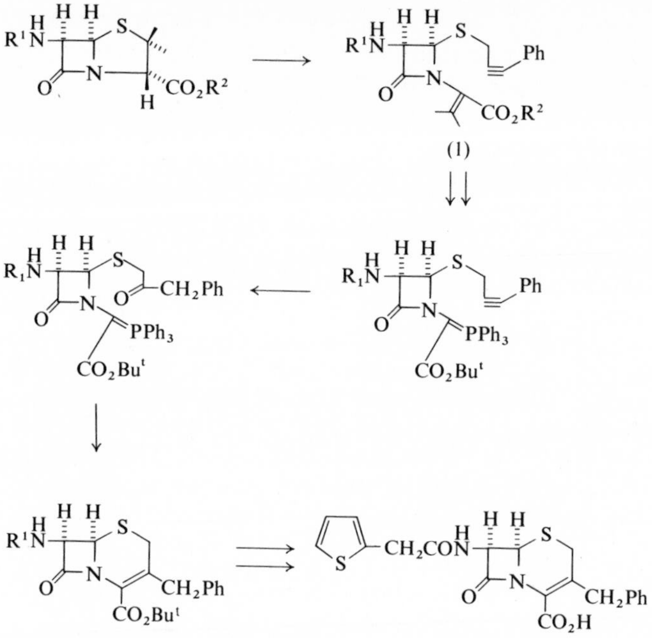

(1)

(2)

$R^1 = Ph_3C-$
$R^2 = PhCH_2-$

Reaction of a penicillin with mercury acetate in acetic acid[84] led to efficient synthesis of the acetoxyazetidinone (3).

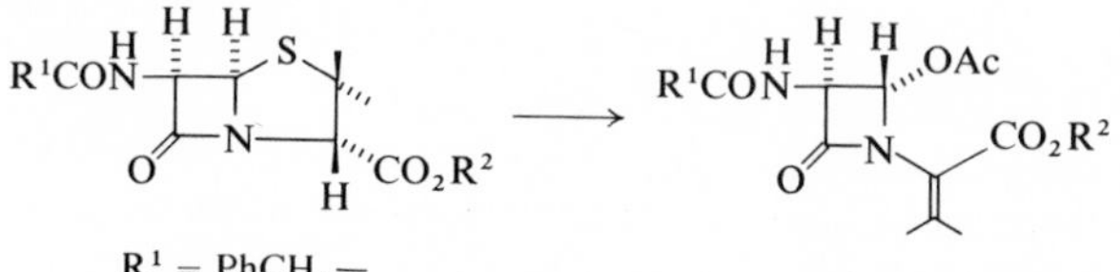

$R^1 = PhCH_2-$
$R^2 = Me$

(3)

A review of the conversion of penicillin to cephalosporin via rearrangement of the sulphoxide has appeared[85]. Further reactions of the sulphenic acid derived from penicillin sulphoxide have been reported, e.g. reaction with sulphuryl chloride[86] gave the sulphinyl chloride (4) from which the deacetoxycephem sulphoxide (5) was obtained in high yield. Use of thionyl chloride[87] gave (6) and (7), presumably via the sulphenyl chloride (8). Treatment of (7) with silver acetate gave the mixture (9), (10), and (11). The conversion (7) → (9) represents the first cepham → penam interconversion.

H H
O
S
R=N
O
N
H
CO_2R^2
SO_2Cl_2
H H
O
S−Cl
R=N
O
N
CO_2R^2
(4)
$SOCl_2$
H H
S−Cl
R=N
O
N
CO_2R^2
(8)
H H
O
S
R=N
O
N
CO_2R^2
(5)
H H
S
R=N
CH_2Cl
O
N
H
CO_2R^2
(6)
+
H H
S
R=N
Cl
O
N
H
CO_2R^2
(7)
AgOAc
H H
S
R=N
O
N
CO_2R^2
(10)
+
H H
S
CH_2OAc
R=N
O
N
H
CO_2R^2
(9)
+
H H
S
R=N
OAc
O
N
H
CO_2R^2
(11)
O
R =
O
R^2 = Me

Reaction of the sulphenic acid with dimethylazodicarboxylate[88] gave deacetoxycephem (12) in 26% yield together with the adduct (13) (14%) and the dimer (14) (11%).

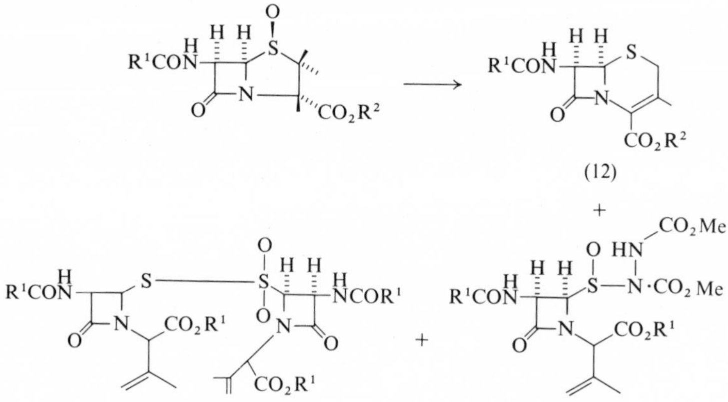

(12)

(14)

(13)

$R \cdot CON(Pr^i) \cdot NH \cdot Pr^i$

(15)

$R^1 = PhOCH_2-$
$R^2 = Me-$

A new ester protecting group, the *N,N'*-di-isopropylhydrazide (15) has been developed for use in penicillin chemistry[89]. Its removal can be effected in high yield by mild oxidative procedures, e.g. $Pb(OAc)_4$, HIO_4, or NBS.

The first conversion of a penicillin to a deacetoxycephem without the intermediacy of the sulphoxide was achieved[90]. This involved treatment of the penicillin with azidoformic ester when the initial sulphur-nitrogen ylide underwent a 6-electron electrocyclic reaction to give (16) (in 12% yield). Treatment with acid furnished a low yield of the deacetoxycephem. Reaction with a carbene instead of the nitrene gave (17)[91], which after conversion to the sulphone (18) could be ring closed to the cephem (19).

Cephalosporin chemistry

An interesting ring contraction of a cephem to a penam has been discovered[92]. The ylide formed by reaction of a deacetoxycephem with N_2CHCO_2Et/Cu underwent a 2,3-sigmatropic rearrangement to give (20).

The 3-exomethylene deacetoxycephem isomer (21) originally isolated by Morin *et al.*[93] from penicillin sulphoxide rearrangement has now been synthesised in reasonably good yields by reduction either electrochemically[94] or by chromous acetate[95]. The methyl ester of this isomer was quantitatively converted to the Δ^3 isomer (ester) with pyridine.

H H
R^1CONH S
O N
H CO_2R^3

$N_3CO_2R^2$ →

H H
R^1CONH S—NH·CO_2R^2
O N CO_2R^3

(16)

↓ H^+

H H
R^1CONH S
O N
CO_2R^3

where $R^1 = PhOCH_2-$
$R^2 = Et$
$R^3 = Me$

↓ $N_2CHCO_2R^2$

H H
R^1CONH S $CH_2CO_2R^2$
O N CO_2R^3

(17)

O O
H H S
R^1CONH $CH_2CO_2R^2$
O N CO_2R^3

(18)

H H
R^1CONH S CO_2R^2
H
O N
H CO_2R^3

(19)

where $R^1 = R^3 = PhCH_2-$
$R^2 = Et$

H H
R^1CONH S
O N
CO_2R^2

→

[$\bar{C}H \cdot CO_2Et$
H H S^+
R^1CONH
O N
CO_2R^2]

↓

$R^1 = PhCH_2-$
$R^2 = Me$

R^1CONH S CO_2Et
H
O N
CO_2R^2

(20)

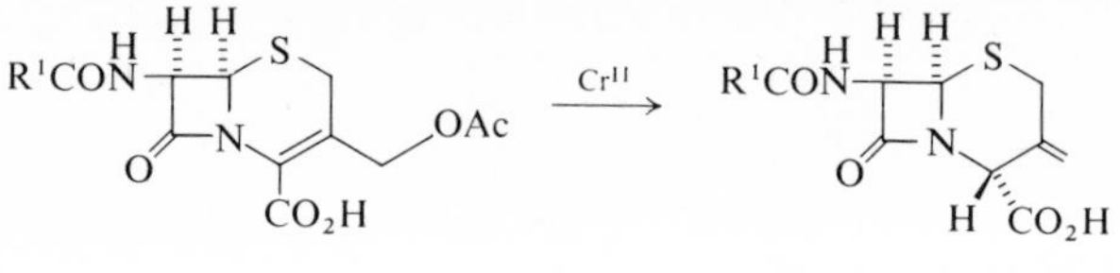

(21)

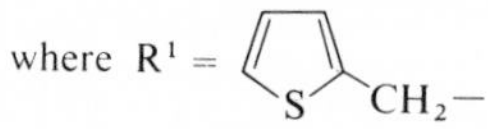

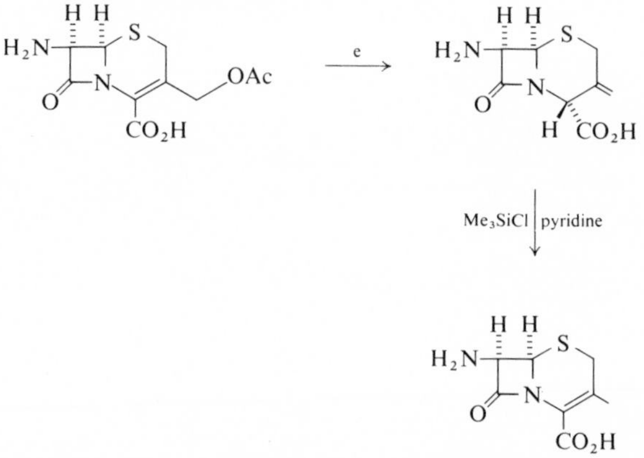

Determination of the stereochemistry at C-4 of the 3-exomethylene derivatives was achieved both chemically[96] and by n.m.r.[97].

β-Lactam functionalisation

This aspect has continued with numerous publications having appeared in the last few months. Applications of the Schiff base alkylation procedure has allowed synthesis of penicillins substituted in the 6-position with the following groups: CH_2CH_2CN[98], CH_2OH[98], CHO[98], CH(OH)Me[98], COMe[98], CO_2Me[98], CO_2Na[98], $CH_2\phi$[99], and SMe[100]. Furthermore, the 6-SMe derivative has shown the added utility of being converted in good yield into the 6-OMe and 6-OEt derivatives[100]. Similar substituted derivatives were also reported for the cephalosporin nucleus. However, the most direct synthesis of 6-methoxypenicillins was by direct oxidation of a penicillin sulphoxide to the acylimine[101] (22) with t-butylhypochlorite. In the presence of methanol, the 6α-methoxy derivative (23) was obtained. X-Ray crystallographic analysis[102] determined the stereochemistry of (24), obtained by this process from anhydropenicillin.

Modification of these reaction conditions[103] has allowed the direct oxidation of both penicillin and cephalosporin esters to their respective 6- and 7-methoxy substituted derivatives in good yield.

$\xrightarrow[\text{base}]{Bu^tOCl}$

(22)

$\downarrow$ MeOH

where $R^1 = PhOCH_2-$
$R^2 = pNO_2 \cdot C_6H_4 \cdot CH_2-$

(23)

(24)

References

1. Morin, R. B., Jackson, B. G., Mueller, R. A., Lavagnino, E. R., Scanlon, W. B. and Andrews, S. L. (1963). *J. Amer. Chem. Soc.*, **85,** 1896; (1969). *J. Amer. Chem. Soc.*, **91,** 1401
2. Nagarajan, R., Boeck, L. D., Gorman, M., Hamill, R. L., Higgens, C. E., Hoehn, M. M., Stark, W. M. and Whitney, J. G. (1971). *J. Amer. Chem. Soc.*, **93,** 2308
3. Morin, R. B. and Jackson, B. G. (1970). *Fortschr. Chem. Org. Naturst.*, **28,** 343
4. Luche, T. L. and Balavoire, G. (1971). *Bull. Soc. Chim. Fr.*, 2733
5. Nomine, G. (1971). *Chim. Ther.*, **6** (1), 53
6. Han, J. P. and Poole, J. W. (1971). *J. Pharm. Sci.*, **60**(4), 503
7. Penicillins and Cephalosporins, Their Chemistry and Biology. (1972). (E. H. Flynn, editor). (New York: Academic Press)
8. Wolfe, S., Godfrey, J. C., Holdrege, C. T. and Perron, Y. G. (1963). *J. Amer. Chem. Soc.*, **85,** 643; (1968). *Can. J. Chem.*, **46,** 2549
9. Naylor, J. H. C., Clayton, J. P., Southgate, R. and Talliday, P. (1971). *Chem. Commun.*, 590
10. Stoodley, R. J. and Ramsey, B. G. (1970). *Chem. Commun.*, 1517; *J. Chem. Soc. C*, 3859; (1971). *J. Chem. Soc. C*, 3864
11. Kukolja, S. (1971). *J. Amer. Chem. Soc.*, **93,** 6267; **93,** 6269
12. Wolfe, S. and Lee, W. S. (1968). *Chem. Commun.*, 242
13. Ramsey, B. G. and Stoodley, R. J. (1971). *Chem. Commun.*, 450
14. Wolfe, S., Lee, W. S. and Misra, R. (1970). *Chem. Commun.*, 1067
15. Clayton, J. P., Naylor, J. H. C., Southgate, R. and Stove, E. R. (1969). *Chem. Commun.*, 129
16. Johnson, D. A., Mania, D., Panetta, C. A. and Silvestin. (1968). *Tetrahedron Lett.*, 1903
17. Jackson, J. R. and Stoodley, R. J. (1971). *Chem. Commun.*, 647
18. Jackson, J. R. and Stoodley, R. J. (1972). *J. Chem. Soc.*, 895
19. Cooper, R. D. G., Demarco, P. V. and Spry, D. O. (1969). *J. Amer. Chem. Soc.*, **91,** 1528
20. Gutowski, G. E. (1970). *Tetrahedron Lett.*, 1779

21. Vlietinck, A., Roets, E., Claes, P. and Vanderhaeghe, H. (1972). *Tetrahedron Lett.*, 285
22. Hauser, D. and Sigg, H. P. (1967). *Helv. Chim. Acta*, **50**, 1327
23. Brunwun, D. M. and Lowe, G. (1972). *J. Chem. Soc.*, 192
24. Cooper, R. D. G., Demarco, P. V., Murphy, C. F. and Spangle, L. A. (1972). *J. Chem. Soc.*, 0000
25. Cooper, R(D. G., unpublished results
26. Kaiser, G. V., Ashbrook, C. W., Goodson, T., Wright, I. G. and Van Heyningen, E. M. (1971). *J. Med. Chem.*, **14**, 426; Wright, I. G., Ashbrook, C. W., Goodson, T., Kaiser, G. V. and Van Heyningen, E. M. (1971). *J. Med. Chem.*, **14**, 420
27. Strominger, J. L. and Tipper, D. J. (1965). *Amer. J. Med.*, **39**, 708
28. Bell, M. R., Oesterlin, R., Clemans, S. D. and Carlson, J. A. (1971). *Abstracts XXIII IUPAC Congress*, Boston, p. 74
29. Bell, M. R., Carlson, J. A. and Oesterlin, R. (1970). *J. Amer. Chem. Soc.*, **92**, 2177
30. Kaiser, G. V., Ashbrook, C. W. and Baldwin, J. E. (1971). *J. Amer. Chem. Soc.*, **93**, 2342
31. Boehme, E. H. W., Applegate, H. E., Toeplitz, B., Dolfini, J. and Gougoutas, J. Z. (1971). *J. Amer. Chem. Soc.*, **93**, 4324; Firestone, R. A., Schelechow, N., Johnston, D. B. R. and Christensen, B. G. (1971). *Tetrahedron Lett.*, 375

Stapley, E. O., Hendlin, D., Hernandez, S., Jackson, M., Mata, J. M., Miller, A. K., Woodruff, H. B., Miller, T. W., Albers-Schanberg, G., Arison, B. H. and Smith, J. L. (1971). *Abstracts XI Interscience Conf. on Antimicrobial Agents and Chemotherapy*, Atlantic City, p. 8

33. Murphy, C. F. and Koehler, R. E. (1972). *Tetrahedron Lett.*, 1585
34. Karady, S., Pines, S. H., Weinstock, L. M., Roberts, F. E., Brenner, G. S., Hainowski, A. M., Cheng, T. Y. and Sletzinger, M. (1972). *J. Amer. Chem. Soc.*, **94**, 1410
35. Cama, L. D., Leanza, W. J., Beattie, T. R. and Christensen, B. G. (1972). *J. Amer. Chem. Soc.*, **94**, 1408
36. Wick, W. E. (1967). *Appl. Microbiol.*, **15**, 4; Wick, W. E. and Boniece, W. S. (1967). *Proc. 6th International Congress Chemotherapy*, Vienna, 717
37. Chauvette, R. R., Pennington, P. A., Ryan, C. W., Cooper, R. D. G., Jose, F. L., Wright, I. G., Van Heyningen, E. M. and Huffman, G. W. (1971). *J. Org. Chem.*, **36**, 1259
38. Hatfield, L. D. (1971). *U.S. Pat.*, 3 591 585
39. Cooper, R. D. G., Demarco, P. V., Cheng, J. C. and Jones, N. D. (1969). *J. Amer. Chem. Soc.*, **91**, 1408
40. Cooper, R. D. G., Demarco, P. V. and Spry, D. O. (1969). *J. Amer. Chem. Soc.*, **91**, 1528
41. Barton, D. H. R., Comer, F. and Sammes, P. G. (1969). *J. Amer. Chem. Soc.*, **91**, 1529
42. Spry, D. O. (1972). *J. Org. Chem.*, **37**, 793
43. Barton, D. H. R., Comer, F., Greig, D. G. T., Lucente, G., Sammes, P. G. and Underwood, W. G. E. (1970). *Chem. Commun.*, 1059
44. Cooper, R. D. G. (1970). *J. Amer. Chem. Soc.*, **92**, 5010
45. Cooper, R. D. G. and Jose, F. L. (1970). *J. Amer. Chem. Soc.*, **92**, 2575
46. Hatfield, L. D., Fisher, J. W., Jose, F. L. and Cooper, R. D. G. (1970). *Tetrahedron Lett.*, 4897
47. Barton, D. H. R., Greig, D. G. T., Lucente, G., Sammes, P. G., Taylor, M. V., Cooper, C. M., Hewitt, G. and Underwood, W. G. E. (1970). *Chem. Commun.*, 1683
48. Ager, I., Barton, D. H. R., Lucente, G. and Sammes, P. G. (1972). *J. Chem. Soc.*, 601
49. Barton, D. H. R., Sammes, P. G., Taylor, M. V., Cooper, C. M., Hewitt, G., Looker, B. E., Underwood, W. G. E. (1971). *Chem. Commun.*, 1137
50. Gutowski, G. E., Daniels, C. J. and Cooper, R. D. G. (1971). *Tetrahedron Lett.*, 3429; Gutowski, G. E., Foster, B. J., Daniels, C. J., Hatfield, L. D. and Fisher, J. W. (1971). *Tetrahedron Lett.*, 3433
51. Hatfield, L. D., unpublished results
52. Cooper, R. D. G. and Spry, D. O. (1972). *Penicillins and Cephalosporins, Their Chemistry and Biology*, Ch. 5. (E. H. Flynn, editor). (New York: Academic Press)
53. Neidleman, S. L., Pan, S. C., Last, J. C. and Dolfini, J. E. (1970). *J. Med. Chem.*, **13**, 386
54. Woodward, R. B., Heusler, K., Gosteli, J., Naegeli, P., Oppolzer, W., Ramage, R., Ranganathan, S. and Vorbruggen, H. (1966). *J. Amer. Chem. Soc.*, **88**, 852
55. Woodward, R. B. (1970). *Hanburg Memorial Lecture before the Pharmaceutical Society of Great Britain*, London; Heusler, K. (1971). *XXIII International Congress Pure and Applied Chem.*, Boston, Vol. 3, 87

56. Scartazzini, R. and Bickel, H. (1972). *Helv. Chim. Acta,* **55,** 423
57. Heusler, K. (1972). *Helv. Chim. Acta,* **55,** 388
58. Sheehan, J. C. and Brandt, K. G. (1966). *J. Amer. Chem. Soc.,* **87,** 5468
59. Bechtig, B., Bickel, H. and Heusler, K. (1972). *Helv. Chim. Acta,* **55,** 417
60. Cooper, R. D. G. and Jose, F. L. (1972). *J. Amer. Chem. Soc.,* **94,** 1021
61. Barton, D. H. R., Greig, D. G. T., Sammes, P. G. and Taylor, M. V. (1971). *Chem. Commun.,* 845
62. Brain, E. G., Eglington, A. J., Nayler, J. H. C., Pearson, M. J. and Southgate, R. (1972). *Chem. Commun.,* 229
63. Spry, D. O. (1970). *J. Amer. Chem. Soc.,* **92,** 5006
64. Webber, J. A., Van Heyningen, E. M. and Vasileff, R. T. (1969). *J. Amer. Chem. Soc.,* **91,** 5764
65. Cooper, R. D. G. and Jose, F. L. (1970). *Abstracts 160th American Chemical Society National Meeting,* Chicago
66. Murphy, C. F. and Koehler, R. E. (1970). *J. Org. Chem.,* **35,** 2429
67. Kaiser, G. V., Cooper, R. D. G., Koehler, R. E., Murphy, C. F., Webber, J. A., Wright, I. G. and Van Heyningen, E. M. (1970). *J. Org. Chem.,* **35,** 2430
68. Cooper, R. D. G. (1972). *J. Amer. Chem. Soc.,* **94,** 1018
69. Ugi, I. and Wischhofer, E. (1962). *Chem. Ber.,* **95,** 136
70. Sjoberg, K. (1970). Ph.D. Thesis, Technical University, Stockholm
71. Wasserman, H. H., Adidses, H. W. and Espejo de Ochoo, O. (1971). *J. Amer. Chem. Soc.,* **93,** 5586
72. Rapoport, H. (1971). *Ger. Offen.* 2 130 730
73. Bose, A. K., Spiegelman, G. and Manhas, M. S. (1968). *J. Amer. Chem. Soc.,* **90,** 4506
74. Bose, A. K., Spiegelman, G. and Manhas, M. S. (1971). *J. Chem. Soc. C,* 2468
75. Bachi, M. D. and Goldberg, O. (1972). *J. Chem. Soc.,* 319
76. Lowe, G. and Parker, J. (1971). *Chem. Commun.,* 577
77. Lowe, G., Brunwin, D. M. and Parker, J. (1971). *Chem. Commun.,* 865
78. Brunwin, D. M., Lowe, G. and Parker, J. (1971). *J. Chem. Soc., C,* 3756
79. Johannson, N. G. and Akermark, B. (1971). *Tetrahedron Lett.,* 4785
80. Bauer, K. (1970). *Z. Naturforsch. B,* **25,** 1125
81. Loder, B. P. and Abraham, E. P. (1971). *Biochem. J.,* **123,** 477
82. Neuss, N., Nash, C. H., Lemke, P. A. and Grutzner, J. B. (1971). *J. Amer. Chem. Soc.,* **93,** 2337
83. Nayler, J. H. C., Pearson, M. J. and Southgate, R. (1973). *Chem. Commun.,* **58**
84. Stoodley, R. J. and Whitehouse, N. R. (1973). *J. Chem. Soc.,* Perkin, **1,** 32
85. Cooper, R. D G., Hatfield, L. D. and Spry, D. O. (1973). *Acc. Chem. Res.,* **6,** 32
86. Kukolja, S. and Lammert, S. R. (1973). *Ang. Chem.,* **1,** 40
87. Kukolja, S. and Lammert, S. R. (1972). *J. Amer. Chem. Soc.,* **94,** 7169
88. Tereo, S., Matsuo, T., Tsushima, S., Matsumoto, N., Mikiwaki, T. and Miyamoto, M. (1973). *Chem. Commun.,* 1304
89. Barton, D. H. R., Girijavallabhan, M. and Sammes, P. G. (1972). *J. Chem. Soc.,* Perkin Trans., 929
90. Numata, M., Imashiro, Y., Minamida, I. and Yamaoka, M. (1972). *Tetrahedron Lett.,* 5097
91. Yoshimoto, M., Ishihara, S., Nakayama, E., Shoji, E., Kuwano, H. and Soma, N. (1972). *Tetrahedron Lett.,* 4387
92. Yoshimoto, M., Ishihara, S., Nakayama, E. and Soma, N. (1972). *Tetrahedron Lett.,* 2923
93. Morin, R. B. (1969). *U.S. Patent* 3 466 275
94. Ochiai, H., Aki, O., Morimoto, A., Okada, T., Shinozaki, K. and Asahi, Y. (1972). *Tetrahedron Lett.,* 2341
95. Ochiai, M., Aki, O., Morimoto, A., Okada, T. and Shimadzu, H. (1972). *Chem. Commun.,* 800
96. Spry, D. O. (1973). *Tetrahedron Lett.,* 165
97. Ochiai, M., Aki, O., Morimoto, A. and Ikada, T. (1972). *Tetrahedron Lett.,* 3241
98. Rasmusson, G. H., Reynolds, G. F. and Arth, G. E. (1973). *Tetrahedron Lett.,* 145
99. Johnson, D. B. R., Schmitt, S. M., Firestone, R. A. and Christensen, B. G. (1972). *Tetrahedron Lett.,* 4917

100. Spitzer, W. A. and Goodson, T. (1973). *Tetrahedron Lett.,* 273
101. Baldwin, J. E., Urbin, F. J., Cooper, R. D. G. and Jose, F. L. (1973). *J. Amer. Chem. Soc.,* **95,** 0000
102. Jones, N. D. and Chaney, M. O., Unpublished results
103. Koppel, G. A. and Koehler, R. E. (1973). *J. Amer. Chem. Soc.,* **95,** 0000